SARS-CoV-2 in Water

In Focus – Special Book Series

SARS-CoV-2 in Water

Editors

Eiji Haramoto, Gertjan Medema, John Scott Meschke and Susan Petterson

IWA PUBLISHING

Published by **IWA Publishing**
Unit 104–105, Export Building
1 Clove Crescent
London E14 2BA, UK
Telephone: +44 (0)20 7654 5500
Fax: +44 (0)20 7654 5555
Email: publications@iwap.co.uk
Web: www.iwapublishing.com

First published 2022

British Library Cataloguing in Publication Data
A CIP catalogue record for this book is available from the British Library

ISBN: 9781789063226

Contents

SARS-CoV-2 in Water doi: 10.2166/wh.2022.001

Editorial: SARS-CoV-2 in water

We are delighted to share with you our special issue on SARS-CoV-2 in water, specifically targeted towards our Journal readership. Cutting-edge studies and reviews on the water-related health aspects of SARS-CoV-2 have been contributed by our research community related to the potential risk of waterborne COVID-19 transmission, household water uses, and hygiene during the pandemic, and surveillance of SARS-CoV-2 RNA in wastewater.

RISK OF WATERBORNE TRANSMISSION

In the early days of the pandemic, when shedding of virus with faeces was identified, the most critical question that needed to be addressed was: Does SARS-CoV-2 in water and wastewater present a risk to human health? Could water-based exposures, for example to plumbing workers and wastewater treatment plant operators, present a transmission pathway that needs to be managed?

While several lines of evidence emerged, indicating that there is likely minimal risk via waterborne exposure (including the lack of epidemiological signal of cases amongst wastewater workers, the lack of evidence of infectious virus in wastewater, and the lack of evidence of SARS-CoV-2 persistence in water or wastewater), this topic is still frequently raised for discussion from various sectors of the community. Included in this special issue is an invited review by Professor Mark Sobsey on the risk of COVID19 via faecal waste, wastewater, and water exposures (Sobsey 2022). We would like to express our gratitude to Mark for taking the time to undertake this important review and document his perspective on the available evidence for the benefit of our readership.

Of course, scientific evidence grows rapidly with ongoing research efforts and there is a need to stay alert and open-minded should there be a need to revisit. With that in mind, we commend to you a study from Dey *et al.* (2022) investigating the potential role of amoebae in the environmental persistence of SARS-CoV-2 virus. Understanding the microbial ecology of water systems and the interaction between different microorganisms is of great value and is important for managing the waterborne risk of several pathogens (from SARS-CoV-2 or other RNA viruses) in the future.

HOUSEHOLD WATER USE AND COMPLIANCE WITH PUBLIC HEALTH ADVICE

Public health interventions to limit person-to-person transmission including stay-at-home orders, social distancing, and handwashing were implemented swiftly and globally at the start of the COVID-19 pandemic, with varying degrees of success. As a result of these interventions, water usage patterns changed both due to stay-at-home orders and an increase in water demand for hand washing. The interaction between water availability and compliance with public health orders became the focus of interesting research.

Kumpel *et al.* (2022) investigated household water usage patterns in Hubballi-Dharwad, India during 2020–2021. One-quarter of all households reported insufficient water for handwashing, suggesting that the increased demand for water was not being met. Their results showed that water insecurity negatively affected a household's ability to adhere to protective public health measures during the COVID-19 pandemic and highlighted the importance of access to uninterrupted, on-premises water during public health emergencies.

When a household water supply is insufficient or interrupted, there is a need to leave the household to obtain an essential water supply. Leaving the home increases person-to-person interactions in the community, and this is exacerbated when needing to visit public water and sanitation facilities where physical distancing is often not possible. Denpetkul *et al.* (2021) report a quantitative risk assessment highlighting the risks of person-to-person spread within public toilet facilities.

Handwashing is a simple and effective hygiene measure. Biswas & Karmakar (2022) analysed data from a recent National Sample Survey for India to assess trends in handwashing across the country. They considered a very broad range of factors including caste, class, monthly expenditure, religion, together with water and sanitation access criteria. They showed clearly that handwashing throughout India is far from universal, and sections of the population with poor handwashing practices are relatively more vulnerable during the COVID-19 pandemic. Identification of these vulnerable sections of the population is of great value for intervention.

WASTEWATER SURVEILLANCE

While disease surveillance via analysis of wastewater is far from new, the rate of growth in the implementation of wastewater surveillance since the beginning of the COVID-19 pandemic has been extraordinary. Wastewater molecular biology has moved into the spotlight of the scientific arena, gaining attention, interest, and critique like never before. The importance of these activities for our research community, as well as for supporting the public health response to the pandemic, is reflected in the large number of papers in this special issue specifically focussing on wastewater surveillance.

Since the beginning of 2020, wastewater surveillance programmes have been implemented in 58 countries (COVID-Poops19 Dashboard|covid19wbec.org). Given this rapid global expansion of predominantly new programmes, Tlhagale *et al.* (2022) reported on the logistics, implementation, advantages, and limitations of national surveillance programmes. A range of countries were included with specific details on the situation in England, the Netherlands, Turkey, and South Africa. In all case studies, the aim of the surveillance programme was to provide scientific evidence regarding the trends in the circulation of SARS-CoV-2 in communities. Wastewater monitoring data has been recognised as an important thread of scientific evidence for public health authorities; however prior to the current pandemic, many decision-makers were unfamiliar with the science behind wastewater surveillance. The unique nature of molecular data coupled with the diversity of methods applied has meant that scientific details are not simple or trivial. This has presented a unique challenge for our scientific community needing to bridge the gap in translating our science for decision-making, accurately, quickly, and effectively. This is particularly addressed by Tlhagale *et al.* (2022) who comment that '*communication and knowledge dissemination have been demanding and time-consuming for the researchers and scientists working on the topic in all four case studies, which shows an area to better prepare for and support in the future*'.

In further support of research translation, we include an excellent review prepared by Hrudey & Conant (2022) drawing on their experience with the Canadian Water Network (CWN) Wastewater Coalition. Hrudey & Conant (2022) argue that appropriate interpretation of wastewater data for public health decision-makers requires an understanding of the importance of different knowns and unknowns. They seek to unpack these details by reviewing the literature regarding the strengths and limitations of wastewater surveillance for SARS-CoV-2 towards the goal of ensuring findings serve needs of public health authorities. In conclusion, it is noted that '*Perhaps one of the enduring benefits from all of the WBS initiatives has been that it required and achieved extensive interdisciplinary collaboration between environmental scientists/engineers and public health professionals.*' This is a welcome and needed avenue of growth for our entire scientific community.

In addition to these broader reviews, this special edition includes a selection of important individual scientific studies from Brazil, Austria, Hungary, Australia, and Japan, with each touching on different key aspects of wastewater surveillance.

Barbosa *et al.* (2022) used their wastewater surveillance programme to assess trends in SARS-CoV-2-RNA in four vulnerable urban communities (informal settlements and low-income neighbourhoods) in metropolitan Sao Paulo. They were able to positively correlate new cases of COVID-19 with N1 viral loads from the two largest communities. This study provides important and detailed information regarding the performance of their analytical methods, a valuable addition to this high-quality scientific study. Barbosa *et al.* (2022) comment that '*Poor sanitation conditions, water scarcity, hyper-dense dwellings, crowded household, lack of health care access, among other factors, make urban slums and other vulnerable communities major COVID-9 hotspots and relevant areas for application of wastewater surveillance.*' Limited studies have been reported in these contexts; however, many programmes are currently ongoing in several LMICs including Pakistan, Nepal, Malawi and Indonesia, and therefore we look forward to future publications in our journal.

Róka *et al.* (2022) demonstrated the use of wastewater surveillance as an efficient tool in tracing emerging SARS-CoV-2 variants of concern in Hungary, with sewage data closely correlated with the clinical emergence of B.1.1.7 cases.

In Tyrol (a federal province of Austria), wastewater monitoring programme was initiated by health authorities. Daleiden *et al.* (2022) report on this programme and show how wastewater surveillance provided value as an early warning system; provided independent confirmation of temporal trends in COVID-19 prevalence; enabled the assessment of the effectiveness of measures, alerted about bursts of disease activity; and provided evidence for the absence of COVID-19.

Camphor *et al.* (2022) report data from Australia and illustrate the interpretation of wastewater data within a more traditional epidemiologic framework used for decision support by public health professionals. They calculate odds ratios (odds of a positive detection given case notifications in the catchment) and evaluate the performance of wastewater testing as a population-level diagnostic tool.

Finally, the case study from an urban community in Japan addresses the important issue of back-calculation, which refers to the process of quantitatively interpreting community disease prevalence from wastewater data. Many studies have sought to estimate the number of cases in the community from the reported number of genome copies in a wastewater sample. Zhu *et al.* (2022) review the uncertainty associated with estimating the number of cases in the catchment area using back-calculation and conclude that '*at the current stage, the uncertainty associated with the wastewater viral load is still a great hindrance to reliable back-calculation*'. Their results, however, demonstrate a strong correlation between virus detection and reported cases, which they argue may guide towards novel wastewater surveillance strategies.

PATH

PATH, with support from the Bill & Melinda Gates Foundation, is delighted to sponsor this very important special issue on SARS-CoV-2 in water, which gathers articles highlighting the role that water plays in transmission, control, and surveillance of SARS-CoV-2 and reflects on the challenges encountered in different settings.

PATH is a global non-profit organisation dedicated to achieving health equity. We believe that wastewater-based surveillance and more broadly environmental surveillance can provide a cost-effective tool for disease surveillance in countries where clinical surveillance is severely hampered by limited resources and where the population get delayed access to vaccines. However, integrating non-sewered areas and the resulting variety of sample types in wastewater surveillance systems increases challenges in implementation with the interpretation of data. With funding from the Bill & Melinda Gates Foundation and the Global Innovation Fund, PATH is supporting groups in Indonesia, Malawi, Nepal, and Pakistan to set up and implement wastewater-based surveillance of SARS-CoV-2 and to estimate the costs and cost-effectiveness of environmental surveillance, which is essential for scale-up and buy-in by public health authorities.

PATH has been supporting water sanitation and hygiene (WASH) with several projects, and acknowledges the topics presented in this issue which highlight the importance of WASH in the control of the transmission of the disease and as a factor of inequities. We are also pleased to see the representation of articles presenting wastewater surveillance initiatives and use cases from high-income countries but also from low-resource settings, where wastewater surveillance could potentially have the greatest impact. Several articles discuss the challenges of interpreting wastewater data, highlighting some of the barriers to wastewater surveillance data-based public health decision-making. This is an issue that will need to be addressed to build momentum from the SARS-CoV2 pandemic.

Guest Editors
Eiji Haramoto IWA
Interdisciplinary Center for River Basin Environment, University of Yamanashi, Japan

Gertjan Medema
KWR Water Research Institute and Delft University of Technology, The Netherlands; Michigan State University, USA

John Scott Meschke IWA
Department of Environmental and Occupational Health Sciences, University of Washington, USA

Susan Petterson
Water & Health Pty Ltd, North Sydney, Australia; School of Medicine and Dentistry, Griffith University, Australia

REFERENCES

Barbosa, M. R. F., Garcia, S. C., Bruni, A. D. C., Machado, F. S., de Oliveira, R. X., Dropa, M., da Costa, A. C., Leal, E., Brandão, C. J., da Silva, R. L. O. & Iko, B. Y. 2022 One-year surveillance of SARS-CoV-2 in wastewater from vulnerable urban communities in metropolitan São Paulo, Brazil. *Journal of Water and Health* **20** (2), 471–490.

Biswas, S. S. & Karmakar, R. 2022 Determinants of hand-hygiene practices in India: reflections from the 76th round National Sample Survey, 2018. *Journal of Water and Health* **20** (1), 68–82.

Camphor, H. S., Nielsen, S., Bradford-Hartke, Z., Wall, K. & Broome, R. 2022 Retrospective epidemiological analysis of SARS-CoV-2 wastewater surveillance and case notifications data – New South Wales, Australia, 2020. *Journal of Water and Health* **20** (1), 103–113.

Daleiden, B., Niederstätter, H., Steinlechner, M., Wildt, S., Kaiser, M., Lass-Flörl, C., Posch, W., Fuchs, S., Pfeifer, B., Huber, A. & Oberacher, H. 2022 Wastewater surveillance of SARS-CoV-2 in Austria: development, implementation, and operation of the Tyrolean wastewater monitoring program. *Journal of Water and Health* **20** (2), 314–328.

Denpetkul, T., Pumkaew, M., Sittipunsakda, O., Mongkolsuk, S. & Sirikanchana, K. 2021 Effects of face masks and ventilation on the risk of SARS-CoV-2 respiratory transmission in public toilets: a quantitative microbial risk assessment. *Journal of Water and Health* **20** (2), 300–313.

Dey, R., Dlusskaya, E. & Ashbolt, N. J. 2022 SARS-CoV-2 surrogate (phi6) environmental persistence within free-living amoebae. *Journal of Water and Health* **20** (1), 83–91.

Hrudey, S. E. & Conant, B. 2022 The devil is in the details: emerging insights on the relevance of wastewater surveillance for SARS-CoV-2 to public health. *Journal of Water and Health* **20** (1), 246–270.

Kumpel, E., Billava, N., Nayak, N. & Ercumen, A. 2022 Water use behaviors and water access in intermittent and continuous water supply areas during the COVID-19 pandemic. *Journal of Water and Health* **20** (1), 139–148.

Róka, E., Déri, D., Khayer, B., Kis, Z., Schuler, E., Magyar, N., Pályi, B., Pándics, T. & Vargha, M. 2022 SARS-CoV-2 variant detection from wastewater: rapid spread of B. 1.1. 7 lineage in Hungary. *Journal of Water and Health* **20** (2), 277–286.

Sobsey, M. D. 2022 Absence of virological and epidemiological evidence that SARS-CoV-2 poses COVID-19 risks from environmental fecal waste, wastewater and water exposures. *Journal of Water and Health* **20** (1), 126–138.

Tlhagale, M., Liphadzi, S., Bhagwan, J., Naidoo, V., Jonas, K., van Vuuren, L., Medema, G., Andrews, L., Béen, F., Ferreira, M. L. & Saatci, A. M. 2022 Establishment of local wastewater-based surveillance programmes in response to the spread and infection of COVID-19 – case studies from South Africa, the Netherlands, Turkey and England. *Journal of Water and Health* **20** (2), 287–299.

Zhu, Y., Oishi, W., Maruo, C., Bandara, S., Lin, M., Saito, M., Kitajima, M. & Sano, D. 2022 COVID-19 case prediction via wastewater surveillance in a low-prevalence urban community: a modeling approach. *Journal of Water and Health* **20** (2), 459–470.

SARS-CoV-2 in Water doi: 10.2166/wh.2021.182

Absence of virological and epidemiological evidence that SARS-CoV-2 poses COVID-19 risks from environmental fecal waste, wastewater and water exposures

Mark D. Sobsey

Department of Environmental Sciences and Engineering, Gillings School of Global Public Health, University of North Carolina, Chapel Hill, NC 27599-7431, USA
E-mail: mark_sobsey@unc.edu

ABSTRACT

This review considers evidence for infectious severe acute respiratory syndrome coronavirus 2 (SARS-CoV-2) presence and COVID-19 infection and illness resulting from exposure to environmental fecal wastes and waters. There is no documented evidence that (1) infectious, replication-capable SARS-CoV-2 is present in environmental fecal wastes, wastewater or water, and (2) well-documented epidemiological evidence of COVID-19 infection, illness or death has never been reported for these exposure media. COVID-19 is transmitted mainly by direct personal contact and respiratory secretions as airborne droplets and aerosols, and less so by respiratory-secreted fomites via contact (touch) exposures. While SARS-CoV-2 often infects the gastrointestinal tract of infected people, its presence as infectious, replication-capable virus in environmental fecal wastes and waters has never been documented. There is only rare and unquantified evidence of infectious, replication-capable SARS-CoV-2 in recently shed feces of COVID-19 hospital patients. The human infectivity dose–response relationship of SARS-CoV-2 is unknown, thereby making it impossible to estimate evidence-based quantitative health effects assessments by quantitative microbial risk assessment methods requiring both known exposure assessment and health effects assessment data. The World Health Organization, Water Environment Federation, US Centers for Disease Control and Prevention and others do not consider environmental fecal wastes and waters as sources of exposure to infectious SARS-CoV-2 causing COVID-19 infection and illness.

Key words: absence, COVID-19, exposure, SARS-CoV-2, wastewater, water

HIGHLIGHTS

- No infectious severe acute respiratory syndrome coronavirus 2 (SARS-CoV-2) has been found in fecal wastes or waters.
- There is no epidemiological evidence of COVID-19 infection, illness or death from exposures via environmental fecal wastes or waters.
- Health risk assessments for COVID-19 by quantitative microbial risk assessment are not possible.
- There is no evidence of infectious SARS-CoV-2 in fecal wastes and waters or COVID-19 infection, illness and death attributable to such exposure media.
- Additional and coordinated efforts are recommended to further seek infectious, replication-capable SARS-CoV-2 in environmental fecal wastes and water using state-of-the-science methods.

GRAPHICAL ABSTRACT

Question:

Is infectious SARS-CoV-2 in wastewater, fecal sludge, recreational water and drinking water?

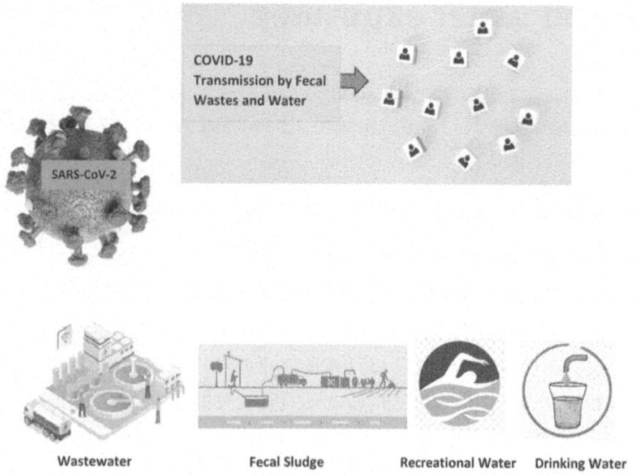

Answer: NO!

INTRODUCTION AND BACKGROUND

The purpose of this report is to identify, review and consider any documented evidence that severe acute respiratory syndrome coronavirus 2 (SARS-CoV-2), the cause of the COVID-19 pandemic, is present in and poses human health risks from exposure to environmental fecal wastes (fecal sludge), wastewaters (sewage), biosolids and fecally contaminated environmental waters used for drinking, irrigation, recreation and other beneficial purposes. Based on an extensive review of the existing literature and my analysis and interpretation of it, no documented evidence exists for infectious SARS-CoV-2 presence in environmental fecal wastes and water or COVID-19 disease causation from these exposure and transmission sources. According to the World Health Organization, there are more than 415,709 scientific publications on SARS-CoV-2 and COVID-19 as of November 26, 2021 (WHO 2021a). Of those many published papers, only two reported efforts to detect infectious replication-capable SARS-CoV-2 in wastewater and water, and such viruses were not found (Rimoldi *et al.* 2020; Albert *et al.* 2021). It is likely that other laboratories, including some known to this author, have been and are continuing to examine wastewater and related fecal wastes for infectious, replication-capable SARS-CoV-2, but none so far have reported positive detection. Furthermore, there are no epidemiologically rigorous studies that attribute COVID-19 infection, illness or death to such environmental fecal waste or water exposure sources. Such key findings are presented in this report.

SARS-COV-2 VIRUS AND ITS EXPOSURE SOURCES: EXPOSURE ASSESSMENT

SARS-CoV-2 is a member of the *Coronaviridae* family and is genetically related to other coronaviruses causing respiratory illnesses in humans and as well as other animal hosts ranging from bats to mammals and birds (Ghai *et al.* 2021). SARS-CoV-2 causes the respiratory disease called COVID-19 (an acronym standing for **co**rona**vi**rus **d**isease of 20**19**). The virus is similar to other coronaviruses causing human infection and illness, including the 2002–2003 severe acute respiratory syndrome (SARS) virus, the Middle East Respiratory Virus (MERS), first recognized in 2012, and several coronaviruses causing upper respiratory illnesses referred to as the common cold and first recognized in the 1960s. Other coronaviruses that cause infection and sometimes illness in livestock, companion animals and laboratory animals were recognized as early as the 1930s. There are also coronaviruses of wild animals such as bats, birds and rodents, with bats being the main reservoir host and source of viruses infecting humans and other animals. SARS-CoV-2 was first detected in humans in Wuhan, China in late 2019, but the origin of the virus and how it first got to humans remain uncertain (WHO 2021b).

Like other coronaviruses, the SARS-CoV-2 virion is about 120 nm in diameter and has an outer lipoprotein envelope (Bar-on *et al.* 2020; Yang *et al.* 2020). It is comprised of an internal nucleic acid genome of single-stranded plus-sense RNA, which is associated with an internal nucleocapsid (N) protein covered by an outermost lipid bilayer (an envelope) containing several proteins or glycoproteins, namely (1) a membrane glycoprotein, (2) a small membrane glycoprotein and (3) a spike (S) glycoprotein that extends outward from the virus surface. The spike protein binds to host cells via the cell's angiotensin-converting enzyme 2 (ACE-2) receptor, resulting in the penetration of host cells. Virus replication then proceeds, with further spread by progressively more infection of cells in various host tissues and organs, primarily the lungs but other tissues and organs as well.

Despite the wealth of data collected and published since the beginning of the pandemic, with more than 415.709 reports by late November 2021, there is still no credible and robust scientific evidence for the presence of infectious and replication-capable SARS-CoV-2 in environmental sources of feces, wastewater, biosolids, fecal sludge or fecally contaminated waters. This assessment is supported by communications from other investigators (Jones *et al.* 2020; Haas *et al.* 2021).

For the known and better-documented routes of SARS-CoV-2 exposure and transmission by direct physical contact, respiratory droplets and aerosols, fomites, urine and possibly ocular (eye) secretions, the quality of available virological and epidemiological evidence is still considered weak (Aiello *et al.* 2020; Brönimann *et al.* 2020; Heneghan *et al.* 2021a, 2021b; Onakpoya *et al.* 2021). Nevertheless, it is clear from available virological and epidemiological evidence that sources of SARS-CoV-2 are primarily from direct personal contact or exposure to virus-containing, environmental media of respiratory origin shed from people infected with COVID-19. These exposure media are (1) airborne droplets and perhaps aerosol particles, (2) fomites containing respiratory secretions from which viruses get transferred via touching (e.g., hand contact) or another direct contact to a susceptible person who then transfers them to their mouth, nose or eye to initiate infection and less frequently, (3) urine and (4) ocular secretions. The fomite exposure route has been considered less common than the airborne particle route (Goldman 2020; CDC 2021; Chen 2021; Lewis 2021; Mondelli *et al.* 2021; Onakpoya *et al.* 2021). However, recent evidence of relatively high concentrations of infectious SARS-CoV-2 virus in and on fomites in hospitalized patient's rooms, including surfaces, door handles and knobs, used nasal tissues and cell phones in patient rooms, suggests that fomite exposure risk may be more important than previously thought (Lin *et al.* 2021).

Recently, Jefferson *et al.* (2021) proposed a 4-tier hierarchical framework of evidence for any proposed exposure and transmission route of a respiratory virus and SARS-CoV-2 in particular. Their analysis is based on systematically reviewing and synthesizing 378 primary studies for an evidence-based update of SARS-CoV-2 modes of transmission. Their analysis revealed significant methodological shortcomings in nearly all reported studies, with a lack of standardization in design, conduct, testing and reporting of SARS-CoV-2 transmission. Further details are provided below.

Most of the evidence for virus presence in clinical and environmental samples is based on detection of viral RNA by reverse transcription (RT)-polymerase chain reaction (PCR) amplification methods (RT-PCR). Few studies report the presence and concentrations of infectious, replication-capable SARS-CoV-2 in such samples. This may be due to the need for such analysis to be done in biosafety level 3 laboratories having high-level containment and other protective measures to contain infectious agents like SARS-CoV-2 that are transmitted through the air to cause potentially lethal infections.

It is important to know that the presence and concentrations of infectious SARS-CoV-2 in human clinical specimens and environmental samples are not predictable by the presence and concentrations of the viral RNA. This is because the ratios of viral RNA concentrations to infectious virus concentrations are very low and highly variable, typically only 0–1 infectious unit per 1,000 to >100,000 RNA genome copies. The high variability in this ratio depends in part on the source and type of clinical or environmental sample, the stage of infection in the human host and the duration of time the sample has been in the environment. It is also important to know that the duration and magnitude of shedding of SARS-CoV-2 RNA by COVID-19-infected people is much longer and higher than that of infectious, replication-capable virus (Widders *et al.* 2020; van Kampen *et al.* 2021). Viral RNA shedding continues typically for days to weeks after apparent cessation (based on lack of detectability) of infectious virus shedding. Also, the longer the period of time the sample is in the environment relative to when it was shed by an infected host, the greater the opportunity for virus die-off or inactivation by environmental stressors.

Based on a systematic review of the literature, of 77 reported primary studies for evidence of infectious SARS-CoV-2 presence in fecal samples from COVID-19-infected humans, little evidence was found (Heneghan *et al.* 2021b). Only six studies attempted virus culture, with only one observing viral cytopathogenic effects (CPE) in cell culture and none reporting or attempting serial viral culture positivity as confirmatory evidence. The available evidence for infectious, replication-competent SARS-CoV-2 presence in feces does not meet the highest hierarchical level of evidence proposed by Jefferson *et al.* (2021) to document fecal presence as an important exposure source of SARS-CoV-2.

Infectious virus concentrations in the few reported virus-positive stool (fecal) samples in the literature have not been reported or quantified (Zapor 2020; Heneghan *et al.* 2021b; Jefferson *et al.* 2021). A systematic review by Heneghan *et al.* (2021b) found only 6 of 77 primary studies that attempted to culture SARS-CoV-2 in feces (Jeong *et al.* 2020; Kim *et al.*

2020; Wang *et al.* 2020; Wölfel *et al.* 2020; Xiao *et al.* 2020; Zhang *et al.* 2020a, 2020b). Of these studies, only three reported evidence of SARS-CoV-2 by cell culture infectivity in some but not all fecal samples tested; the other three studies detected no infectious virus by cell culture. Only one study observed viral CPE in cell culture, and none reported or attempted serial viral culture positivity as confirmatory evidence.

To further elaborate on these reviewed primary studies, Wang *et al.* (2020) reported fecal sample positivity for SARS-CoV-2 infectivity in cell culture in two out of four samples, with visual confirmation only by electron microscopy. In the report by Xiao *et al.* (2020), the stool of a single COVID-19 patient was cell culture-positive, with virus presence further supported by visible cytopathic effects after second-round cell culture passage. Cell culture supernatant contained coronavirus-like particles by electron microscopy as well as full-length SARS-CoV-2 RNA detected by sequencing. In the same study, fecal samples from two additional patients were also positive for infectious virus by cell culture analysis. Zhang *et al.* (2020a) reported that stools of only one patient from among an unreported number of patients was positive for infectious virus. The presence of SARS CoV-2 was further confirmed by electron microscopic observation of coronavirus particles and by full length genome sequencing of viral RNA. Although Jeong *et al.* (2020) were unable to directly infect cell cultures with a fecal sample, they successfully infected ferrets by nasal inoculation, which resulted in signs and symptoms of illness, as well as evidence of virus infection in nasal washes collected 2–6 days post-infection based on increasing viral RNA concentrations and then successful cell culture detection of SARS-CoV-2 from these nasal washes.

The rarity of detecting infectious, replication-capable SARS-CoV-2 despite high levels of viral RNA in shed feces has a plausible explanation, which is the rapid inactivation of virus infectivity in the colon (lower intestines). Zang *et al.* (2020) observed SARS-CoV-2 replication in the small intestine of humans. However, they also reported using a recombinant SARS-CoV-2 mNeonGreen reporter virus, to document experimentally that most of the viruses (99%) are inactivated in 1 h by simulated colonic fluid. This evidence helps to explain the low frequency of infectious, replication-capable SARS-CoV-2 detection in fecal samples.

Other studies for the presence of SARS-COV-2 in feces of COVID-19-infected people also lack evidence for infectious, replication-capable virus. Two such studies report histological and immunofluorescence evidence of SARS-CoV-2 virus antigen presence in biopsied intestinal tissues. However, no confirmation of the presence of infectious, replication-competent virus in cell culture was done (Qian *et al.* 2020; Xiao *et al.* 2020). Xiao *et al.* (2020) reported evidence of SARS-CoV-2 replication in intestinal tissue only by histologic and immunofluorescent staining, using SARS-CoV-2 nucleoprotein antibodies for detection by microscopic analyses. Qian *et al.* (2020) reported evidence of SARS-CoV-2 in rectal tissues by positivity for SARS-CoV-2 RNA using RT-PCR. They also reported the apparent observation of coronavirus particles in rectal tissues by electron microscopy and the presence of SARS-CoV-2 nucleoprotein antigen in rectal tissue specimens using immunofluorescence microscopy.

Despite the rare and unquantified presence of infectious, replication-capable SARS-CoV-2 in fecal wastes of people infected with COVID-19, the possibility of such virus presence has generated interest and concern by some for its presence in environmental fecal wastes, such as wastewater, biosolids, fecal sludge, and environmental waters and wastes used beneficially by humans. The focus has been on possible virus presence in drinking, recreational and irrigation waters and in wastewater and fecal solids used in food crop agriculture.

Interest in the potential presence of infectious SAR-CoV-2 in environmental waters and fecal wastes has been motivated by the documented presence of SARS-CoV-2 RNA in feces, sewage, biosolids, latrine samples and fecally contaminated waters. This is because COVID-19 infections in community members result in viral RNA presence in shed bodily wastes, such as saliva, sputum, feces, urine and ocular discharges of the population that occur in community wastewaters and fecal sludges. This finding has led to several reports speculating on fecal-associated and water-related environmental exposures to infectious, replication-capable SARS-CoV-2 (Kang *et al.* 2020; Kitajima *et al.* 2020; La Rosa *et al.* 2020; Abdelodun *et al.* 2021; Del Brutto *et al.* 2021; Giacobbo *et al.* 2021; Gwenzi 2021; Mohan *et al.* 2021; Thakur *et al.* 2021; Tran *et al.* 2020). However, no infectious, replication-capable SARS-CoV-2 has yet to be reported, quantified and further confirmed from such environmental exposure sources.

Although rare and unlikely, it is important to recognize that if infectious, replication SARS-COV-2 was shed fecally or was otherwise present in shed human wastes, it would be expected to decline relatively rapidly and more rapidly than viral RNA levels in the environment (Bivins *et al.* 2020b) This is because wastewater, biosolids and fecal sludge treatment processes, and various environmental antagonists, including elevated temperatures, sunlight, chemical antagonists such as surfactants, metals, enzymes (proteases, nucleases amylases and lipases) and disinfectants, as well as microbial predation, will all contribute to relatively rapid loss of infectivity of infectious SARS-COV-2 virions released into the aquatic environment (Bosch *et al.* 2006; Pinon & Vialette 2018; Paul *et al.* 2021). Indeed, enveloped viruses such as SARS-CoV-2 are considered the least resistant of all microbes, compared to non-enveloped viruses, vegetative bacteria, protozoan parasites, mycotic agents and bacteria spores. The infectivity of these viruses would decline relatively rapidly (Brisolara *et al.* 2021; Maal-Bared *et al.* 2021). Antiviral environmental stressors are a likely contributing reason why all reported efforts to detect and quantify infectious, replication-capable SARS-CoV-2 in fecal wastes and water have been unsuccessful (Huraimel *et al.* 2020; Rimoldi *et al.* 2020; Tran *et al.* 2020; Gwenzi 2021; Kumar *et al.* 2021; Westhaus *et al.* 2021). Because such evidence of virus inactivation has been well documented in previous reviews, it will not be covered in further detail here (Maal-Barad *et al.* 2021; Brisolara *et al.* 2021; Nassri *et al.* 2021; Chen *et al.* 2021a).

COVID-19 INFECTION, DISEASE AND HEALTH RISKS: HEALTH EFFECTS ASSESSMENT

COVID-19 infection and disease occur mainly in the respiratory tract, with abundant virus presence in cells and tissues of the lungs, trachea, nose, throat and mouth, often resulting in cellular and tissue damage and inflammation. The disease occurs in both the upper and lower respiratory tract, with pneumonia in the lower respiratory tract as the more severe health effect. The adverse effects of pneumonia are greater in more susceptible hosts, such as immune-deficient people and those with other existing health conditions (e.g., obesity, heart disease and chronic lung disease) (Bar-on *et al.* 2020; Yang *et al.* 2020). The virus can also be present in other organs and tissues, including the intestinal tract (the gut), liver, gallbladder and pancreas, the eye, the genito-urinary system and skin. The respiratory tract is the main site of infection and illness and the major source of virus release in saliva and mucus that becomes airborne as droplets and smaller particles (aerosols). Such released virus then results in virus exposures to others to cause further transmission by infection of other susceptible human hosts.

COVID-19 infection can produce human disease of differing severity (Rod *et al.* 2020). It is also noteworthy that some COVID-19-infected people, between 4 and 94% depending on geographic location and population demographics (e.g., age and preexisting health conditions), remain asymptomatic but still shed the virus and its RNA as a source of virus exposure to others (Heneghan *et al.* 2020; Lauer *et al.* 2020; Lee *et al.* 2020; WHO 2020a; Gao *et al.* 2021; Tan *et al.* 2021; Wilmes *et al.* 2021). For asymptomatic COVID-19 patients that are positive for SARS-CoV-2 by RNA analysis, the time period to when some developed symptoms of illness averaged 15 days in one reported study (Lee *et al.* 2020). Other studies report typical presymptomatic durations of only few days, but up to 14 days in some cases (Lauer *et al.* 2020; WHO 2020a; Tan *et al.* 2021). In the study by Lee *et al.* 2020, both symptomatic and asymptomatic patients had comparable RNA levels in respiratory specimens. Based on viral RNA, viral loads of asymptomatic patients decreased more slowly than those of symptomatic (including presymptomatic) patients. In the same study, viral RNA Ct values from upper and lower respiratory tract specimens of asymptomatic and symptomatic (including presymptomatic) patients were not different. However, some recent studies indicate lower risks of transmission by asymptomatic than by symptomatic patients (Miller 2021; Widders *et al.* 2020). Yet other evidence suggests that asymptomatic cases pose high risks of transmission because both they and others who are susceptible to infection are unaware that they are contagious, thus unwittingly resulting in more risky interactions between them. Overall, the extent to which symptomatic and asymptomatic people with COVID-19 contribute to further transmission to other people still remains uncertain.

COVID-19 disease can range from mild to moderate illness resembling seasonal influenza and other common respiratory illnesses (a 'cold'). However, when the disease progresses to more serious illness as severe pneumonia, it requires urgent medical care and often can be life-threatening. Those with pneumonia have difficulty breathing, chest pain, shortness of breath, hemodynamic instability and sometimes loss of taste or smell, nausea or vomiting and diarrhea; they typically require oxygenation or mechanical ventilation therapy to survive.

Virus and viral nucleic acid (RNA) shed in respiratory releases of COVID-19-infected people by actions, such as coughing, sneezing, spitting, talking and singing, begins a few days (typically 2–5 days) after initial infection. Shed viral RNA can be detected for up to 2 months post-infection and sometimes even longer in some high-risk COVID-19 hosts (Zapor 2020). However, infectious virus is detectable in respiratory secretions only for up to 1.5–2 weeks post-infection (van Kampen *et al.* 2021). As was previously noted above, viruses released from the respiratory tract can also be deposited on surfaces or other fomites, such as tissues, tables, doorknobs and handles, cell phones and outer garments. These deposited viruses can potentially be transferred to others who touch or otherwise come in contact with such materials and surfaces and then transfer the viruses to their mouth, nose or eyes to initiate infection at these sites.

COVID-19 infections in many (from 30 to 60%) but not all infected people also occur extensively in the gastrointestinal tract, sometimes producing diarrhea and related enteric symptoms. The duration and magnitude of virus RNA shedding from the gut and presence in feces is often somewhat longer than in respiratory secretions. Only rarely (in about 5% of infections) is the virus and its RNA found in urine, as another but less common source of virus shedding (Jones *et al.* 2020; Zapor 2020). The estimated concentrations of viral RNA as gene copies (GC) per ml are about 10^5–10^{11} in respiratory fluids, 10^2–10^7 in feces and 10^2–10^5 in urine. These data further document respiratory secretions as the major transmission source of virus shed from infected people (Jones *et al.* 2020; Zapor 2020); Chen *et al.* 2021a.

SARS-CoV-2 TRANSMISSION, EPIDEMIOLOGY AND HUMAN INFECTIVITY FOR RISK CHARACTERIZATION

Based on the extensive clinical and epidemiological evidence gathered so far, COVID-19 transmission can be described by the established process known as the chain of infection and its six elements (Ahmad *et al.* 2020). The first element is the infectious agent, SARS-COV-2, which causes the disease COVID-19. The second element is the reservoir or the place where the infectious agent, SARS-CoV-2, is present, which is primarily COVID-19-infected human hosts. The third element is how the agent exits the infected COVID-19 host, which occurs primarily through the mouth and nose when a person coughs or sneezes to release SARS-CoV-2 in respiratory secretions. The fourth element is the mode of transmission, or how SARS-CoV-2 gets from one person to another. For those with COVID-19 infection or disease, SARS-CoV-2 release is as respiratory secretions from direct contact (e.g., kissing), or as airborne particles (coughs, sneezes, etc.), or by indirect contact such as deposition on fomites that another person can touch to acquire the virus. The fifth element in the chain of COVID-19 infection is the portal of entry, or where SARS-CoV-2 enters another person's body. This is essentially the same as the portal of exit. Virus entry into susceptible human hosts is via the mouth, nose or eye and is the sixth element in the chain. The clinical and epidemiological evidence in support of this infection chain for COVID-19 transmission is well documented for the respiratory transmission scenario. However, there is no conclusive virological or epidemiological evidence documenting other transmission routes or pathways for SARS-CoV-2 virus transmission that result in COVID-19 infection or disease, such as from environmental fecal wasters and waters.

Like other viruses as well as other pathogens, the ability of a person infected with SARS-CoV-2 to transmit the virus to others is quantified as the number of people they successfully infect with COVID-19. This is referred to as its reproduction number or R_0 value. The estimated R_0 for COVID-19 ranges from 1.9 to 6.5 based on 20 reported studies, with 13 of these studies in the range of 2–3 (Spencer *et al.* 2020). This R_0 value is typical of many respiratory viruses, except for the measles virus, for which the R_0 is estimated to be >10. It is noteworthy that the R_0 value of SARS-CoV-2 differs among the several known variants of the virus. The recent so-called Delta variant, which has become prevalent globally, has a higher R_0 value and observed greater transmissibility than earlier variants (Liu & Rocklöv 2021).

The transmission of COVID-19 through direct human-to-human contact is most commonly reported among health-care employees and primary caregivers of diseased patients who must come in direct contact with them. However, the human transmission of COVID-19 is also considered to occur from sources such as respiratory secretions released as airborne droplets and smaller airborne particles and possibly from contact with fomites that results in observed clusters or outbreaks of COVID-19 among people in other settings. These settings include households, nursing homes, colleges and schools, crowded bars and restaurants, and other commercial and public settings such as food production facilities and retail commercial enterprises. Such indoor locations can have large numbers of people in poorly ventilated and often crowded spaces (Chen *et al.* 2021b).

To reduce COVID-19 transmission risks, public health agencies and other concerned stakeholders invoke and promote public health and social measures of prevention and control. These measures include: (1) keeping physical distance from other people (usually 1–2 m), (2) encouraging frequent hand hygiene, (3) rapidly identifying people with COVID-19 and encouraging or requiring their isolation, (4) encouraging or requiring timely quarantining (confinement) of people who have been in recent contact with others who have COVID-19, infection or illness, (5) encouraging or requiring the wearing of face masks or respirators by both those with COVID-19 infection or illness (source control) and others who may become exposed to them (susceptibles) and (6) disinfecting various surfaces and objects to inactivate any SARS-CoV-2 on them (Ahmad *et al.* 2020). In some communities and countries, preventing or reducing COVID-19 transmission is further addressed by encouraging or requiring people to stay home and not venture out, except for essential reasons (e.g., buying groceries); this is referred to as ‘lockdown’. Some countries restrict the entrance of people from other countries having high levels of COVID-19 disease.

COVID-19 transmission by environmental fecal wastes and waters has been raised as a possibility, and there is no credible virological or epidemiological evidence documenting this transmission source or chain of infection. The World Health Organization and other health agencies such as the US Centers for Disease Control and Prevention (CDC) have developed formal and rigorous categorical criteria to determine if an exposure medium such as water or wastewater is a plausible and documentable cause of infectious disease transmission by a pathogen considered wastewater- and water-borne (Tillett *et al.* 1998; Eisenberg *et al.* 2001; Bartram & Hunter 2015; CDC 2019). The quality or strength of evidence is based on documenting the presence of the pathogen in the exposure medium at the time of human exposure using culture-based and molecular methods and then temporally linking the estimated exposure to ill people who ingested such contaminated water or fecal wastes. The volume of water ingested is also a key criterion to establish and quantify risk, because the more water consumed the higher the potential risk of becoming infected and ill (i.e., a dose–response relationship).

While there are no virological or epidemiological data documenting infectious, replication-capable SARS-CoV-2 presence or epidemiologically documented COVID-19 infection and disease attributable water- and fecal waste-related exposures and transmission routes, there is one reported COVID-19 disease incident tentatively and only weakly linked to fecal matter. A cluster of nine COVID-19 cases among three families was speculated, based on only circumstantial evidence, to have been caused by airborne exposure to fecal droplets from plumbing in a high-rise building in Guangzhou, China (Kang *et al.* 2020). However, no infectious, replication-capable virus was found in this fecal waste or in environmental fomite samples, although some environmental samples in bathrooms were positive for viral RNA. Furthermore, other possible routes of exposure for transmission such as fomites could not be ruled out. This report is similar to an earlier outbreak of another coronavirus disease, SARS, in Hong Kong in 2004. That outbreak was tentatively attributed to entrained fecal droplets as the airborne exposure medium for SARS disease transmission in a high-rise apartment complex (Yu *et al.* 2004). However, the epidemiological evidence was only circumstantial, and no infectious, replication-capable virus or viral RNA was looked for in the suspected human fecal waste source (faulty toilets) or in other environmental samples.

The extent to which infectious, replication-capable SARS-CoV-2 would pose a risk of human infection from environmental exposures, such as by ingestion of fecally contaminated drinking or recreational water or wastewater-irrigated produce, is also dependent on the magnitude and duration of exposure. That is, an environmental exposure would have to be high enough and long enough for there to be a high probability of resulting in infection in a susceptible human host. A sufficient number or quantity of infectious, replication-capable SARS-CoV-2 would need to enter a target site in the body, such as the mouth, nose, throat or intestinal tract, to cause infection. It is also noteworthy that, to this author's knowledge, no documented epidemiological evidence of infection, illness or death has been reported for people who are occupationally exposed regularly to environmental fecal wastes, such as those in contact with wastewater and biosolids at wastewater treatment and management facilities.

The presence of readily detectable SARS-CoV-2 RNA in wastewater by RT-PCR has led to the development and implementation of monitoring and surveillance efforts for COVID-19 based on SARS-CoV-2 RNA detection in wastewater samples. Such surveillance is becoming increasingly used as an alert and tracking system for the presence of COVID-19 infections in communities and more specific settings, such as college campuses, hospitals and other workplaces (Bivins *et al.* 2020a; Hill *et al.* 2021; Lundy *et al.* 2021; Mackul'ak *et al.* 2021; Panchal *et al.* 2021). Such wastewater-based epidemiology (WBE) has become a complementary approach to inform public health systems, programs and resources of SARS-CoV-2 presence. This wastewater-based epidemiological approach complements conventional public health surveillance based on identifying COVID-19-infected people by testing individual community members for evidence of infection. Such community-based testing typically uses RT-PCR analysis of nasopharyngeal clinical specimens for the presence of SARS-CoV-2 RNA genome targets to identify COVID-19-infected individuals. In some countries, regions and communities, WBE has become a supporting approach to (1) monitor and track the magnitude of presence of COVID-19 infections in communities and other specific locations, (2) identify ‘hotspots’ of high infection prevalence in geographic areas and their populations that require more effective COVID-19 vaccination coverage and (3) confirm virus absence, presence and magnitude of presence in specific locations (Bivins *et al.* 2020a; Fuschi *et al.* 2021; Smith *et al.* 2021).

It is also noteworthy that reported concentrations of SARS-CoV-2 RNA in wastewater samples are typically relatively low based on estimated genome copies or CT values, compared to their concentrations in saliva, sputum and fresh feces of infected humans. Hence, the typical concentrations of SARS-CoV-2 RNA in wastewaters and fecally contaminated waters are usually much too low and variable to be indicative of the presence and concentrations of infectious and replication-capable SARS-CoV-2.

An important data gap in estimating the potential transmission risks of COVID-19 infection, illness and death from exposure to environmental fecal, wastewater and water contamination is the absence of a documented human infectivity dose–response relationship for infectious, replication-capable SARS-CoV-2. There are no reported studies in which human volunteers were dosed with different known quantities of infectious SARS-CoV-2 by a specific exposure route to develop dose–response relationships (Karimzadeh *et al.* 2021). In addition, there are no documented epidemiological data for COVID-19 infection, illness or death from human exposure to environmental fecal wastes or waters in which infectious, replication-capable was detected and quantified.

There are limited and poorly quantified data on the infectivity of SARS-CoV-2 in experimental animal models that are susceptible to infection, such as ferrets, Syrian hamsters, minks and genetically modified mice (Karimzadeh *et al.* 2021). However, these few animal studies rarely used a range of infectious SARS-CoV-2-graded doses, most doses administered were very high and an insufficient number of replicate animals were challenged per dose to develop reliable quantitative dose–response relationships. The one exception is the study by Rosenke *et al.* (2020) who challenged Syrian hamsters with infectious SARS-CoV-2 at graded doses from 1 to 100,000 TCID-50. They estimated the 50% infectious dose of SARS-CoV-2 to be about five infectious virions. However, the infectivity dose–response data were not presented in the published paper, and therefore, it is impossible to know how this dose–response relationship result was determined and if it is accurate. Furthermore, whether such animal infectivity dose–response data can be assumed to represent human infectivity dose–response relationships is unknown, highly uncertain and cannot be verified directly until human infectivity dose–response studies are done. It is noteworthy that a human volunteer dose–response study was approved and recently began in the United Kingdom (Akst 2021).

Despite the absence of quantitative human infectivity dose–response data for SARS-CoV-2, there have been unverifiable assumptions about what the infectivity dose–response relationship might be (Karimzadeh *et al.* 2021). Such attempts have been based on the use of human infectivity dose–response data for other viruses in quantitative microbial risk assessments (QMRA) (Zaneti *et al.* 2020; Karimzadeh *et al.* 2021). Such an analysis was done to quantify the risks from a sewage drainage system in multi-unit apartment buildings as a transmission route of SARS-CoV-2 (Shi *et al.* 2020). Such estimated COVID-19 infection risks not based on human infectivity dose–response data for infectious SARS-CoV-2 have been based on either (1) concentrations of viral RNA and unverifiable assumptions of the ratio RNA GC to infectious viruses (Kumar *et al.* 2021) or (2) they have been based on the use of human infectivity dose–response relationships for other viruses such as SARS (Zaneti *et al.* 2020). Therefore, the validity of these QMRAs as being representative of COVID-19 infection risks from infectious SARS-CoV-2 is uncertain and highly questionable (Haas *et al.* 2021).

In summary, there is an absence of documented evidence for the presence and concentrations of infectious, replication-capable SARS-CoV-2 in environmental fecal wastes and waters. There is also an absence of documented epidemiological evidence that environmental fecal wastes and waters have ever caused human cases, clusters or outbreaks of COVID-19 infection, illness or death. For these reasons, there is no basis for establishing a chain of infection or providing evidence of human health risk for COVID-19 transmission from environmental fecal waste and water exposures. To address this absence of evidence, greater and better-coordinated efforts are recommended to (1) determine if infectious, replication-capable SARS-CoV-2 is present in environmental fecal wastes and waters and at what concentrations using state-of-the-science detection methods and (2) improve and increase rigorous epidemiological efforts to determine if COVID-19 infection, illness and death can be attributed to such environmental fecal waste and water exposures.

CONCLUSIONS

Based on the information reviewed and analyzed in this study as well as reviews and analyses by expert authorities such as the World Health Organization, the US Centers for Disease Control and Prevention, Water Environment Federation and their many expert scientific and technical staff and advisers, as well as recent reviews by others, there is no credible evidence for the presence of infectious, replication-capable SARS-COV-2 in fecal wastes and waters such as those used for recreation, agricultural irrigation or drinking. Furthermore, there is no documented epidemiological evidence of human infections, clusters of cases or outbreaks due to such environmental exposure sources of fecal wastes and waters (CDC 2020; WHO 2020b; WEF 2021). These conclusions are also supported by other investigators, such as Jones *et al.* (2020), who stated that '… the likelihood of infection due to contact with sewage-contaminated water (e.g., swimming, surfing, angling) or food (e.g., salads, shellfish) is extremely low or negligible based on very low predicted abundances and limited environmental survival of SARS-CoV-2', and that '... exposure to feces or wastewater has never been implicated as a transmission vector'. Overall, it is concluded that there is no documented evidence for the presence of infectious, replication-capable SARS-CoV-2 in fecally related environmental media or of epidemiologically documented COVID-19 human health risks from environmental water and waste exposures or that COVID-19 disease has resulted from such exposure.

Greater and better-coordinated investigation is recommended to determine if infectious, replication-capable SARS-CoV-2 is present in environmental fecal wastes and waters by the use of state-of-the-science methods for recovery, detection and quantitative analysis and to determine if such presence results in epidemiologically documented COVID-19 infection, illness or death.

DATA AVAILABILITY STATEMENT

All relevant data are available from an online repository or repositories.

REFERENCES

Abdelodun, B., Ajibade, F. O., Tiamiyu, A. G. O., Nwogwu, N. A., Ibrahim, R. G., Kumar, P., Kumar, V., Odey, G., Yadav, K. K., Khan, A. H., Cabral-Pinto, M. M. S., Kareem, K. Y., Bakare, H. O., Ajibade, T. F., Naveed, Q. N., Islam, S., Fadare, O. O. & Choia, K. S. 2021 Monitoring the presence and persistence of SARS-CoV-2 in water-food-environmental compartments: state of the knowledge and research needs. *Environmental Research* 111373. doi:10.1016/j.envres.2021.111373.

Ahmad, S., Usman, M. R. M., Baviskar, K. D. & Patil, T. P. 2020 Break the chain of coronavirus disease (Covid-19) infection: a review. *International Journal of Pharmaceutical Sciences Review and Research* **64** (2), 187–191. doi:10.47583/ijpsrr.2020.v64i02.030.

Aiello, F., Afflitto, G. G., Mancino, R., Cesareo, M., Giannini, C. & Nucci, C. 2020 Coronavirus disease 2019 (SARS-CoV-2) and colonization of ocular tissues and secretions: a systematic review. *Eye* **34**, 1206–1211. https://doi.org/10.1038/s41433-020-0926-9.

Akst, J. 2021 Q&A: human challenge studies of COVID-19 underway in UK. *The Scientist*, June 18, 2021. Available from: https://www.the-scientist.com/news-opinion/q-a-human-challenge-studies-of-covid-19-underway-in-uk-68908.

Albert, S., Ruíz, A., Pemán, J., Sakavert, M. & Domingo-Calap, P. 2021 Lack of evidence for infectious SARS-CoV-2 in feces and sewage. *European Journal of Clinical Microbiology and Infectious Disease*. https://doi.org/10.1007/s10096-021-04304-4

Bar-on, Y. M., Flamholz, A., Phillips, R. & Milo, R. 2020 SARS-CoV-2 (COVID-19) by the numbers. *eLife* **9**, e57309. 15 pp. https://doi.org/10.7554/eLife.57309.

Bartram, J. & Hunter, P. 2015 Bradley classification of disease transmission routes for water-related hazards. In: *Routledge Handbook of Water and Health* (Bartram, J., Baum, R., Coclanis, P. A., Gute, D. M., Kay, D., McFadyen, S., Pond, K., Robertson, W. & Rouse, M. J., eds). Routledge, London and New York. ISBN: 9781138910072. https://doi.org/10.4324/9781315693606.ch03.

Bivins, A., North, D., Ahmad, A., Ahmed, W., Alm, E., Been, F., Bhattacharya, P., Bijlsma, L., Boehm, A. B., Brown, J., Buttiglieri, G., Calabro, V., Carducci, A., Castiglioni, S., Gurol, C. Z., Chakraborty, S., Costa, F., Curcio, S., de los Reyes III, F. L., Vela, J. D., Farkas, K., Fernandez-Casi, X., Gerba, C., Gerrity, D., Girones, R., Gonzalez, R., Haramoto, E., Harris, A., Holden, P. A., Islam Md, I., Jones, D. T., Kasprzyk-Hordern B, L., Kitajima, M., Kotlarz, N., Kumar, M., Kuroda, K., La Rosa, G., Malpei, F., Mautus, M., McLellan, S. L., Medema, G., Meschke, J. S., Mueller, J., Newton, R. J., Nilsson, D., Noble, R. T., van Nuijs, A., Peccia, J., Perkins, T. A. & Pickering, A. J. 2020a Wastewater-based epidemiology: global collaborative to maximize contributions in the fight against COVID-19. *Environmental Science and Technology* **54**, 7754–7757. doi:10.1021/acs.est.0c02388..

Bivins, A., Greaves, J., Fischer, R., Yinda, K. C., Ahmed, W., Kitajima, M., Munster, V. J. & Bibby, K. 2020b Persistence of SARS-CoV-2 in water and wastewater. *Environmental Science & Technology Letters* **7** (12), 937–942. https://doi.org/10.1021/acs.estlett.0c00730.

Bosch, A., Pintó, R. M. & Abad, F. X. 2006 Survival and transport of enteric viruses in the environment. *Viruses in Foods* **2006**, 151–187. doi:10.1007/0-387-29251-9_6.

Brisolara, K. F., Maal-Bared, R., Sobsey, M. D., Reimers, R. S., Rubin, A., Bastian, R. K., Gerba, C., Smith, J. E., Bibby, K., Kester, G. & Brown, S. 2021 Assessing and managing SARS-CoV-2 occupational health risk to workers handling residuals and biosolids. *Science of the Total Environment*. **774**, 145732. https://doi.org/10.1016/j.scitotenv.2021.145732.

Brönimann, S., Rebhan, K., Lemberger, U., Misrai, V., Shariat, S. F. & Praderea, B. 2020 Secretion of severe acute respiratory syndrome coronavirus 2 in urine. *Current Opinion in Urology*. doi:10.1097/MOU.0000000000000808.

Centers for Disease Control and Prevention (CDC) 2019 *Strength-of-Evidence Classification for Waterborne Disease and Outbreaks*. CDC, Atlanta. Available from: https://www.cdc.gov/healthywater/surveillance/outbreak-classifications.html.

CDC 2020 *Information for Sanitation and Wastewater Workers on COVID-19*. Available from: https://www.cdc.gov/coronavirus/2019ncov/community/sanitation-wastewater-workers.html#print.

CDC 2021 *Science Brief: SARS-CoV-2 and Surface (Fomite) Transmission for Indoor Community Environments*. p. 5. Available from: https://www.cdc.gov/coronavirus/2019-ncov/more/science-and-research/surface-transmission.html#print.

Chen, T. 2021 *Fomites and the COVID-19 Pandemic: An Evidence Review on Its Role in Viral Transmission*. National Collaborating Centre for Environmental Health, Vancouver, BC. ISBN 978-1-988234-56 4. Available from: https://ncceh.ca/documents/evidence-review/fomites-and-covid-19-pandemic-evidence-review-its-role-viral-transmission.

Chen, C., Hayward, K., Khan, S. J., Örmeci, B., Pillay, S., Rose, J. B., Thanikal, J. V. & Zhang, T. 2021a Role of wastewater treatment in COVID-19 control. *Water Quality Research Journal* **56** (2), 68–82. https://doi.org/10.2166/wqrj.2020.025.

Chen, P. Z., Bobrovitz, N., Premji, Z., Koopmans, M., Fisman, D. N. & Gu, F. X. 2021b Heterogeneity in transmissibility and shedding SARS-CoV-2 via droplets and aerosols. *eLife* **10**, e65774. doi:10.7554/eLife.65774.

Del Brutto, O. H., Costa, A. F., Mera, R. M., Andrade-Molina, D., Recalde, B. Y., Garcia, H. H. & Femandez-Cadena, J. C. 2021 SARS-CoV-2 RNA in swabbed samples from latrines and flushing toilets: a case-control study in a rural Latin American setting. *American Journal of Tropical Medicine and Hygiene* **104** (3), 1045–1047. doi:10.4269/ajtmh20-1380.

Eisenberg, J. N. S., Bartram, J. & Hunter, P. A. 2001 A public health perspective for establishing water-related guidelines and standards. In: *World Health Organization (WHO). Water Quality: Guidelines, Standards a Health* (Fewtrell, L. & Bartram, J., eds). IWA Publishing, London, UK, chapter 11. ISBN: 1 900222 28 0.

Fuschi, C., Pu, H., Negri, M., Colwell, R. & Chen, J. 2021 Wastewater-based epidemiology for managing the COVID-19 pandemic. *ACS EST Water* **1** (6), 1352–1362. https://doi.org/10.1021/acsestwater.1c00050.

Gao, Z., Xu, Y., Sun, C., Wang, X., Guo, Y., Qiu, S. & Ma, K. 2021 A systematic review of asymptomatic infections with COVID-19. *Journal of Microbiology, Immunology and Infection* **54** (1), 12–16. https://doi.org/10.1016/j.jmii.2020.05.001.

Ghai, R. R., Carpenter, A., Liew, A. Y., Martin, K. B., Herring, M. K., Gerber, S. I., Hall, A. J., Sleeman, J. M., VonDobschuetz, S. & Barton Behravesh, C. 2021 Animal reservoirs and hosts for emerging alphacoronaviruses and betacoronaviruses. *Emerging Infectious Diseases* **27** (4), 1015–1022. doi:10.3201/eid2704.203945.

Giacobbo, A., Rodrigues, M. A. S., Ferreira, J. Z., Bernardes, A. M. & de Pinhob, M. N. 2021 A critical review on SARS-CoV-2 infectivity in water and wastewater. What do we know? *Science of the Total Environment* **774**, 145721. doi:10.1016/j.scitotenv.2021.145721.

Goldman, E. 2020 Exaggerated risk of transmission of COVID-19 by fomites. *The Lancet* **20**, 893. https://doi.org/10.1016/S1473-3099(20)30561-2.

Gwenzi, W. 2021 Leaving no stone unturned in light of the COVID-19 faecal-oral hypothesis? A water, sanitation and hygiene (WASH) perspective targeting low-income countries. *Science of the Total Environment* **753**, 141751. doi:10.1016/j.scitotenv.2020.141751.

Haas, C. N., Bivins, A., Ahmed, W., Hamilton, K. A. & Khan, S. J. 2021 Discussion on 'Potential discharge, attenuation and exposure risk of SARS-CoV-2 in natural water bodies receiving treated wastewater'. *npj Clean Water* **4** (32). https://doi.org/10.1038/s41545-021-00123-4

Heneghan, C., Brassey, J. & Jefferson, T. 2020 *COVID-19: What Proportion Are Asymptomatic?*. The Center for Evidence-Based Medicine, April 6, 2020. Available from: https://www.cebm.net/covid-19/covid-19-what-proportion-are-asymptomatic/.

Heneghan, C. J., Spencer, E. A., Brassey, J., Plüddemann, A., Onakpoya, I. J., Evans, D. H., Conly, J. M. & Jefferson, T. 2021a SARS-CoV-2 and the role of airborne transmission: a systematic review [version 2; peer review: 1 approved with reservations, 2 not approved]. *F1000Research* **10**, 232. https://doi.org/10.12688/f1000research.52091.2.

Heneghan, C. J., Spencer, E. A., Brassey, J., Plüddemann, A., Onajpoya, I. J., Evans, D. A., Conly, J. M. & Jefferson, T. 2021b SARS-CoV-2 and the role of orofecal transmission: a systematic review [version 1; peer review: 2 approved with reservations]. *F1000Research* **10**, 231. https://doi.org/10.12688/f1000research.51592.1.

Hill, K., Zamyadi, A., Deere, D., Vanrolleghem, P. A. & Crosbie, N. D. 2021 SARS-CoV-2 known and unknowns, implications for the water sector and wastewater-based epidemiology to support national responses worldwide: early review of global experiences with the COVID-19 pandemic. *Water Quality Research Journal* **56** (2), 57–67. https://doi.org/10.2166/wqrj.2020.100.

Huraimel, K. A., Alhosani, M., Kunhabdulla, S. & Stietiya, M. H. 2020 SARS-CoV-2 in the environment: modes of transmission, early detection and potential role of pollutions. *Science of the Total Environment* **744**, 140946. doi:10.1016/j.scitotenv.2020.140946.

Jefferson, T., Heneghan, C., Spencer, E. A., Brassey, J., Plüddemann, A., Onakpoya, I., Evans, D. & Conly, J. M. 2021 *A Hierarchical Framework for Assessing Transmission Causality of Respiratory Viruses*. Preprints (www.preprints.org). Not peer-reviewed. doi:10.20944/preprints202104.0633.v1.

Jeong, H. W., Kim, S. M., Kim, H. S., Kim, Y. I., Kim, J. H., Cho, J. Y., Kim, S. H., Kang, H., Kim, S. G., Park, S. J., Kim, E. H. & Choi, Y. K. 2020 Viable SARS-CoV-2 in various specimens from COVID-19 patients. *Clinical Microbiology & Infectious* **26** (11), 1520–1524. doi:10.1016/j.cmi.2020.07.020.

Jones, D. L., Baluja, M. Q., Graham, D. W., Corbishley, A., McDonald, J. E., Malham, S. K., Hillary, L. S., Connor, T. R., Gaze, W. H., Moura, I. B., Wilcox, M. H. & Farkasa, K. 2020 Shedding of SARS-CoV-2 in feces and urine and its potential role in person-to-person transmission and the environment-based spread of COVID-19. *Science of the Total Environment* **749**, 141364. doi:10.1016/j.scitotenv.2020.141364.

Kang, M., Wei, J., Yuan, J., Guo, J. & Zhang, Y. 2020 Probable evidence of fecal aerosol transmission of SARS-CoV-2 in a high-rise building. *Annals of Internal Medicine* **173** (12), 974–981. https://doi.org/10.7326/M20-0928.

Karimzadeh, S., Bhopal, R. & Nguyen, T. H. 2021 Review of infective dose, routes of transmission and outcome of COVID-19 caused by the SARS-COV-2: comparison with other respiratory viruses. *Epidemiology and Infection* **149** (e96), 1–8. https://doi.org/10.1017/S0950268821000790.

Kim, J.-M., Kim, H. M., Lee, E. J., Jo, H. J., Yoon, Y., Lee, N.-J., Son, J., Lee, Y.-J., Kim, M. S., Lee, Y.-P., Chae, S.-J., Park, K. R., Cho, S.-R., Par, S., Kim, S. J., Wang, E., Woo, S. H., Lim, A., Park, S.-J., Jang, J. H., Chung, Y.-S., Chin, B. S., Lee, J.-S., Lim, D., Han, M.-G. & Yooe, C. K. 2020

Detection and isolation of SARS-CoV-2 in serum, urine, and stool specimens of COVID-19 patients from the Republic of Korea. *Osong Public Health and Research Perspectives* **11** (3), 112–117. doi:10.24171/j.phrp.2020.11.3.02.

Kitajima, M., Ahmed, W., Bibby, K., Carducci, A., Gerba, C. P., Hamilton, K. A., Haramoto, E. & Rose, J. B. 2020 SARS-CoV-2 in wastewater: state of the knowledge and research needs. *Science of the Total Environment* **739**, 139076. doi:10.1016/j.scitotenv.2020.139076.

Kumar, M., Alamin, M. & Kuroda, K. 2021 Potential discharge, attenuation and exposure risk of SARS-CoV-2 in natural water bodies receiving treated wastewater. *npj Clean Water* **4**, 8. https://doi.org/10.1038/s41545-021-00098-2.

La Rosa, G., Bonadonna, L., Lucentini, L., Kenmoe, S. & Suffredini, E. 2020 Coronavirus in water environments: occurrence, persistence and concentration methods – a scoping review. *Water Research* **179**, 115899. doi:10.1016/j.watres.2020.115899.

Lauer, S. A., Grantz, K. H., Bi, Q., Jones, F. K., Zheng, Q., Meredith, H. R., Azman, A. S., Reich, N. G. & Lessler, J. 2020 The incubation period of coronavirus disease 2019 (COVID-19) from publicly reported confirmed cases: estimation and application. *Annals of Internal Medicine* **172** (9). https://doi.org/10.7326/M20-0504

Lee, S., Kim, T., Lee, E., Lee, C., Kim, H., Rhee, H., Park, S. Y., Son, H.-J., Yu, S., Park, J. W., Choo, E. J., Suyeon Park, S., Mark Loeb, M. & Kim, T. H. 2020 Clinical course and molecular viral shedding among asymptomatic and symptomatic patients with SARS-CoV-2 infection in a community treatment center in the Republic of Korea. *JAMA Internal Medicine* **180** (11), 1447–1452. doi:10.1001/jamainternmed.2020.3862.

Lewis, D. 2021 COVID-19 rarely spreads through surfaces. So why are we still deep cleaning?. *Nature* **590**, 26–28. https://doi.org/10.1038/d41586-021-00251-4.

Lin, Y.-C., Malott, R. J., Ward, l., Kiplagat, L., Pabbaraju, K., Gil, K., Berenger, B. M., Hu, J., Fonseca, K., Noyce, R., Louie, T., Evans, D. H. & Conly, J. M. 2021 *Detection and Quantification of Infectious Severe Acute Respiratory Coronavirus-2 in Diverse Clinical and Environmental Samples from Infected Patients: Evidence to Support Respiratory Droplet, and Direct and Indirect Contact as Significant Modes of Transmission*. https://doi.org/10.1101/2021.07.08.21259744.

Liu, Y. & Rocklöv, J. 2021 The reproductive number of the Delta variant of SARS-CoV-2 is far higher compared to the ancestral SARS-CoV-2 virus. *Journal of Travel Medicine* **28** (7), taab124. doi:10.1093/jtm/taab124.

Lundy, L., Fatta-Kassinos, D., Slobodnik, J., Karaolia, P., Lubos Cirka, l., Kreuzinger, N., Castiglioni, S., Bijlsma, L., Dulio, V., Deviller, G., Lai, F. Y., Alygizaki, N., Barneo, M., Baz-Lomba, J. A., Béen, F., Cíchová, M., Conde-Pérez, K., Covaci, A., Donner, E., Ficek, A., Hassard, F., Hedström, A., Hernandez, F., Jansk, V., Jellison, K., Hofman, J., Hill, K., Hong, P.-Y., Kasprzyk-Hordern, B., Kolarević, S., Krahule, J., Lambropoulou, D., de Llanos, R., Mackuľak, T., Martinez-García, L., Martíne, F., Medema, G., Micsinai, A., Myrmel, M., Nasser, M., Niederstätter, H., Nozal, L., Oberacher, H., Očenášková, V., Ogorzaly, L., Papadopoulos, D., Peinad, B., Pitkänen, T., Poz, M. & Soraya Rumbo-Fea, S. 2021 Making waves: collaboration in the time of SARS-CoV-2 – rapid development of an international co-operation and wastewater surveillance database to support public health decision-making. *Water Research* **199**, 117–167. doi:10.1016/j.watres.2021.117167.

Maal-Bared, R., Brisolara, K., Munakata, N., Bibby, K., Gerba, C., Sobsey, M., Schaefer, S., Swift, J., Gary, L., Sherchan, S., Babatola, A., Bastian, R., Olabode, L., Reimers, R. & Rubin, A. 2021 Implications of SARS-CoV-2 on current and future operation and management of wastewater systems. *Water Environment Research* **93** (4), 502–515. doi:10.1002/wer.1446.

Mackuľak, T., Gál, M., Špalková, V., Fehér, M., Briestenská, K., Mikušová, M., Tomčíková, K., Tamáš, M. & Škulcová, A. B. 2021 Wastewater-based epidemiology as an early warning system for the spreading of SARS-CoV-2 and its mutations in the population. *International Journal of Environmental Research and Public Health* **18** (11), 5629. doi:10.3390/ijerph18115629.

Miller, C. P. 2021 Do asymptomatic carriers of SARS-COV-2 transmit the virus? *The Lancet Regional Health Commentary* **4**, 100082. https://doi.org/10.1016/j.lanepe.2021.100082.

Mohan, S. V., Hemalatha, M., Kopperi, H., Ranjith, I. & Kumar, A. K. 2021 SARS-CoV-2 in environmental perspective: occurrence, persistence, surveillance, inactivation and challenges. *Chemical Engineering Journal* **405**, 126893. doi:10.1016/j.cej.2020.126893.

Mondelli, M. U., Colaneri, M., Seminari, E. M., Baldanti, F. & Bruno, R. 2021 Low risk of SARS-CoV-2 transmission by fomites in real-life conditions. *The Lancet Infectious Diseases* **21** (5), E112. https://doi.org/10.1016/S1473-3099(20)30678-2.

Nassri, I., Khazraji, M., Mouhir, I. & Fekhaoui, M. 2021 *International Journal of Development Research* **11** (08), 49175–49181. https://doi.org/10.37118/ijdr.22541.08.2021.

Onakpoya, I. J., Heneghan, C. J., Spencer, E. A., Brassey, J., Annette Plüddemann, A., David, H., Evans, D. H., John, M., Conly, J. M. & Jefferson, T. 2021 SARS-CoV-2 and the role of fomite transmission: a systematic review. *F1000Research* **10**, 233.

Panchal, D., Prakas, O., Bobde, P. & Pal, S. 2021 SARS-CoV-2: sewage surveillance as an early warning system and challenges in developing countries. *Environmental Science and Pollution Research International* **28** (18), 22221–22240. doi:10.1007/s11356-021-13170-8.

Paul, D., Kolar, P. & Hall, S. G. 2021 A review of the impact of environmental factors on the fate and transport of coronaviruses in aqueous environments. *npj Clean Water* **4**. Article number: 7. https://doi.org/10.1038/s41545-020-00096-w.

Pinon, A. & Vialette, M. 2018 Survival of viruses in water. *Intervirology* **2018** (61), 214–222. doi:10.1159/000484899.

Qian, Q., Fan, L., Liu, W., Li, J., Yue, J., Wang, M., Ke, X., Yin, Y., Chen, Q. & Jiang, C. 2020 Direct evidence of active SARS-CoV-2 replication in the intestine. *Clinical Infectious Diseases*. doi:10.1093/cid/ciaa925.

Rimoldi, S. G., Stefani, F., Gigantiello, A., Polesello, S., Comandatore, F., Mileto, D., Maresca, M., Longobardi, C., Mancon, A., Romeri, F., Pagani, C., Cappelli, F., Roscioli, C., Moja, L., Gismondo, M. R. & Salerno, F. 2020 Presence and infectivity of SARS-CoV-2 virus in wastewaters and rivers. *Science of the Total Environment* **744**, 140911. doi:10.1016/j.scitotenv.2020.140911.

Rod, J. E., Oviedo-Trespalacios, O. & Cortes-Ramirez, J. 2020 A brief-review of the risk factors for covid-19 severity. *Revista De Saúde Pública* **54**, 60. https://doi.org/10.11606/s1518-8787.2020054002481.

Rosenke, K., Meade-White, K., Letko, M., Clancy, C., Hansen, F., Liu, Y., Okumura, A., Tang-Huau, T.-L., Li, R., Saturday, G., Feldmann, F., Scott, D., Wang, Z., Munster, V., Jarvis, M. A. & Feldmann, H. 2020 Defining the Syrian hamster as a highly susceptible preclinical model for SARS-CoV-2 infection. *Emerging Microbes & Infections* **9** (1). https://doi.org/10.1080/22221751.2020.1858177.

Shi, K.-W., Huang, Y.-S., Quon, H., Ou-Yang, Z. L., Wang, C. & Jiang, S. C. 2020 Quantifying the risk of indoor drainage system in multi-unit apartment building as a transmission route of SARS-CoV-2. *Science of the Total Environment* **762**, 143056. doi:10.1016/j.scitotenv.2020.143056.

Spencer, J. A., Shutt, D. P., Moser, S. K., Clegg, H., Helen, J., Wearing, H. J., Mukundan, H. & Manore, C. A. 2020 Epidemiological parameter review and comparative dynamics of influenza, respiratory syncytial virus, rhinovirus, human coronavirus, and adenovirus. medRxiv preprint. 29 pp. https://doi.org/10.1101/2020.02.04.20020404.

Tan, J. K., Leong, D., Munusamy, H., Ariffin, N. H. Z., Kori, N., Hod, R. & Periyasamy, P. 2021 The prevalence and clinical significance of presymptomatic COVID-19 patients: how we can be one step ahead in mitigating a deadly pandemic. *BMC Infectious Diseases* **21**, 249. https://doi.org/10.1186/s12879-021-05849-7.

Thakur, A. K., Sathyamurthy, R., Velraj, R., Lynch, I., Saidur, R., Pandey, A., Sharshir, S. W., Kabeel, A. E., Hwang, J.-Y. & GaneshKumar, P. 2021 Secondary transmission of SARS-CoV-2 through wastewater: concerns and tactics for treatment to effectively control the pandemic. *Journal of Environmental Management* **290**, 112668. doi:10.1016/j.jenvman.2021.112668.

Tillett, H. E., de Louvois, J. & Wall, P. 1998 Surveillance of outbreaks of waterborne infectious disease: categorizing levels of evidence. *Epidemiology and Infection* **120**, 37–42.

Tran, H. N., Le, G. T., Nguyen, D. T., Juang, R.-S., Rinklebe, J., Bhatnagar, A., Lima, E. C., Iqbal, H. M. N., Sarmah, A. K. & Chaom, H.-P. 2020 SARS-CoV-2 coronavirus in water and wastewater: a critical review about presence and concern. *Environmental Research* **193**, 110265. doi:10.1016/j.envres.2020.110265..

van Kampen, J. J. A., van de Vijver, D. A. M. C., Fraaij, P. L. A., Haagmans, B. L., Lamers, M. M., Okba, N., van den Akker, J. P. C., Endeman, H., Gommers, D. A. M. P., Cornelissen, J. J., Hoek, R. A. S., van der Eerden, M. M., Hesselink, D. A., Metselaar, H. J., Verbon, A., de Steenwinkel, J. E. M., Aron, G. I., van Gorp, E. C. M., van Boheemen, S., Voermans, J. C., Boucher, C. A. B., Molenkamp, R., Koopmans, M. P. G., Geurtsvankessel, C. & van der Eijk, A. A. 2021 Duration and key determinants of infectious virus shedding in hospitalized patients with coronavirus disease-2019 (COVID-19). *Nature Communications* **12**. Article number: 267. https://doi.org/10.1038/s41467-020-20568-4.

Wang, W., Xu, Y., Gao, R., Lu, R., Han, K., Wu, G. & Tan, W. 2020 Detection of SARS-CoV-2 in different types of clinical specimens. *JAMA* **323** (18), 1843–1844. doi:10.1001/jama.2020.3786.

WEF 2021 *Current Priority: Coronavirus*. Available from: https://www.wef.org/coronavirus.

Westhaus, S., Weber, F.-A., Schiwy, S., Linnemann, V., Brinkmann, M., Widera, M., Greve, C., Janke, A., Hollert, H., Wintgens, T. & Cieseka, S. 2021 Detection of SARS-CoV-2 in raw and treated wastewater in Germany – suitability for COVID-19 surveillance and potential transmission risks. *Science of the Total Environment* **751**, 141750. doi:10.1016/j.scitotenv.2020.141750.

WHO 2020a *Coronavirus Disease 2019 (COVID-19)*. Situation Report – 73, 13 pp. Available from: https://www.who.int/docs/default-source/coronaviruse/situation-reports/20200402-sitrep-73-covid-19.pdf?sfvrsn=5ae25bc7_2#:~:text=The%20incubation%20period%20for%20COVID,occur%20before%20symptom%20onset.

WHO 2020b *WHO Water, Sanitation, Hygiene, and Waste Management for SARS-CoV-2, the Virus That Causes COVID-19*. Interim guidance. July 29, 2020. 15 pp. WHO reference number: WHO/2019-nCoV/IPC_WASH/2020.4. Available from: https://www.who.int/publications/i/item/WHO-2019-nCoV-IPC-WASH-2020.4.

WHO 2021a *COVID-19 Global Literature on Coronavirus Disease*. As of November 26, 2021. Available from: https://search.bvsalud.org/global-literature-on-novel-coronavirus-2019-ncov/.

WHO 2021b *WHO-Convened Global Study of the Origins of SARS-CoV-2 (including annexes)*. China Part. Joint WHO-China Study. January 14–February 10, 2021. Joint Report. 120 pp. Available from: https://www.who.int/health-topics/coronavirus/origins-of-the-virus.

Widders, A., Broom, A. & Broom, J. 2020 SARS-CoV-2: the viral shedding vs infectivity dilemma. *Infection, Disease & Health* **25** (3), 210–215. doi:10.1016/j.idh.2020.05.002.

Wilmes, P., Zimmer, J., Schulz, J., Glod, F., Veiber, L., Mombaerts, L., Rodriguez, B., Aalto, A., Pastore, J., Snoeck, C. J., Ollert, M., Fagherazzi, G., Mossong, J., Goncalves, J., Skupin, A. & Nehrbass, U. 2021 SARS-CoV-2 transmission risk from asymptomatic carriers: results from a mass screening programme in Luxembourg. *The Lancet Regional Health* **4**, 1000056. https://doi.org/10.1016/j.lanepe.2021.100056.

Wölfel, R., Corman, V. M., Guggemos, W., Seilmaier, M., Zange, S., Müller, M. A., Niemeyer, D., Jones, T. C., Vollmar, P., Rothe, C., Hoelscher, M., Bleicker, T., Brünink, S., Schneider, J., Ehmann, R., Zwirglmaier, K., Drosten, C. & Wendtner, C. 2020 Virological assessment of hospitalized patients with COVID-2019. *Nature* **581** (7809), 465–469. doi:10.1038/s41586-020-2196-x.

Xiao, F., Sun, J., Xu, Y., Li, F., Huang, X., Li, H., Zhao, J., Huang, J. & Zhao, J. 2020 Infectious SARS-CoV-2 in feces of patient with severe COVID-19. *Emerging Infectious Diseases* **26** (8), 1920–1922. Available from: www.cdc.gov/eid.

Yang, Y., Xiao, Z., Ye, X., He, X., Sun, B., Qin, Z., Yu, J., Yao, J., Wu, Q., Bao, Z. & Zhao, W. 2020 SARS-CoV-2: characteristics and current advances in research. *Virology Journal* **17**. Article number: 117. https://doi.org/10.1186/s12985-020-01369-z.

Yu, I. T. S., Li, Y., Wong, T. W., Tam, W., Chan, A. T., Lee, J. H. W., Leung, D. Y. C. & Ho, T. 2004 Evidence of airborne transmission of the severe acute respiratory syndrome virus. *New England Journal of Medicine* **350** (17), 1731–1739. doi:10.1056/NEJMoa032867.

Zaneti, R. N., Girardi, V., Spilki, F. R., Mena, K., Westphalen, A. P. C., da Costa Colares, E. R., Pozzebon, A. G. & Etchepare, R. G. 2020 Quantitative microbial risk assessment of SARS-CoV-2 for workers in wastewater treatment plants. *Science of the Total Environment* **754** (92021), 142163. https://doi.org/10.1016/j.scitotenv.2020.142163.

Zang, R., Maria Florencia Gomez Castro, M. F. G., McCune, B. T., Zeng, Q., Rothlauf, P. W., Sonnek, N. M., Liu, Z., Brulois, K. F., Wang Greenberg, H. B., Diamond, M. S., Ciorba, M. A., Whelan, S. P. J. & Ding, S. 2020 TMPRSS2 and TMPRSS4 promote SARS-CoV-2 infection of human small intestinal enterocytes. *Science Immunology* **5** (47), eabc3582. 18 pp. doi:10.1126/sciimmunol.abc3582.

Zapor, M. 2020 Persistent detection and infectious potential of SARS-CoV-2 virus in clinical specimens from COVID-19 patients. *Viruses* **12** (12), 1384. https://doi.org/10.3390/v12121384.

Zhang, W., Du, R. H., Li, B., Zheng, X. S., Yang, X. L., Hu, B., Wang, Y. Y., Xiao, G. F., Yan, B., Shi, Z. L. & Zhou, P. 2020a Molecular and serological investigation of 2019-nCoV infected patients: implication of multiple shedding routes. *Emerging Microbes and Infections* **9**, 386–389. doi:10.1080/22221751.2020.1729071.

Zhang, Y., Chen, C., Zhu, S., Shu, C., Wang, D., Song, J., Song, Y., Zhen, W., Feng, Z., Wu, G., Xu, J. & Xu, W. 2020b Isolation of 2019-nCoV from a stool specimen of a laboratory-confirmed case of the coronavirus disease 2019 (COVID-19). *China CDC Weekly* **2** (8), 123–124. doi:10.462 +34/ccdcw2020.033.

First received 28 July 2021; accepted in revised form 28 November 2021. Available online 10 December 2021

SARS-CoV-2 in Water doi: 10.2166/wh.2021.167

SARS-CoV-2 surrogate (*Phi6*) environmental persistence within free-living amoebae

Rafik Dey [a,*], Elena Dlusskaya [b] and Nicholas J. Ashbolt [c]

[a] School of Public Health, University of Alberta, Edmonton, AB, Canada
[b] Faculty of Agricultural, Life & Environmental Sciences, University of Alberta, Edmonton, AB, Canada
[c] Faculty of Science and Engineering, Southern Cross University, Gold Coast, QLD, Australia
*Corresponding author. E-mail: rafik@ualberta.ca

RD, 0000-0002-4532-1751; ED, 0000-0003-4027-9221; NJA, 0000-0002-3853-0096

ABSTRACT

The reported persistence of SARS-CoV-2 virions in aquatic environments highlights the need to better understand potential mechanisms that may prolong its dissemination. We evaluated the possibility that amoebae might serve as transport hosts by studying the interaction of the enveloped bacteriophage Phi6, as a potential surrogated along with one of the most common amoebae in engineered aquatic environments, *Vermamoeba vermiformis*. Using microscopy, imaging flow cytometry and bacteriophage cell culture, our results imply that the SARS-CoV-2 surrogate triggers amoebic mitochondria and induced apoptosis to promote viral persistence in trophozoites. Furthermore, virus-infected amoebae were still infectious after 2 months within FLA cysts. These results suggest that amoebae could contribute to the environmental persistence of SARS-CoV-2, including disinfection processes. In addition, amoebae could be a successful model system for understanding respiratory virus-eukaryotic biology at the cellular and molecular levels.

Key words: apoptosis, bacteriophage Phi6, environmental persistence, free-living amoebae, mitochondria, SARS-CoV-2, transmission

HIGHLIGHTS

- Amoebic mitochondria serve as a support organelle platform for the SARS-CoV-2 surrogate persistence.
- Long-term stability of RNA enveloped viruses in amoeba-forming cysts.
- Free-living amoebae as a potential vector and environmental reservoir for RNA enveloped viruses.

INTRODUCTION

Free-living amoebae (FLA) are ubiquitous in natural and engineered aquatic and moist environments (Jadin 1973; Rodríguez-Zaragoza 1994; Lau & Ashbolt 2009; Samba-Louaka *et al.* 2019), and known natural environmental reservoirs for a plethora of respiratory pathogens and a vector of human water-associated infections (Rowbotham 1986; Essig *et al.* 1997; la Scola *et al.* 2003; Lamoth & Greub 2010; Bousbia *et al.* 2013; Santos-Montañez *et al.* 2015; Taylor-Mulneix *et al.* 2017). Increasing evidence suggests that FLA are potential environmental reservoirs for human viruses (enteric and respiratory) (Danes & Cerva 1981; la Scola *et al.* 2005; Mattana *et al.* 2006; Scheid & Schwarzenberger 2012; Boratto *et al.* 2014; Hsueh & Gibson 2015; Atanasova *et al.* 2018; Folkins *et al.* 2020), but how the virus infect and survive in amoeba hosts is unclear. To date, there is also no epidemiologic evidence for waterborne transmission of COVID-19, unlike SARS-CoV-1 demonstrated via wastewater aerosols (Yu *et al.* 2014). However, we have recently reported that enveloped human respiratory syncytial virus (RSV) persists within various FLA genera and could be transmitted via extracellular amoebal-vesicles in the respiratory size range (Dey *et al.* 2021).

While the environmental persistence underlying severe acute respiratory syndrome coronavirus 2 (SARS-CoV-2-induced COVID-19) is subject to intense investigation (Chin *et al.* 2020; Fukuta *et al.* 2021; Sala-Comorera *et al.* 2021), any potential role played by FLA is unreported. Since viral RNA of SARS-CoV-2 have been detected in various waste/water systems (Ahmed *et al.* 2020; Bivins *et al.* 2020; Haramoto *et al.* 2020; la Rosa *et al.* 2020; Rimoldi *et al.* 2020; Wu *et al.* 2020a; de Oliveira *et al.* 2021), and the presence of biota increased the aquatic persistence of infectious SARS-CoV-2 (Sala-Comorera *et al.* 2021), we considered the possibility that amoebae might serve as natural hosts and increase its persistence for environmental dispersal.

We tested the amoeba-host hypothesis by studying the interaction of the enveloped bacteriophage Phi6 surrogate, a dsRNA phage of the *Cystoviridae* family that has been a popular model for studying enveloped RNA viruses including coronaviruses SARS-CoV-1 and SARS-CoV-2 (Casanova & Weaver 2015; Aquino De Carvalho *et al.* 2017; Prussin *et al.* 2018; Silverman & Boehm 2020; di Novo *et al.* 2021; Ma *et al.* 2021; Pino *et al.* 2021; Vatter *et al.* 2021); and its association with one of the most common amoeba in human aquatic environments *Vermamoeba vermiformis* (Jadin 1973; de Moura *et al.* 1985; Bradbury 2014; Delafont *et al.* 2018).

MATERIALS AND METHODS

Propagation and purification of bacteriophages

Bacteriophages Phi6 (HER-102) and its host strain *Pseudomonas syringae* HB1OY#3 (HER-1102) were purchased from the Félix d'Hérelle Reference Center for Bacterial Viruses (Université Laval, QC, Canada). Bacteriophage Phi6 was propagated in the host strain; the host culture was grown in BBL Trypticase Soy Broth (BD, Ref# 211768) at 26 °C and 250 rpm. The overnight stock culture was added (20% v/v) to fresh Tryptic Soy Broth that also contained 5 mM of each $CaCl_2$ and $MgCl_2$ and incubated for 2 h. Then, the exponentially growing host was infected with the Phi6 bacteriophage, and the incubation was continued overnight without shaking. Overnight lysates were centrifuged at 4,000 g for 30 min to precipitate host cell debris and supernatant was filtered through a 0.22 mm syringe filter (Merck Millipore, REF # SLGS033SS) into a sterile Amicon Ultra 100 K centrifugal filter device (Merck Millipore, REF # UFC910024). The Ultra 100 K device was centrifuged again at 4,000 g for 20 min to remove growth media. The bacteriophage-containing retentate that remained on the filter part of the device (about 250 μL) was treated with DNAse I (Roche Diagnostics, cat # 10104159001) to remove residual host DNA. Twenty-five microliters of 10× DNAse I buffer (100 mM Tris HCl, pH 7.5, 25 mM $MgCl_2$, and 5 mM $CaCl_2$ in MQ water) and 2 μL of 2.5 mg/mL DNAse I, dissolved in storage buffer (20% glycerol in 75 mM NaCl), were added and the retentate was incubated for 45 min at 37 °C. All chemicals were purchased from Sigma unless stated otherwise. After the incubation, bacteriophage was rinsed with 10 mL of 1× HyClone PBS (HyClone Laboratories, REF #SH30256.02) and resuspended in PBS to the original volume.

Amoebae growth conditions

The amoeba used in this study was *V. vermiformis* (ATCC 50237) isolated from a hospital cooling tower drain. Amoebae were grown in tissue culture flasks in SCGYEM (Serum-Casein-Glucose-Yeast-Extract-Medium: ATCC medium 1021) at 25 °C in a 5% CO_2 incubator. The trophozoites were maintained in exponential growth phase by sub-culturing every 3–4 days in fresh SCGYEM, and then harvested by tapping the flasks to dislodge surface-adhered cells and washed three times with sterile distilled water with centrifugation to remove carried-over nutrients in the supernatants. Experiments were carried out in 15 mL screw-cap tubes, using 10 mL of medium.

Phi6 infectivity assays

V. vermiformis and Phi6 were co-cultured at a ratio of 1:100 in conical Falcon tubes, vortexed to favour phages interaction with amoebae and then transferred to 6-well culture plates (Fisher Scientific 130185). After 72 h incubation at 30 °C, Phi6 phages were released from amoeba trophozoites by lysing the FLA by passing the sample through a sterile 20-gauge syringe needles (BD, Franklin lakes, USA 304827) five to six times. To assess the number of internalized phages, the viral titer (plaque-forming units per mL, PFU mL^{-1}) was employed and determined by double-layer agar assay as described by Fedorenko *et al.* (2020) with minor modification. Briefly, solid and soft trypticase soy agar was prepared from BBL Trypticase Soy Broth with the addition of 1.5 and 0.6% agar, respectively. Then, 0.015% of Triphenyltetrazolium Chloride was added to the sloppy agar to increase the contrast of the plaques against the host lawn due to the formation of pink-coloured formazan by the actively growing bacteria. The concentration of internalized phages was compared with a control of Phi6-only (without *V. vermiformis*). The total number of Phi6 phage (with or without FLA) was established by performing serial 10-fold dilutions of the co-culture medium containing lysed amoebae and phages that were subsequently spread in triplicate.

Data were analyzed by means of Student's *t*-test and are means ± SEM of three independent experiments.

Induction of cyst state in *V. vermiformis*

After incubating the phages with amoeba trophozoites for 24 h, amoebae cells were washed with PBS to remove extracellular and surface-bound phages and resuspended in 5 mL sterile distilled water in culture flasks and stored at 4 °C. After 2 months, aliquots from the amoeba cysts were collected for TEM analysis and the other aliquots from the same experimental samples were resuspended in tissue culture flasks in fresh SCGYEM medium at 25 °C for 72 h for a full trophozoites state recovery. The intracellular phage concentrations (PFU mL^{-1}) were estimated with the culture method as previously described.

Imaging flow cytometry analysis

Since staining with Acridine Orange (AO) is a reliable method for identifying RNA within cells by showing an intense red fluorescence (Keeble & Jay 1962; Doynikova & Vekshin 2019), Phi6 phages were pre-stained with AO (ThermoFisher A3568) for 10 min, after staining the phages were washed and filtered through an Amicon-Ultra-15 Ultracel™ 100 K centrifugal filter unit (EMD Millipore UFC910024) several times to remove residual AO. Unexposed amoebae to AO-Phi6 provided as a control to assess if any presence of autofluorescence in the AO channel (see Supplementary Material, Figure S1).

ImageStream® cytometry analysis and the instrument gating strategy for FLA was performed as previously described (Dey *et al.* 2019). Briefly, *V. vermiformis* trophozoites were infected for 24 h with AO-Phi6 at MOI of 100, washed and resuspended in PBS prior to processing through the ImageStream®X Mark II (Millipore Sigma). Cells were examined at 60× magnification. Analysis was performed using the IDEAS software (Amnis®, Seattle) and cells (fluorescent phages and amoebae) were identified based on bright-field morphology, size, and red fluorescence signal.

Transmission electron microscopy

V. vermiformis trophozoites were co-cultured with Phi6 at an MOI of 100 on Thermonax® cover slips (Thermo Fisher 174985). After decanting the medium, amoebae were fixed at room temperature with 2.5% glutaraldehyde and 0.1M sodium cacodylate buffer (Electron Microscopy Sciences 15960). The samples were submitted for processing at the imaging core at the University of Alberta, Faculty of Biological sciences. Sectioned and carbon-coated samples were observed with a Hitachi H-7650 transmission electron microscope.

RESULTS AND DISCUSSION

AO-Phi6 phages were shown by ImageStream® flow cytometry to be internalised within the amoeba (Figure 1(a)) and further examined by transmission electron microscopy (TEM). The phage particles were observed attachment at the amoeba cell surface and present within numerous *V. vermiformis* replicative phagosomes (Figure 1(b)). Like human RSV, an enveloped

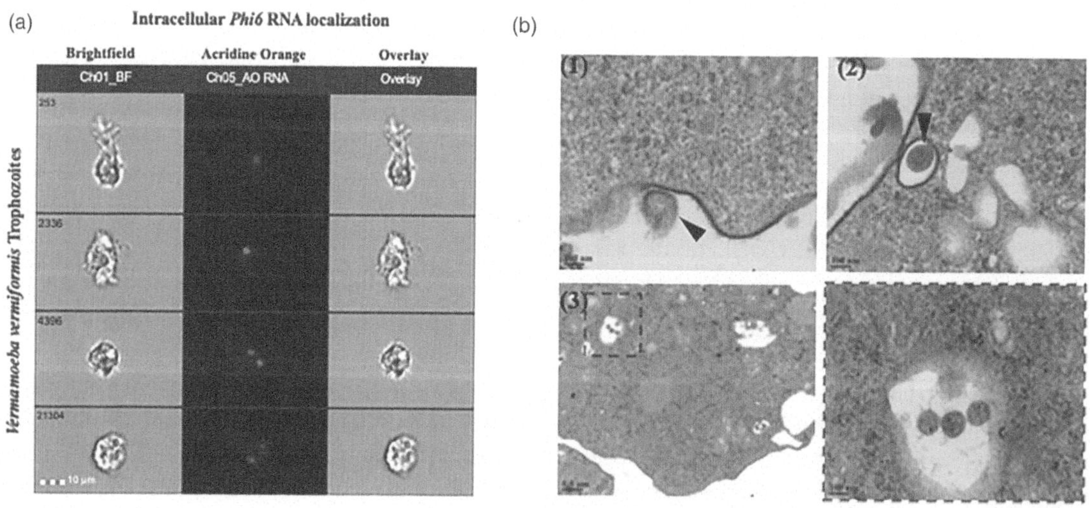

Figure 1 | Bacteriophage Phi6 internalization. (a) Localization of internalized AO-Phi6 RNA particles in live amoebae using ImageStream® flow cytometry. (b) Transmission electron micrographs of thin sections of *V. vermiformis* cells incubated with Phi6 phages for 6 h. Phi6 particles (black arrowheads) are seen at the cell surface (1), within cytoplasmic vesicle (2), and within food vacuoles (3).

human pathogen we have shown to subvert amoebal innate immunity (Dey *et al.* 2021), Phi6 altered the host mitochondrial ultrastructure at 24 h post-infection (Figure 2(a)), with swollen mitochondria and cristae disappearance by 48 h (Figure 2(b)). At 72 h post-infection, amoebae cells showed depletion of all organelles with characteristics of apoptotic cells (Figure 2(c)). Also present in Figure 2(c), are intracellular clusters of phages within apoptotic amoeba cytosol. Of note, similar results were obtained with the FLA *Willaertia magna* and *Acanthamoeba polyphaga*, indicating that the observed phenomenon was not limited to a single amoebic species (see Supplementary Material, Figure S2). To determine if internalized phages remained infectious, we performed a quantitative analysis of the intracellular phage population at 72 h post-infection. The intracellular phages were released by lysing the amoebae and then used to infect their host bacteria for quantification of the recovered phages by titration analysis. The results revealed that the phages retained their infectious capacity during the 72 h incubation period with titres peaking at $\sim 10^6$ PFU mL^{-1}, at a similar infectivity to Phi6-only controls (Figure 2(d)).

Hence, the surrogate Phi6 appears to rapidly induce amoeba apoptosis by targeting their mitochondria in order to remain infectious within 'ghost' host cells. Given the bacterial origin of mitochondria, the high mutation rates and expanded host range of the RNA-Phi6 phage (Duffy *et al.* 2006; Ferris *et al.* 2007; Ford *et al.* 2014); our results imply that the amoebic mitochondria served as a support structure for Phi6 phage persistence. Mitochondria are dynamic multifunctional organelles involved in a variety of cellular and metabolic processes including ATP production, Ca^{2+} homeostasis, and programmed cell death (Twig *et al.* 2008). An increasing number of pathogens, including viruses and bacteria, have been shown to directly or indirectly alter mitochondrial function and dynamics leading to cell death (James & Martinou 2008; Ohta & Nishiyama 2011; Kim *et al.* 2013; Ding *et al.* 2017). Interestingly and like Phi6 phage, SARS-CoV-1 and SARS-CoV-2 also target mitochondrial functions to proliferate and subsequently disseminate by killing host cells (Yan *et al.* 2004; Bordi *et al.* 2006; Shi *et al.* 2014; Gordon *et al.* 2020; Singh *et al.* 2020; Wu *et al.* 2020b; Zhu *et al.* 2020; Flynn *et al.* 2021). Additionally, evidence of evolutionary affinity between cystoviruses (*Pseudomonas* bacteriophages) and the dsRNA viruses infecting diverse

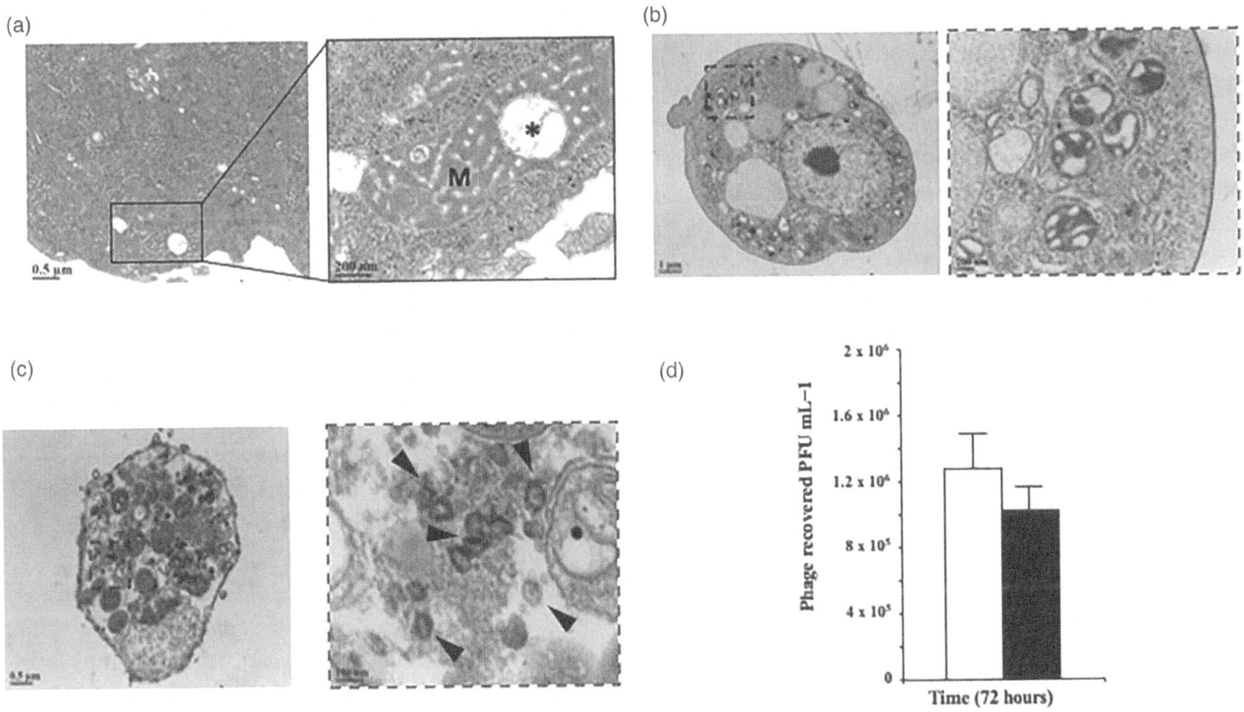

Figure 2 | *Phi6* infection induces mitochondrial alteration and cell apoptosis. Ultrastructure analysis of Phi6-infected *V. vermiformis* cells showing: (a) at 24 h post-infection, TEM of altered mitochondria (M) with dilated cristae and separation of the mitochondrial outer membrane (star). (b) The ultrastructure of various altered mitochondria and abnormal organelles after 48 h post-infection. (c) At 72 h post-infection, TEM showing apoptotic amoeba with depletion of all organelles. In the zoomed images, altered and swollen mitochondria with loss of cristae in infected cells are shown in (a and b). Phi6 particles (black arrowheads) within apoptotic amoeba cell is also shown in (c). (d) Quantification of internalized Phi6 phages from *V. vermiformis* trophozoites. Amoebae were incubated with phages for 72 h. Control Phi6 alone (white bar) and Phi6-containing amoeba trophozoites (black bar) analyzed by plaque assay.

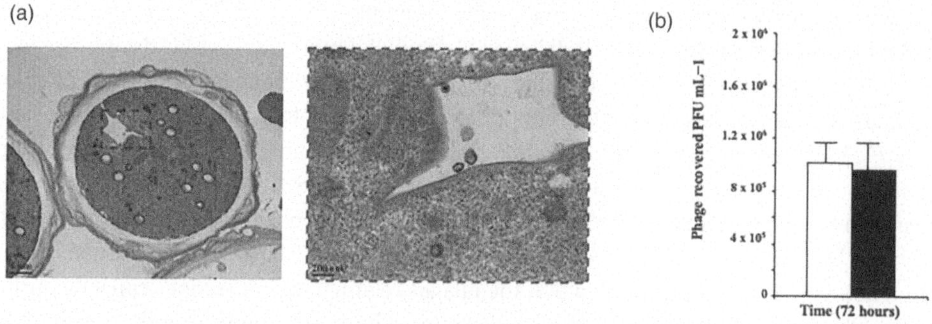

Figure 3 | Ultrastructure visualization and persistence of Phi6-containing amoeba cysts. (a) Transmission electron microscopy (TEM) of Phi6 particles contained within *V. vermiformis* cyst phagosome after 2 months of co culture. (b) Infectivity assay of Phi6 phages from *V. vermiformis* cysts. Amoebae were incubated with phages for 2 months. Control Phi6 alone (white bar) and Phi6-containing amoeba cysts (black bar).

eukaryotic hosts has been presented (Poranen & Bamford 2012; el Omari *et al.* 2013; Pflug *et al.* 2014; Koonin *et al.* 2015; Wolf *et al.* 2018). Thus, complementary phylogenic and comparative genomic analysis with human coronaviruses against their persistence within amoebae, and in particular, SARS-CoV-2 is needed. This subversion of amoebae innate immunity by Phi6 may have important implications for a better understanding of RNA enveloped viruses disease pathogenesis, which could make amoebae a suitable model system for understanding respiratory virus-eukaryotic biology at the cellular and molecular levels. Further studies are required to elucidate the precise mechanisms by which RNA viruses affect amoebic mitochondrial dynamics, initiate and regulate apoptotic cell death.

Since amoebic cysts are highly resistant to various harsh environmental stressors, thereby protecting their intracellular microbes and readily aerosolized to enhance dispersal from engineered water systems (Aksozek *et al.* 2002; Lambrecht *et al.* 2017; Borges *et al.* 2019), we also investigated the stability of Phi6 phage inside *V. vermiformis* cysts. TEM analysis showed that at 2 months after the infection vacuoles still contained infectious viruses in *V. vermiformis* cysts (Figure 3(a) and 3(b)). It has been shown that during viral respiratory diseases investigations, FLA cysts were isolated from the air where patients being nursed and from patients' upper respiratory tract (Wang & Feldman 1967; Kingston & Warhurst 1969). Consequently, it is reasonable to hypostatize that amoeba cysts may also be a vehicle and an intermediate environmental host for SARS-CoV-2 transmission through inhalation, and possibly also within the nasal-phalangeal track where similar amoeba have long been known to reside (Skocil *et al.* 1972; Cabello-Vílchez *et al.* 2014). It is, therefore, conceivable that amoebae cysts containing infectious RNA viruses could be protected against disinfecting agents/conditions and easily dispersed via aerosols. Thus, future studies are necessary to explore these possibilities.

CONCLUSION

Here, we successfully demonstrate the persistence of an infectious SARS-CoV-2 surrogate (Phi6 phage) within active amoebal trophozoites and dormant cysts. Prospectively, viral respiratory pathogens persistence and stability could be maintained by colonizing amoebae in water systems. However, further research is required to confirm the comparability of the fate of Phi6 within amoebae with the viruses of interest (i.e., SARS-CoV-1 and SARS-CoV-2).

It is noteworthy that our work sheds light on the environmental fate of FLA-associated enveloped viruses and raises new questions regarding the persistence of human-pathogenic RNA viruses in nature. Our study also exemplifies the utility of including amoebae as an environmental reservoir when conducting risk assessment for emerging RNA enveloped viruses in aquatic environments.

ACKNOWLEDGEMENTS

These studies were supported by Alberta Innovates (grant #201300490), Alberta, and the Canadian Institute for Health Research (CIHR grant TGEHIPR 150713), Canada. We would like to thank Arlene Oatway for help with transmission electron microscopy (Microscopy Facility Biological Sciences, University of Alberta).

DATA AVAILABILITY STATEMENT

All relevant data are included in the paper or its Supplementary Information.

REFERENCES

Ahmed, W., Angel, N., Edson, J., Bibby, K., Bivins, A., O'Brien, J. W., Choi, P. M., Kitajima, M., Simpson, S. L., Li, J., Tscharke, B., Verhagen, R., Smith, W. J. M., Zaugg, J., Dierens, L., Hugenholtz, P., Thomas, K. v. & Mueller, J. F. 2020 First confirmed detection of SARS-CoV-2 in untreated wastewater in Australia: a proof of concept for the wastewater surveillance of COVID-19 in the community. *Science of the Total Environment* **728**. https://doi.org/10.1016/j.scitotenv.2020.138764.

Aksozek, A., McClellan, K., Howard, K., Niederkorn, J. Y. & Alizadeh, H. 2002 Resistance of *Acanthamoeba castellanii* cysts to physical, chemical, and radiological conditions. *Journal of Parasitology* **88**. https://doi.org/10.1645/0022-3395(2002)088[0621:roacct]2.0.co;2.

Aquino De Carvalho, N., Stachler, E. N., Cimabue, N. & Bibby, K. 2017 Evaluation of Phi6 persistence and suitability as an enveloped virus surrogate. *Environmental Science and Technology* **51**. https://doi.org/10.1021/acs.est.7b01296.

Atanasova, N. D., Dey, R., Scott, C., Li, Q., Pang, X. L. & Ashbolt, N. J. 2018 Persistence of infectious enterovirus within free-living amoebae – a novel waterborne risk pathway? *Water Research* **144**. https://doi.org/10.1016/j.watres.2018.07.023.

Bivins, A., Greaves, J., Fischer, R., Yinda, K. C., Ahmed, W., Kitajima, M., Munster, V. J. & Bibby, K. 2020 Persistence of SARS-CoV-2 in water and wastewater. *Environmental Science and Technology Letters* **7**. https://doi.org/10.1021/acs.estlett.0c00730.

Boratto, P. V. M., Dornas, F. P., Andrade, K. R., Rodrigues, R., Peixoto, F., Silva, L. C. F., la Scola, B., Costa, A. O., de Almeida, G. M. F., Kroon, E. G. & Abrahão, J. S. 2014 Amoebas as mimivirus bunkers: increased resistance to UV light, heat and chemical biocides when viruses are carried by amoeba hosts. *Archives of Virology* **159**. https://doi.org/10.1007/s00705-013-1924-z.

Bordi, L., Castilletti, C., Falasca, L., Ciccosanti, F., Calcaterra, S., Rozera, G., di Caro, A., Zaniratti, S., Rinaldi, A., Ippolito, G., Piacentini, M. & Capobianchi, M. R. 2006 Bcl-2 inhibits the caspase-dependent apoptosis induced by SARS-CoV without affecting virus replication kinetics. *Archives of Virology* **151**. https://doi.org/10.1007/s00705-005-0632-8.

Borges, I., Rodrigues, R. A. L., Dornas, F. P., Almeida, G., Aquino, I., Bonjardim, C. A., Kroon, E. G., la Scola, B. & Abrahão, J. S. 2019 Trapping the enemy: *Vermamoeba vermiformis* circumvents *Faustovirus Mariensis* dissemination by enclosing viral progeny inside cysts. *Journal of Virology* **93**. https://doi.org/10.1128/jvi.00312-19.

Bousbia, S., Papazian, L., Saux, P., Forel, J. M., Auffray, J. P., Martin, C., Raoult, D. & la Scola, B. 2013 Serologic prevalence of amoeba-associated microorganisms in intensive care unit pneumonia patients. *PLoS ONE* **8**. https://doi.org/10.1371/journal.pone.0058111.

Bradbury, R. S. 2014 Free-living amoebae recovered from human stool samples in Strongyloides agar culture. *Journal of Clinical Microbiology*. https://doi.org/10.1128/JCM.02738-13.

Cabello-Vílchez, A. M., Mena, R., Zuñiga, J., Cermeño, P., Martín-Navarro, C. M., González, A. C., López-Arencibia, A., Reyes-Batlle, M., Piñero, J. E., Valladares, B. & Lorenzo-Morales, J. 2014 Endosymbiotic *Mycobacterium chelonae* in a *Vermamoeba vermiformis* strain isolated from the nasal mucosa of an HIV patient in Lima, Peru. *Experimental Parasitology* **145**. https://doi.org/10.1016/j.exppara.2014.02.014.

Casanova, L. M. & Weaver, S. R. 2015 Inactivation of an enveloped surrogate virus in human sewage. *Environmental Science and Technology Letters* **2**. https://doi.org/10.1021/acs.estlett.5b00029.

Chin, A. W. H., Chu, J. T. S., Perera, M. R. A., Hui, K. P. Y., Yen, H.-L., Chan, M. C. W., Peiris, M. & Poon, L. L. M. 2020 Stability of SARS-CoV-2 in different environmental conditions. *The Lancet Microbe* **1**. https://doi.org/10.1016/s2666-5247(20)30003-3.

Danes, L. & Cerva, L. 1981 Survival of polioviruses and echoviruses in *Acanthamoeba castellanii* cultivated in vitro. *Journal of Hygiene Epidemiology Microbiology and Immunology* **25**, 169–74.

de Moura, H., Salazar, H. C., Fernandes, O., Lisboa, D. C. & de Carvalho, F. G. 1985 Free-living amoebae in human bowel: evidence of parasitism. *Revista do Instituto de Medicina Tropical de Sao Paulo* **27**. https://doi.org/10.1590/s0036-46651985000300007.

Delafont, V., Rodier, M. H., Maisonneuve, E. & Cateau, E. 2018 *Vermamoeba vermiformis*: a free-living amoeba of interest. *Microbial Ecology*. https://doi.org/10.1007/s00248-018-1199-8.

de Oliveira, L. C., Torres-Franco, A. F., Lopes, B. C., Santos, B. S. Á. d. S., Costa, E. A., Costa, M. S., Reis, M. T. P., Melo, M. C., Polizzi, R. B., Teixeira, M. M. & Mota, C. R. 2021 Viability of SARS-CoV-2 in river water and wastewater at different temperatures and solids content. *Water Research* **195**. https://doi.org/10.1016/j.watres.2021.117002.

Dey, R., Rieger, A. M., Stephens, C. & Ashbolt, N. J. 2019 Interactions of *Pseudomonas aeruginosa* with *Acanthamoeba polyphaga* observed by imaging flow cytometry. *Cytometry Part A* **95**. https://doi.org/10.1002/cyto.a.23768.

Dey, R., Folkins, M. A. & Ashbolt, N. J. 2021 Extracellular amoebal-vesicles: potential transmission vehicles for respiratory viruses. *NPJ Biofilms and Microbiomes* **7**. https://doi.org/10.1038/s41522-021-00201-y.

Ding, L., Buckwalter, J. A. & Martin, J. A. 2017 DAMPs synergize with cytokines or fibronectin fragment on inducing chondrolysis but lose effect when acting alone. *Mediators of Inflammation* **2017**. https://doi.org/10.1155/2017/2642549.

di Novo, N. G., Carotenuto, A. R., Mensitieri, G., Fraldi, M. & Pugno, N. M. 2021 Modeling of virus survival time in respiratory droplets on surfaces: a new rational approach for antivirus strategies. *Frontiers in Materials* **8**. https://doi.org/10.3389/fmats.2021.631723.

Doynikova, A. N. & Vekshin, N. L. 2019 Detection of RNA hydrolysis with binase by acridine orange fluorescence. *Applied Biochemistry and Microbiology* **55**. https://doi.org/10.1134/S0003683819050028.

Duffy, S., Turner, P. E. & Burch, C. L. 2006 Pleiotropic costs of niche expansion in the RNA bacteriophage Φ6. *Genetics* **172**. https://doi.org/10.1534/genetics.105.051136.

el Omari, K., Sutton, G., Ravantti, J. J., Zhang, H., Walter, T. S., Grimes, J. M., Bamford, D. H., Stuart, D. I. & Mancini, E. J. 2013 Plate tectonics of virus shell assembly and reorganization in phage φ8, a distant relative of mammalian reoviruses. *Structure* **21**. https://doi.org/10.1016/j.str.2013.06.017.

Essig, A., Heinemann, M., Simnacher, U. & Marre, R. 1997 Infection of *Acanthamoeba castellanii* by *Chlamydia pneumoniae*. *Applied and Environmental Microbiology* **63**. https://doi.org/10.1128/aem.63.4.1396-1399.1997.

Fedorenko, A., Grinberg, M., Orevi, T. & Kashtan, N. 2020 Survival of the enveloped bacteriophage Phi6 (a surrogate for SARS-CoV-2) in evaporated saliva microdroplets deposited on glass surfaces. *Scientific Reports* **10**. https://doi.org/10.1038/s41598-020-79625-z.

Ferris, M. T., Joyce, P. & Burch, C. L. 2007 High frequency of mutations that expand the host range of an RNA virus. *Genetics* **176**. https://doi.org/10.1534/genetics.106.064634.

Flynn, R. A., Belk, J. A., Qi, Y., Yasumoto, Y., Wei, J., Alfajaro, M. M., Shi, Q., Mumbach, M. R., Limaye, A., DeWeirdt, P. C., Schmitz, C. O., Parker, K. R., Woo, E., Chang, H. Y., Horvath, T. L., Carette, J. E., Bertozzi, C. R., Wilen, C. B. & Satpathy, A. T. 2021 Discovery and functional interrogation of SARS-CoV-2 RNA-host protein interactions. *Cell* **184**. https://doi.org/10.1016/j.cell.2021.03.012.

Folkins, M. A., Dey, R. & Ashbolt, N. J. 2020 Interactions between human reovirus and free-living amoebae: implications for enteric virus disinfection and aquatic persistence. *Environmental Science and Technology* **54**. https://doi.org/10.1021/acs.est.0c02896.

Ford, B. E., Sun, B., Carpino, J., Chapler, E. S., Ching, J., Choi, Y., Jhun, K., Kim, J. D., Lallos, G. G., Morgenstern, R., Singh, S., Theja, S. & Dennehy, J. J. 2014 Frequency and fitness consequences of bacteriophage 6 host range mutations. *PLoS ONE* **9**. https://doi.org/10.1371/journal.pone.0113078.

Fukuta, M., Mao, Z. Q., Morita, K. & Moi, M. L. 2021 Stability and infectivity of SARS-CoV-2 and viral RNA in water, commercial beverages, and bodily fluids. *Frontiers in Microbiology* **12**. https://doi.org/10.3389/fmicb.2021.667956.

Gordon, D. E., Jang, G. M., Bouhaddou, M., Xu, J., Obernier, K., White, K. M., O'Meara, M. J., Rezelj, V. v., Guo, J. Z., Swaney, D. L., Tummino, T. A., Hüttenhain, R., Kaake, R. M., Richards, A. L., Tutuncuoglu, B., Foussard, H., Batra, J., Haas, K., Modak, M., Kim, M., Haas, P., Polacco, B. J., Braberg, H., Fabius, J. M., Eckhardt, M., Soucheray, M., Bennett, M. J., Cakir, M., McGregor, M. J., Li, Q., Meyer, B., Roesch, F., Vallet, T., mac Kain, A., Miorin, L., Moreno, E., Naing, Z. Z. C., Zhou, Y., Peng, S., Shi, Y., Zhang, Z., Shen, W., Kirby, I. T., Melnyk, J. E., Chorba, J. S., Lou, K., Dai, S. A., Barrio-Hernandez, I., Memon, D., Hernandez-Armenta, C., Lyu, J., Mathy, C. J. P., Perica, T., Pilla, K. B., Ganesan, S. J., Saltzberg, D. J., Rakesh, R., Liu, X., Rosenthal, S. B., Calviello, L., Venkataramanan, S., Liboy-Lugo, J., Lin, Y., Huang, X. P., Liu, Y. F., Wankowicz, S. A., Bohn, M., Safari, M., Ugur, F. S., Koh, C., Savar, N. S., Tran, Q. D., Shengjuler, D., Fletcher, S. J., O'Neal, M. C., Cai, Y., Chang, J. C. J., Broadhurst, D. J., Klippsten, S., Sharp, P. P., Wenzell, N. A., Kuzuoglu-Ozturk, D., Wang, H. Y., Trenker, R., Young, J. M., Cavero, D. A., Hiatt, J., Roth, T. L., Rathore, U., Subramanian, A., Noack, J., Hubert, M., Stroud, R. M., Frankel, A. D., Rosenberg, O. S., Verba, K. A., Agard, D. A., Ott, M., Emerman, M., Jura, N., von Zastrow, M., Verdin, E., Ashworth, A., Schwartz, O., d'Enfert, C., Mukherjee, S., Jacobson, M., Malik, H. S., Fujimori, D. G., Ideker, T., Craik, C. S., Floor, S. N., Fraser, J. S., Gross, J. D., Sali, A., Roth, B. L., Ruggero, D., Taunton, J., Kortemme, T., Beltrao, P., Vignuzzi, M., García-Sastre, A., Shokat, K. M., Shoichet, B. K. & Krogan, N. J. 2020 A SARS-CoV-2 protein interaction map reveals targets for drug repurposing. *Nature* **583**. https://doi.org/10.1038/s41586-020-2286-9.

Haramoto, E., Malla, B., Thakali, O. & Kitajima, M. 2020 First environmental surveillance for the presence of SARS-CoV-2 RNA in wastewater and river water in Japan. *Science of the Total Environment* **737**. https://doi.org/10.1016/j.scitotenv.2020.140405.

Hsueh, T. Y. & Gibson, K. E. 2015 Interactions between human norovirus surrogates and *Acanthamoeba* spp. *Applied and Environmental Microbiology* **81**. https://doi.org/10.1128/AEM.00649-15.

Jadin, J. B. 1973 Hypothesis about adaptation to man and animals of amoebae of the limax group. *Parasite* **48**. https://doi.org/10.1051/parasite/1973482199.

James, D. I. & Martinou, J. C. 2008 Mitochondrial dynamics and apoptosis: a painful separation. *Developmental Cell*. https://doi.org/10.1016/j.devcel.2008.08.011.

Keeble, S. A. & Jay, R. F. 1962 Fluorescent staining for the differentiation of intracellular ribonucleic acid and deoxyribonucleic acid. *Nature* **193**. https://doi.org/10.1038/193695a0.

Kim, S. J., Khan, M., Quan, J., Till, A., Subramani, S. & Siddiqui, A. 2013 Hepatitis B virus disrupts mitochondrial dynamics: induces fission and mitophagy to attenuate apoptosis. *PLoS Pathogens* **9**. https://doi.org/10.1371/journal.ppat.1003722.

Kingston, D. & Warhurst, D. C. 1969 Isolation of amoebae from the air. *Journal of Medical Microbiology* **2**. https://doi.org/10.1099/00222615-2-1-27.

Koonin, E. v., Dolja, V. v. & Krupovic, M. 2015 Origins and evolution of viruses of eukaryotes: the ultimate modularity. *Virology*. https://doi.org/10.1016/j.virol.2015.02.039.

Lambrecht, E., Baré, J., Sabbe, K. & Houf, K. 2017 Impact of *Acanthamoeba* cysts on stress resistance of *Salmonella enterica* Serovar Typhimurium, *Yersinia enterocolitica* 4/O:3, *Listeria monocytogenes* 1/2a, and *Escherichia coli* O:26. *Applied and Environmental Microbiology* **83**. https://doi.org/10.1128/AEM.00754-17.

Lamoth, F. & Greub, G. 2010 Amoebal pathogens as emerging causal agents of pneumonia. *FEMS Microbiology Reviews*. https://doi.org/10.1111/j.1574-6976.2009.00207.x.

la Rosa, G., Iaconelli, M., Mancini, P., Bonanno Ferraro, G., Veneri, C., Bonadonna, L., Lucentini, L. & Suffredini, E. 2020 First detection of SARS-CoV-2 in untreated wastewaters in Italy. *Science of the Total Environment* **736**. https://doi.org/10.1016/j.scitotenv.2020.139652.

la Scola, B., Boyadjiev, I., Greub, G., Khamis, A., Martin, C. & Raoult, D. 2003 Amoeba-resisting bacteria and ventilator-associated pneumonia. *Emerging Infectious Diseases* **9**. https://doi.org/10.3201/eid0907.030065.

la Scola, B., Marrie, T. J., Auffray, J. P. & Raoult, D. 2005 Mimivirus in pneumonia patients. *Emerging Infectious Diseases* **11**. https://doi.org/10.3201/eid1103.040538.

Lau, H. Y. & Ashbolt, N. J. 2009 The role of biofilms and protozoa in legionella pathogenesis: implications for drinking water. *Journal of Applied Microbiology*. https://doi.org/10.1111/j.1365-2672.2009.04208.x.

Ma, B., Linden, Y. S., Gundy, P. M., Gerba, C. P., Sobsey, M. D. & Linden, K. G. 2021 Inactivation of coronaviruses and phage phi6 from irradiation across UVC wavelengths. *Environmental Science and Technology Letters* **8**. https://doi.org/10.1021/acs.estlett.1c00178.

Mattana, A., Serra, C., Mariotti, E., Delogu, G., Fiori, P. L. & Cappuccinelli, P. 2006 *Acanthamoeba castellanii* promotion of in vitro survival and transmission of coxsackie B3 viruses. *Eukaryotic Cell* **5**. https://doi.org/10.1128/EC.5.4.665-671.2006.

Ohta, A. & Nishiyama, Y. 2011 Mitochondria and viruses. *Mitochondrion*. https://doi.org/10.1016/j.mito.2010.08.006.

Pflug, A., Guilligay, D., Reich, S. & Cusack, S. 2014 Structure of influenza A polymerase bound to the viral RNA promoter. *Nature* **516**. https://doi.org/10.1038/nature14008.

Pino, N. J., Rodriguez, D. C., Cano, L. C. & Rodriguez, A. 2021 Detection of SARS-CoV-2 in wastewater is influenced by sampling time, concentration method, and target analyzed. *Journal of Water and Health*. https://doi.org/10.2166/wh.2021.133.

Poranen, M. M. & Bamford, D. H. 2012 Assembly of large icosahedral double-stranded RNA viruses. *Advances in Experimental Medicine and Biology* **726**. https://doi.org/10.1007/978-1-4614-0980-9_17.

Prussin, A. J., Schwake, D. O., Lin, K., Gallagher, D. L., Buttling, L. & Marr, L. C. 2018 Survival of the enveloped virus phi6 in droplets as a function of relative humidity, absolute humidity, and temperature. *Applied and Environmental Microbiology* **84**. https://doi.org/10.1128/AEM.00551-18.

Rimoldi, S. G., Stefani, F., Gigantiello, A., Polesello, S., Comandatore, F., Mileto, D., Maresca, M., Longobardi, C., Mancon, A., Romeri, F., Pagani, C., Cappelli, F., Roscioli, C., Moja, L., Gismondo, M. R. & Salerno, F. 2020 Presence and infectivity of SARS-CoV-2 virus in wastewaters and rivers. *Science of the Total Environment* **744**. https://doi.org/10.1016/j.scitotenv.2020.140911.

Rodríguez-Zaragoza, S. 1994 Ecology of free-living amoebae. *Critical Reviews in Microbiology* **20**. https://doi.org/10.3109/10408419409114556.

Rowbotham, T. J. 1986 Current views on the relationships between amoebae, legionellae and man. *Israel Journal of Medical Sciences* **22**, 678–689.

Sala-Comorera, L., Reynolds, L. J., Martin, N. A., O'Sullivan, J. J., Meijer, W. G. & Fletcher, N. F. 2021 Decay of infectious SARS-CoV-2 and surrogates in aquatic environments. *Water Research* **201**. https://doi.org/10.1016/j.watres.2021.117090.

Samba-Louaka, A., Delafont, V., Rodier, M. H., Cateau, E. & Héchard, Y. 2019 Free-living amoebae and squatters in the wild: ecological and molecular features. *FEMS Microbiology Reviews*. https://doi.org/10.1093/femsre/fuz011.

Santos-Montañez, J., Benavides-Montaño, J. A., Hinz, A. K. & Vadyvaloo, V. 2015 *Yersinia pseudotuberculosis* IP32953 survives and replicates in trophozoites and persists in cysts of *Acanthamoeba castellanii*. *FEMS Microbiology Letters* **362**. https://doi.org/10.1093/femsle/fnv091.

Scheid, P. & Schwarzenberger, R. 2012 *Acanthamoeba* spp. as vehicle and reservoir of adenoviruses. *Parasitology Research* **111**. https://doi.org/10.1007/s00436-012-2828-7.

Shi, C.-S., Qi, H.-Y., Boularan, C., Huang, N.-N., Abu-Asab, M., Shelhamer, J. H. & Kehrl, J. H. 2014 SARS-coronavirus open reading frame-9b suppresses innate immunity by targeting mitochondria and the MAVS/TRAF3/TRAF6 signalosome. *The Journal of Immunology* **193**. https://doi.org/10.4049/jimmunol.1303196.

Silverman, A. I. & Boehm, A. B. 2020 Systematic review and meta-analysis of the persistence and disinfection of human coronaviruses and their viral surrogates in water and wastewater. *Environmental Science and Technology Letters*. https://doi.org/10.1021/acs.estlett.0c00313.

Singh, K. K., Chaubey, G., Chen, J. Y. & Suravajhala, P. 2020 Decoding SARS-CoV-2 hijacking of host mitochondria in Covid-19 pathogenesis. *American Journal of Physiology – Cell Physiology*. https://doi.org/10.1152/ajpcell.00224.2020.

Skocil, V., Dvorák, R., Stěrba, J., Slajs, J., Serbus, C. & Cerva, L. 1972 Epidemiological study of the incidence of amoebas of the limax group in military communities. V. Relation between the presence of amoebas of the limax group in nasal swabs and a pathological finding in nasal mucosa. *Journal of Hygiene Epidemiology Microbiology and Immunology* **16**, 101–106.

Taylor-Mulneix, D. L., Soumana, I. H., Linz, B. & Harvill, E. T. 2017 Evolution of Bordetellae from environmental microbes to human respiratory pathogens: amoebae as a missing link. *Frontiers in Cellular and Infection Microbiology* **7**. https://doi.org/10.3389/fcimb.2017.00510.

Twig, G., Elorza, A., Molina, A. J. A., Mohamed, H., Wikstrom, J. D., Walzer, G., Stiles, L., Haigh, S. E., Katz, S., Las, G., Alroy, J., Wu, M., Py, B. F., Yuan, J., Deeney, J. T., Corkey, B. E. & Shirihai, O. S. 2008 Fission and selective fusion govern mitochondrial segregation and elimination by autophagy. *EMBO Journal* **27**. https://doi.org/10.1038/sj.emboj.7601963.

Vatter, P., Hoenes, K. & Hessling, M. 2021 Photoinactivation of the coronavirus surrogate phi6 by visible light. *Photochemistry and Photobiology* **97**. https://doi.org/10.1111/php.13352.

Wang, S. S. & Feldman, H. A. 1967 Isolation of *Hartmannella* species from human throats. *New England Journal of Medicine* **277**. https://doi.org/10.1056/nejm196711302772204.

Wolf, Y. I., Kazlauskas, D., Iranzo, J., Lucía-Sanz, A., Kuhn, J. H., Krupovic, M., Dolja, V. v. & Koonin, E. v. 2018 Origins and evolution of the global RNA virome. *mBio* **9**. https://doi.org/10.1128/mBio.02329-18.

Wu, F., Zhang, J., Xiao, A., Gu, X., Lee, W. L., Armas, F., Kauffman, K., Hanage, W., Matus, M., Ghaeli, N., Endo, N., Duvallet, C., Poyet, M., Moniz, K., Washburne, A. D., Erickson, T. B., Chai, P. R., Thompson, J. & Alm, E. J. 2020a SARS-CoV-2 titers in wastewater are higher than expected from clinically confirmed cases. *mSystems* **5**. https://doi.org/10.1128/msystems.00614-20.

Wu, K. E., Fazal, F. M., Parker, K. R., Zou, J. & Chang, H. Y. 2020b RNA-GPS Predicts SARS-CoV-2 RNA residency to host mitochondria and nucleolus. *Cell Systems* **11**. https://doi.org/10.1016/j.cels.2020.06.008.

Yan, H., Xiao, G., Zhang, J., Hu, Y., Yuan, F., Cole, D. K., Zheng, C. & Gao, G. F. 2004 SARS coronavirus induces apoptosis in Vero E6 cells. *Journal of Medical Virology* **73**. https://doi.org/10.1002/jmv.20094.

Yu, I. T. S., Qiu, H., Tse, L. A. & Wong, T. W. 2014 Severe acute respiratory syndrome beyond Amoy Gardens: completing the incomplete legacy. *Clinical Infectious Diseases* **58**. https://doi.org/10.1093/cid/cit797.

Zhu, N., Wang, W., Liu, Z., Liang, C., Wang, W., Ye, F., Huang, B., Zhao, L., Wang, H., Zhou, W., Deng, Y., Mao, L., Su, C., Qiang, G., Jiang, T., Zhao, J., Wu, G., Song, J. & Tan, W. 2020 Morphogenesis and cytopathic effect of SARS-CoV-2 infection in human airway epithelial cells. *Nature Communications* **11**. https://doi.org/10.1038/s41467-020-17796-z.

First received 21 July 2021; accepted in revised form 12 September 2021. Available online 24 September 2021

SARS-CoV-2 in Water doi: 10.2166/wh.2021.184

Water use behaviors and water access in intermittent and continuous water supply areas during the COVID-19 pandemic

Emily Kumpel[a,*], Nayaran Billava[b], Nayanatara Nayak[b] and Ayse Ercumen[c]

[a] Department of Civil and Environmental Engineering, University of Massachusetts Amherst, 130 Natural Resources Road, Amherst, MA 01003, USA
[b] Centre for Multi-Disciplinary Development Research, Dr B.R. Ambedkarnagar, Near Yalakki Shettar Colony, Dharwad- 580 004 Karnataka, India
[c] Department of Forestry and Environmental Resources, North Carolina State University, Raleigh, NC 27695, USA
*Corresponding author. E-mail: ekumpel@umass.edu

EK, 0000-0003-0138-8441; NN, 0000-0002-3580-9614; AE, 0000-0001-6002-1514

ABSTRACT

More than one billion people worldwide receive intermittent water supply (IWS), in which water is delivered through a pipe network for fewer than 24 h/day, limiting the quantity and accessibility of water. During the COVID-19 pandemic, stay-at-home orders and efforts to limit contact with others can affect water access for those with unreliable home water supplies. We explored whether water service delivery and household water-use behaviors changed during the COVID-19 pandemic in Hubballi–Dharwad, India, and whether they differed if households had IWS or continuous (24×7) water supply through a longitudinal household survey in 2020–2021. We found few perceived differences in water service delivery or water access, although one-quarter of all households reported insufficient water for handwashing, suggesting an increased demand for water that was not satisfied. Many households with 24×7 supply reported water outages, necessitating the use of alternative water sources. These findings suggest that water demand at home increased and households with IWS and 24×7 both lacked access to sufficient water. Our findings indicate that water insecurity negatively affected households' ability to adhere to protective public health measures during the COVID-19 pandemic and highlight the importance of access to uninterrupted, on-premise water during public health emergencies.

Key words: COVID-19 pandemic, drinking water, handwashing, intermittent water supply, piped water, water access

HIGHLIGHTS

- Water access in households with intermittent and continuous piped water supply was compared during the COVID-19 pandemic.
- Households reported using more water during the pandemic and having insufficient water for handwashing.
- Social distancing while using public water points and toilets was often infeasible.
- Access to uninterrupted on-premise water should be ensured during public health emergencies.

INTRODUCTION

Insufficient or unreliable water supply, which can exacerbate poor sanitation and inadequate hygiene practices, can fuel the spread of disease in densely populated areas. In settings with access to piped water, available water can be restricted by intermittent water supply (IWS), in which water is delivered through a piped network for a limited number of hours per day or per week. IWS can compromise the safety of delivered water, lead to increased risk of waterborne infections, and result in high household expenditures on coping costs (Andey & Kelkar 2009; Fan *et al.* 2014; Ercumen *et al.* 2015; Guragai *et al.* 2017; Gurung *et al.* 2017; Kumpel *et al.* 2017). Previous studies have also found that households with IWS often have very restricted volumes of water available (Kumpel *et al.* 2017). Furthermore, households in low-income urban communities often access water from off-premise locations, requiring that they leave home to collect water (Elliott *et al.* 2019; WHO/UNICEF 2021).

During the COVID-19 pandemic, governments around the world required residents to stay at home to reduce transmission and promoted hygiene measures such as handwashing. Therefore, households experiencing water insecurity from IWS during the COVID-19 pandemic faced increased demand for water for handwashing and hygiene and from increased time spent at home due to stay-at-home restrictions, while they were at the same time often restricted from leaving their homes to collect additional water. Handwashing, requiring water, is a key mechanism for reducing transmission of SARS-CoV-2; however, 2 billion people in the world lack access to adequate handwashing facilities, a challenge further compounded by the inadequate or unreliable water access provided by IWS (Brauer *et al.* 2020; Freeman & Caruso 2020; Ray 2020; Rohilla 2020; Sayeed *et al.* 2021). While governments around the world extended free water supply to citizens, deferred payments of water bills, or issued moratoria on water disconnections during the public health emergency (Lakhani 2020; NDTV 2020), in many of these places, unpaid bills entered into arrears and households faced large bills when collection resumed (Walton 2020; Rajwi 2021).

While more than one billion people in the world are served by IWS systems, there is little understanding of the impact of IWS on households during public health emergencies as compared to continuous water supply, including the impact of stay-at-home orders on household water access and the effects of IWS on COVID-19 mitigation behaviors such as handwashing and social distancing (e.g., while queuing to collect water). Given the widespread use of IWS worldwide, it is important to know whether households with IWS may have experienced differential changes in water access and water-use behaviors during public health emergencies such as the COVID-19 pandemic and whether IWS affected their ability to adhere to mitigation measures. This information can inform future policies or interventions to improve water access during public health emergencies, such as whether additional measures are needed to ensure water access, and help prioritize efforts to upgrade IWS systems to provide continuous supply.

In this study, we seek to understand water access and use during the COVID-19 pandemic among households with intermittent and continuous water supply. Specifically, the objective was to compare household water access, water usage, and adherence to disease transmission mitigation behaviors during and after the stay-at-home orders in Hubballi–Dharwad, India, in households with IWS and those with continuous ('24×7') water service.

MATERIALS AND METHODS

Study area

Hubballi–Dharwad are twin cities in northern Karnataka, India, with a population of more than one million. In 2006 and 2007, approximately 10% of the population were changed to 24×7 water supply on a pilot scale, while the rest of the city received IWS. Details about Hubballi–Dharwad, the water system, the conversion from IWS to 24×7 supply, and the impact of the conversion on child health, household economics, and water quality are documented elsewhere (Kumpel & Nelson 2013, 2014; Burt & Ray 2014; Ercumen *et al.* 2015; Kumpel *et al.* 2017; Burt *et al.* 2018; Ray *et al.* 2018). The wards that were upgraded to 24×7 in this initial pilot conversion are referred to as '24×7 demonstration wards' hereinafter. Since the initial conversion, 24×7 water supply has been introduced to additional areas of Hubballi–Dharwad; as of 2018, 19 of the cities' 67 wards were reported to be served fully by 24×7 supply, 20 wards were partially served by 24×7 supply, and 28 wards remained on IWS. However, notably, our household survey in 2018 found that, even in wards classified by the water authority as 'fully covered' by 24×7 service, some households continued to have IWS. We focused the current study on the 16 wards where at least 50% of households had 24×7 supply.

COVID-19 in Hubballi–Dharwad

In India, country-wide stay-at-home orders were implemented from March 22–June 1, 2020, with the gradual lifting of restrictions through June and July 2020. In the state of Karnataka, lockdown was ended on July 22, 2020 (ET Bureau 2020). Hubballi–Dharwad reported its first confirmed positive case of COVID-19 on March 22, 2020 (Menasinakai 2020). As of July 14, 2021, Dharwad District reported 60,275 positive cases of COVID-19 and 1,264 deaths related to COVID-19, out of 2,876,587 total cases in the state of Karnataka (GoK 2021).

Study design

We used a genetic matching algorithm (Diamond & Sekhon 2012) to select 16 IWS wards as matched controls for the 16 wards with >50% 24×7 service. Wards were matched on socioeconomic indicators, sanitation conditions, and pre-upgrade water service using a representative dataset collected before the conversion. As described in Ercumen *et al.* (2015), the matching variables included percentages of pukka structures (concrete or reinforced cement concrete), low-income, one-room, and self-identified slum households; illiterate females; households with own tap, receiving water less often than every 5 days, with own latrine, and with a designated garbage disposal location provided by the municipality and garbage collection service; and monthly household health expenses. During field inspection, one of the selected wards was observed to primarily contain commercial units and was dropped.

We enrolled households from the remaining 31 wards in Hubballi–Dharwad, India, with a target of approximately 20 households per ward. The 31 wards had a total of 86,627 households with 432,676 residents in 2011 (the last available census data), with 18.7% of the population designated as residing in a slum area. For each selected ward, participants were identified through snowball sampling, with the initial list generated through personal contacts at the Center for Multi-disciplinary Development Research (CMDR) in Dharwad, India. Further participants were identified by asking these contacts to provide phone numbers of their personal contacts from different wards and from different socioeconomic segments of each ward until the sample size was reached. The sample size was selected based on feasibility with the snowball sampling method, which the study team estimated as 500–1,000 households total. Due to lockdown and public health protection, participants were recruited and surveyed via phone.

We enrolled 606 households from 31 wards (326 with IWS and 280 with 24×7) with a mean of 19.5 households per ward (range: 17–23). Round 1 (R1) data were collected from July 9–August 30, 2020 (after the national lockdown had lifted), Round 2 (R2) data were collected from November 15–December 30, 2020, and Round 3 (R3) data were collected from May 1–June 10, 2021 (Figure 1). Out of 606 households enrolled originally in R1, 580 households (301 with IWS and 279 with 24×7) remained in the study at the time of R2 and 560 remained in R3 (286 in IWS and 274 in 24×7). Reasons for loss to follow up included participants being out of service/unreachable and moving.

Data collection

Trained enumerators from CMDR used a structured questionnaire to assess whether the household had IWS or 24×7 service and collect data on the reported frequency of water delivery, frequency and duration of any water outages, user-perceived water pressure and water quantities, water bills, and reported hygiene behaviors including handwashing, during the early stages of the pandemic. They also asked how these compared to their experience during lockdown. The survey included very limited questions on socioeconomic indicators because of the personal relationships between participants due to the snowball sampling method and potential participant discomfort disclosing socioeconomic information on a telephone survey. Each enrolled household was surveyed a total of three times over the course of approximately 1 year. The study protocol was reviewed and approved by the Institutional Review Board of the University of Massachusetts, Amherst. We also collected information on revenue collection and water service delivery from water service providers (Karnataka Urban Water Supply & Drainage Board and the Hubballi–Dharwad Municipal Corporation) in Hubballi–Dharwad.

Data analysis

Descriptive statistics, including tabulated percentages (for binary variables) and means, medians, and standard deviations (for continuous variables), were generated using R (R Core Team 2020). We compared self-reported household perceptions on water service and water use between the three rounds of data collection and between households with IWS vs. 24×7 service. Significance testing for quantitative data on water delivery durations and days between supplies was performed using the non-parametric Kruskal–Wallis (KW) test. Significance testing for binary values comparing IWS and 24×7 groups was performed using χ^2 tests. The *p*-values were considered significant at the $p \leq 0.05$ level.

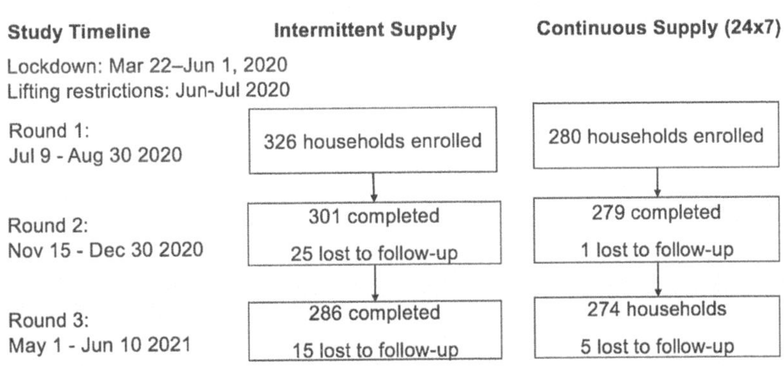

Figure 1 | Number of households enrolled and lost to follow-up across study arms and data collection rounds.

RESULTS

Respondent characteristics

Respondents with IWS and 24×7 water service were 44 and 43% female and mean of 42 vs. 41 years old, respectively. The primary occupation of respondents' head of household was also similarly distributed between both groups: in IWS and 24×7, respectively, 41 vs. 37% held private, government, or NGO jobs, 23 and 26% were business owners, 13 and 19% were day laborers or street vendors, 7 and 5% were retired, and 6 and 5% worked in agriculture.

Water service

Households with IWS received water every 5 days for a mean of 4 h (Table 1). In R1, most households perceived that this was the same number of days between delivery (94%, $n = 326$) and duration of delivery (91%, $n = 326$) as compared to before the COVID-19 pandemic or the lockdown period. The differences in days between deliveries from R1 to R2 was 2.4 h, and from R1 and R3 was 18.8 h. The difference in supply duration from R1 to R2 was 18 min, and from R1 to R3 was 24 min. Differences in the days between deliveries were significant between the rounds (KW: $p < 0.05$), and differences in delivery duration were only significant between R1 and R3 (KW: $p < 0.05$) (Table 1). In R1, the majority of households self-reported that they perceived the same water pressure during the reporting period as prior to lockdown (79% same, 11% lower, 10% higher, $n =$ 325). Delivery frequency and duration varied between wards; some households in the 24×7-demonstration wards and 24×7-fully covered wards still had IWS (Figure 2), consistent with our previous household survey in these wards in 2018 (unpublished).

During R1, households with 24×7 reported a mean of 0.5 outages in the last 2 weeks, with 60% of households reporting no outages, 31% experiencing one outage, and 8% experiencing two ($n = 262$) (Table 1). Of the households reporting an outage, households reported that, during their last outage, water was off for a mean of 2.1 h. All of the 24×7 households (100%, $n =$ 262) said their experience with the number of outages was the same as before pandemic and most (94%, $n = 272$) said it was the same during lockdown, and most households (93%, $n = 274$) perceived a similar outage duration as before the pandemic. During R2, households reported more outages in 24×7 and with longer durations: 91% of households reported an outage in the prior 2 weeks, with an overall mean of 2.3 outages among the 24×7 households ($n = 274$) lasting a mean of 11.8 h ($n =$ 173). During R3, outages reduced from R2 but were higher than R1, with 75% of households reporting an outage in the prior 2 weeks ($n = 271$) for a mean of 5.8 h ($n = 168$). Differences in outage frequency and duration were significant between all of the rounds (KW: $p < 0.001$) (Figure 2). More than half of the 24×7 households (56%, $n = 248$ in R2; 60%, $n = 253$ in R3) reported that the utility gave advance notice of the outages.

Water use

During R1, approximately one-third of both 24×7 and IWS households perceived using more water than before the pandemic (35 and 31%, respectively) (Table 2). Of 24×7 households, 31% perceived higher water bills in R1 than prior to the pandemic,

Table 1 | Reported service delivery among respondents with IWS and 24×7 in R1 (July–August 2020), R2 (November–December 2020), and R3 (May–June 2021)

			R1	R2	R3	R1 compared to pre-pandemic (%)		
		Unit	Mean			Less	Same	More
IWS	Delivery frequency	days	4.6 ($n = 326$)	4.7 ($n = 300$)	5.3 ($n = 286$)	4	94	2
	Duration of supply	h	4.1 ($n = 323$)	4.4 ($n = 300$)	4.5 ($n = 286$)	5	91	4
	Perception of water pressure ($n = 325$)					11	79	10
24×7	Outages in the last 2 weeks	% of hh	40 ($n = 262$)	91 ($n = 274$)	75 ($n = 271$)	0	100	0
		times	0.5	2.3	1.5			
	Duration of outages	h	2.1 ($n = 116$)	11.8 ($n = 173$)	5.8 ($n = 168$)	5	93	2
						R1 compared to lockdown (%)		
IWS	Delivery frequency ($n = 353$)					2	94	4
24×7	Outage frequency ($n = 272$)					3	94	3

Lockdown period refers to March 22–May 31, 2020.

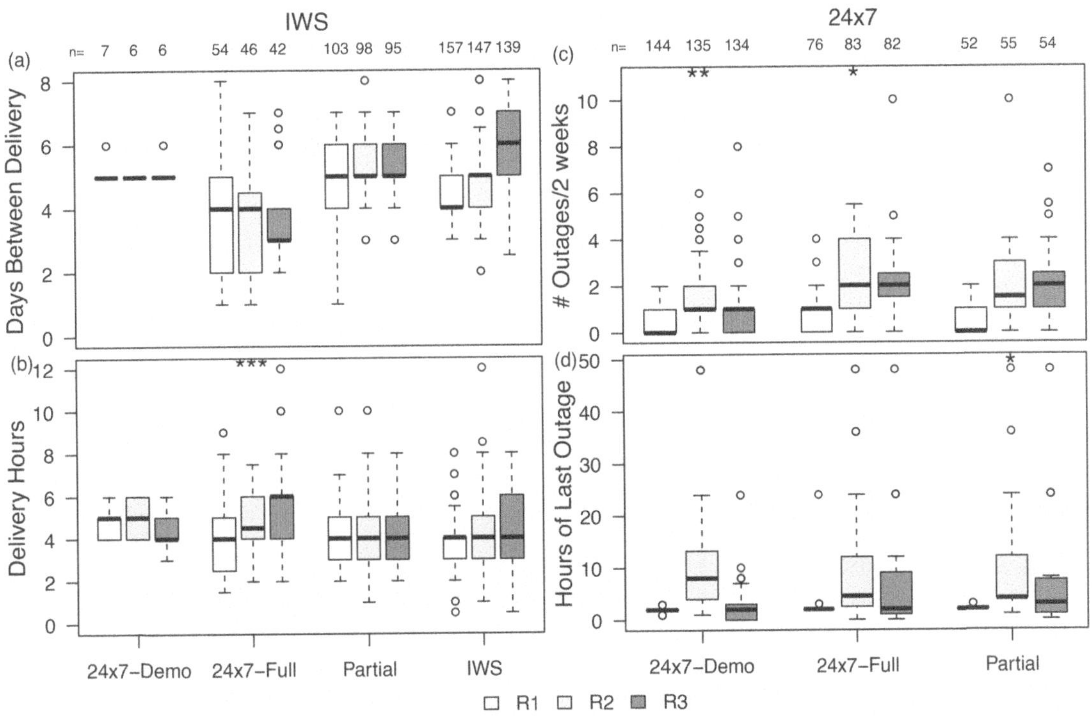

Figure 2 | Boxplots showing household-reported IWS and 24×7 service delivery. (a) Days between water deliveries in IWS. (b) Duration of water delivery in IWS; *** three outliers at 30 h not shown. (c) Number of outages in 24×7; * one outlier at 14 outages not shown; ** two outliers at 24 outages not shown. (d) Duration of the last outage in 24×7; * one outlier at 72 h not shown. Colors represent data collection rounds. 24×7-Demo refers to the eight wards that were converted to 24×7 supply in the pilot conversion in 2007. 24×7-Full refers to additional wards where the water authority reports having extended 24×7 service to all households in the ward since the pilot conversion (however, we note that many households in these wards lacked 24×7 service). Partial refers to wards where the water authority reports having extended 24×7 service to some households. Please refer to the online version of this paper to see this figure in colour: http://dx.doi. 10.2166/wh.2021.184.

Table 2 | Water quantity and water bill perceptions

Variables	Round	Supply type	*n*	Frequency compared to pre-pandemic (%)		
				Less	Same	More
Quantity of water used	R1	IWS	326	1	68	31
		24×7	279	3	62	35
Water bill	R1	IWS	310	1	82	17
		24×7	275	1	68	31
	R2[a]	IWS	241	–	–	40
		24×7	244	–	–	41
	R3	IWS	301	0	72	29
		24×7	279	<1	69	30

[a]Question was phrased as a Yes/No in R2.

while only 17% of IWS households perceived an increase in water bills. However, during R2 and R3, similar percentages between IWS and 24×7 perceived higher water bills, although this may be an artifact of the change in the question from less/same/more to a yes/no.

Use of alternative water sources

In R1, 26% ($n = 326$) of IWS households used a non-municipal water source compared to only 4% ($n = 280$) of 24×7 households (Table 3). Of those IWS households using another source, the most common was borewell water (groundwater) from a piped system (34%), a public non-piped water point (33%), or trucked water (e.g., a vehicle equipped with a water tanker)

Table 3 | Alternative water sources, handwashing, and sanitation access

Variables	Supply type	R1 *n*	R1 Yes (%)	R1 No (%)	R1 *p*[a]	R2 *n*	R2 Yes (%)	R2 No (%)	R2 *p*[a]	R3 *n*	R3 Yes (%)	R3 No (%)	R3 *p*[a]
Rooftop storage tank	IWS	326	71	29	<0.01*	–	–	–	–	–	–	–	–
	24×7	280	55	45		–	–	–		–	–	–	
Roof unable to support a tank[b]	IWS	95	8	92	0.40	–	–	–	–	–	–	–	–
	24×7	124	13	87		–	–	–		–	–	–	
Use any other water[c]	IWS	326	26	74	<0.01*	–	–	–	–	–	–	–	–
	24×7	280	4	96		254	85	15		206	96	4	
Borewell – private	IWS	86	5	95	<0.01*	–	–	–	–	–	–	–	–
	24×7	10	40	60		216	2	98		198	1	99	
Borewell – piped	IWS	86	34	66	0.08	–	–	–	–	–	–	–	–
	24×7	10	0	100		216	2	98		198	1	99	
Borewell – shared tank	IWS	86	1	99	1.00	–	–	–	–	–	–	–	–
	24×7	10	0	100		216	1	99		198	1	99	
Borewell – neighbor	IWS	86	7	93	0.86	–	–	–	–	–	–	–	–
	24×7	10	0	100		216	4	96		198	2	98	
Public water point	IWS	86	33	67	0.45	–	–	–	–	–	–	–	–
	24×7	10	50	50		216	2	98		198	4	96	
Trucked	IWS	86	27	73	0.44	–	–	–	–	–	–	–	–
	24×7	10	10	90		216	91	9		198	89	11	
Bottled	IWS	86	1	99	1.00	–	–	–	–	–	–	–	–
	24×7	10	0	100		216	0	100		198	3	97	
Open well	IWS	86	1	99	1.00	–	–	–	–	–	–	–	–
	24×7	10	0	100		216	0	100		198	0	100	
Use off-premise water sources	IWS	326	12	88	<0.01*	301	7	93	<0.01*	286	4	96	0.9
	24×7	280	5	95		279	1	99		274	3	97	
Social distance at water points[c]	IWS	39	54	46	0.87	–	–	–		–	–	–	
	24×7	13	46	54		3	0	100		9	33	67	
Sufficient water for handwashing	IWS	326	75	25	0.58	301	75	25	0.11	286	80	20	0.06
	24×7	280	77	23		279	69	31		274	73	27	
Use of hand sanitizer	IWS	326	95	5	0.65	301	89	11	0.75	286	80	20	0.20
	24×7	280	94	6		279	88	12		274	76	24	
Use of public toilets	IWS	326	3	97	1.00	301	3	97	0.94	286	0	100	0.23
	24×7	280	3	97		279	3	97		274	1	99	
Social distance at public toilets	IWS	10	10	90	0.04*	12	8	92	0.45	–	–	–	
	24×7	9	67	33		10	30	70		3	67	33	

[a]χ^2 test for difference between IWS and 24×7.
[b]Roof unable to support a tank' refers to whether the respondents lived in a house or an apartment that could not structurally support the weight and/or size of a roof tank.
[c]In R2 and R3, the questions were asked only of 24×7 households and referred to during water outages.
*Significant at $p < 0.01$.

(27%) ($n = 86$). In R1, few households used off-premise water in either IWS or 24×7 (12 and 5%, respectively), and fewer still during R2 (7 and 1%) and R3 (4 and 3%); these were likely a mix of alternative sources and public piped water points. Of those that used off-premise water during R1, approximately half in both water supply schemes said they were able to maintain 2 m (social distance) at the water point. In R2 and R3, only CWS households were asked about their ability to maintain social distance; of the few households that used off-premise water, less than a third reported challenges in doing so (Table 3). In open-ended questions, 17 IWS households and one 24×7 household said they wore a mask and tried to maintain social distancing when collecting other sources, and 10 IWS households and no 24×7 households reported that they stopped using another source or lacked access to another source due to the pandemic.

When 24×7 households were asked whether they accessed other water sources when they had an outage during R2 and R3, many households (85% in R2 and 96% in R3) reported using another water source, primarily trucked water (91 and 89%) (Table 3). A few used borewell water from neighbors (4 and 2% in R2 and R3) or public water points (2 and 4% in R2 and R3). Few households in either IWS or 24×7 reported that the water utility had provided for water during piped water shortages (3 and 4% of IWS households in R2 and R3, and <1 and 3% of 24×7 households in R2 and R3). Seventy-one (71%, $n = 326$) of IWS households in R1 had a rooftop storage tank, while only 55% of households with 24×7 had a rooftop storage tank; accepting large volumes of water, such as from a truck delivering to a home, would require households to have a large amount of household storage.

All households were asked an open-ended question during R1 of how the COVID-19 pandemic lockdown and stay-at-home orders affected their water and sanitation access. When asked during R1 whether the COVID-19 pandemic affected any of their other water and sanitation-related practices, many households in IWS and 24×7 perceived no effect (37 and 41%), one-third perceived cleaning more (27 and 34%,), and one-fifth reported drinking treated water (20 and 18%) ($n = 326$ and $n = 280$). Compared to 24×7 households, IWS households reported storing more water (13 vs. 5%), using more water (7 vs. 1%), and using water more carefully (5 vs. 1%). Other responses from households included obtaining a private water source to use (e.g., neighbor's borewell), storing more treated water, drinking hot water, and washing clothes and vegetables more often. Several houses requested more frequent delivery of water or reported problems with the 24×7 delivery or water pressure.

Hand hygiene and sanitation

Approximately three-quarters of both IWS and 24×7 households perceived having sufficient water for handwashing in R1 (75 and 77%, respectively), with no significant difference between IWS and 24×7 households (Table 3). During R2 and R3, the numbers were similar for IWS households. In R2, the percent reporting sufficient water for handwashing among 24×7 households dropped somewhat, potentially related to the more frequent outages experienced during R2. Most households in both IWS and 24×7 groups reported using hand sanitizers in R1 (95 and 94%, respectively), with a slight drop during R2 (89 and 88%) and a further drop in R3 (80 and 76%); there were no significant differences between IWS and 24×7 households. In an open-ended question of how the COVID-19 pandemic affected handwashing practices in R1, almost all households (91% in IWS and 97% in 24×7) reported washing their hands more frequently, while few households reported no change (7 and 4% in 24×7 and IWS). Few of the households used public toilets (3% in both IWS and 24×7 during both R1 and R2 and <1% in R3); during R1, two-thirds of those using public toilets in 24×7 areas were able to maintain social distance at public toilets, while most of those in IWS (90%) were not ($n = 10$). During R2, only one-third of 24×7 households using public toilets were able to maintain social distancing, while the percentage remained similar for IWS households; however, few households reported using public toilets.

Water utility impacts

The Water Board in Hubballi–Dharwad reported initial challenges in ensuring that workers were outfitted with proper personal protective equipment and identification. They provided alternative water sources to increase water availability at COVID-19 quarantine centers and public facilities (e.g., schools, hostels, community halls, and public places), including tanker trucks and drinking water barrels and bottles to low-income populations during lockdown. They did not report changes in supplying water nor unusual problems with the 24×7 area supplies but did report pipe leak issues in IWS areas. They worked at half staffing capacity in Water Board offices through lockdown.

Water billing and revenue collection were reported to be challenging during the COVID-19 pandemic. This included delays in issuing water bills and, when finally issued, that the bills then reflected 6 months of consumption. Households frequently prohibited meter readers from accessing the household to read meters. Only 10% of revenue was collected during the lockdown period (March 22–May 31, 2020), and only 40% of revenue was collected in the 2020–2021 financial year (April 2020–March 2021). This was attributed to lack of online bill payment options in 24×7 areas, while online bill payment was available in IWS areas.

DISCUSSION

Households with IWS perceived few differences in water delivery during the early months of the pandemic or during lockdown in Hubballi–Dharwad, India. The Water Board and other government offices reported no changes to their water production or delivery. Among households with IWS, days between subsequent water deliveries varied by 2–19 h and the duration of delivery varied by 18–24 min across three rounds of data collection over the course of 1 year. Since households, particularly low-income households, often wait at home to collect and store water, varying water timings can make a substantial difference for households with members working outside of the home (Burt *et al.* 2018; Kumar *et al.* 2018). Similarly, available water volumes in an IWS system can be proportional to the total amount of time the supply is on; a difference in minutes in the duration of supply can directly and substantially affect water volumes available to households (Kumpel *et al.* 2017). Notably, in areas with 24×7 supply, there were increases in outages during the pandemic. While we do not know the reason for this, it did pose potential challenges for households with expected 24×7 water access, especially since only half of 24×7 households had an overhead storage tank (larger in-home storage would be needed to accept larger water deliveries from trucked water). Previous studies in Hubballi–Dharwad have noted that public borewells were removed from 24×7 areas as part of the conversion to 24×7 supply, potentially threatening household water security during water outages (Burt *et al.* 2018). The variations in IWS service delivery and in water outages in 24×7 supply areas are consistent with seasonal changes in piped water availability observed during earlier studies of water access in Hubballi–Dharwad (Kumpel & Nelson 2013). Our first round of data collection was during the monsoon season where water supplies at the source (and electricity supplies for consistently operating the water system) tend to be higher, while the second and third rounds, collected during dryer seasons, showed more outages and potential shortages of supply. At the height of supply scarcity during the second data collection round, households with 24×7 supply reported a mean of 2.3 outages in the past 2 weeks for 11.8 h in duration (309 h of supply/336 total hours in a week, or supply on 92% of the time). During the same period, IWS households received water for 4.4 h every 4.7 days (13 h of supply/336 total hours in a week, or supply on 4% of the time).

Approximately one-quarter of households in both 24×7 and IWS areas perceived insufficient water for adequate handwashing. This finding is of notable concern due to the importance of handwashing to mitigate the spread of the COVID-19 pandemic. Households with IWS and 24×7 both reported increasing their frequency of handwashing in response to the pandemic, suggesting that changes in human behavior in response to perceived risks led to an increase in demand for water that was not satisfied by either supply scheme. These findings support previous calls for the provision of handwashing stations to interrupt SARS-CoV2 transmission in low-income communities (Freeman & Caruso 2020; Ray 2020). Notably, households in this study did adjust their reported handwashing behavior and reported dedicating scarce water resources towards handwashing.

Households also largely perceived using more water overall during the pandemic. Both IWS and 24×7 households perceived higher water bills starting during the second round of data collection. 24×7 households are metered based on consumption. Some IWS households are metered based on consumption, but many are not and are charged a flat rate no matter the consumption; perceived higher water bills may have been due to higher metered consumption or due to utility-reported delays in billing.

Of note, many residents with 24×7 and IWS collect piped water in their yard and alternative sources, such as their own or a neighbor's borewell, were available on premise or close to home. However, those households that did need to venture out for alternative water sources did report issues with social distancing at public water sources, particularly later in 2020, a finding consistent with a study of water access in a town with IWS in Zimbabwe (Zvobgo & Do 2020). These findings indicate that water insecurity negatively impacted individuals' ability to adhere to protective public health measures (Stoler *et al.* 2020) and highlight the importance of on-premise, uninterrupted water access in reducing the need for collection, and therefore potential exposure, from off-premise public water points during the COVID-19 pandemic. Notably, many households at our study site reported using sources that did not require queuing, such as trucked water delivered to home or neighbors' private sources. While water sharing has been noted as a widespread and important mechanism of ensuring household water security (Stoler *et al.* 2019), it has also been suggested as a mechanism for SARS-CoV2 transmission through close interaction between neighbors (Stoler *et al.* 2020). In particular, households with IWS frequently relied on multiple types of alternative water sources; while they appeared to have continued practices existing prior to the pandemic, it was likely further strained as water use increased and households needed to more actively manage their water supplies and use. Additionally, some households did report having to abandon the use of some water sources due to the pandemic. A notably large portion of household with 24×7 supply reported relying on trucked water due to the numerous outages in water supply experienced in rounds 2 and 3.

This study had several limitations. First, the study population was drawn from a non-random sample; due to restrictions imposed by the COVID-19 pandemic, it was not possible to travel to recruit a random sample of households from the study areas. Therefore, we relied on snowball sampling to recruit participants. We were also unable to collect detailed socioeconomic information from participants via the phone survey approach necessitated by the pandemic, and we therefore report analyses that are not adjusted for socioeconomic covariates. However, our study design relied on multivariate matching to generate well-balanced study groups (Arnold *et al.* 2010) and, in our prior work in Hubballi–Dharwad, matching balanced socioeconomic covariates between households with IWS and 24×7 supply (Ercumen *et al.* 2015). All information was self-reported, and the information about water delivery details during lockdown relied on respondents' memories, subject to the risk of inaccurate recall.

CONCLUSION

Safe water, sanitation, and hygienic conditions are essential to protecting human health during all infectious disease outbreaks, including the COVID-19 pandemic (Stoler *et al.* 2020; Tortajada & Biswas 2020; Desye 2021). Our study represents the only assessment, to our knowledge, comparing water access during the COVID-19 pandemic under intermittent and continuous water supply schemes. We did not find large differences in key metrics of water service delivery and water-related behaviors between households with IWS and 24×7 supply during the COVID-19 pandemic in Hubballi–Dharwad, India. Many households had systems for accessing water, including alternative sources, that persisted during lockdown. While few households reported relying on off-premise sources, among the few households that did, maintaining social distancing while collecting water was often not feasible. Most households reported using more water overall during the pandemic, likely straining households' water management capacities and leading to higher bills. Households with IWS and 24×7 both struggled to access sufficient quantities of water for handwashing, suggesting that water insecurity can limit their ability to adhere to COVID-19 mitigation measures. In particular, households expecting 24×7 water service but facing outages may not have been prepared to cope with the outages. Our findings highlight the role of ensuring adequate supplies of uninterrupted, reliable water during public health emergencies; and ensuring alternative supplies and sufficient household storage for households not otherwise prepared for outages.

ACKNOWLEDGEMENTS

We are thankful to households in Hubballi–Dharwad for their participation and the enumerator team for collecting the data.

FUNDING

Funding was provided by the UMass Amherst Department of Civil and Environmental Engineering.

DATA AVAILABILITY STATEMENT

Data cannot be made publicly available; readers should contact the corresponding author for details.

REFERENCES

Andey, S. P. & Kelkar, P. S. 2009 Influence of intermittent and continuous modes of water supply on domestic water consumption. *Water Resources Management* **23**, 2555–2566.

Arnold, B. F., Khush, R. S., Ramaswamy, P., London, A. G., Rajkumar, P., Ramaprabha, P., Durairaj, N., Hubbard, A. E., Balakrishnan, K. & Colford, J. M. 2010 Causal inference methods to study nonrandomized, preexisting development interventions. *Proceedings of the National Academy of Sciences* **107**, 22605–22610. https://doi.org/10.1073/pnas.1008944107.

Brauer, M., Zhao, J. T., Bennitt, F. B. & Stanaway, J. D. 2020 Global access to handwashing: implications for COVID-19 control in low-income countries. *Environmental Health Perspectives* **128**, 057005. https://doi.org/10.1289/EHP7200.

Burt, Z. & Ray, I. 2014 Storage and non-payment: persistent informalities within the formal water supply of Hubli-Dharwad, India. *Water Alternatives* **7**, 106–120.

Burt, Z., Ercümen, A., Billava, N. & Ray, I. 2018 From intermittent to continuous service: costs, benefits, equity and sustainability of water system reforms in Hubli-Dharwad, India. *World Development* **109**, 121–133. https://doi.org/10.1016/j.worlddev.2018.04.011.

Desye, B. 2021 COVID-19 pandemic and water, sanitation, and hygiene: impacts, challenges, and mitigation strategies. *Environmental Health Insights* **15**. https://doi.org/10.1177/11786302211029447.

Diamond, A. & Sekhon, J. S. 2012 Genetic matching for estimating causal effects: a general multivariate matching method for achieving balance in observational studies. *The Review of Economics and Statistics* **95**, 932–945. https://doi.org/10.1162/REST_a_00318.

Elliott, M., Foster, T., MacDonald, M. C., Harris, A. R., Schwab, K. J. & Hadwen, W. L. 2019 Addressing how multiple household water sources and uses build water resilience and support sustainable development. *npj Clean Water* **2**, 6. https://doi.org/10.1038/s41545-019-0031-4.

Ercumen, A., Arnold, B. F., Kumpel, E., Burt, Z., Ray, I., Nelson, K. & Colford Jr., J. M. 2015 Upgrading a piped water supply from intermittent to continuous delivery and association with waterborne illness: a matched cohort study in urban India. *PLoS Medicine* **12**, e1001892. https://doi.org/10.1371/journal.pmed.1001892.

ET Bureau 2020 *Karnataka Lockdown News: No More Lockdown Anywhere in Karnataka from Wednesday, Says CM BS Yediyurappa. The Economic Times.*

Fan, L., Liu, G., Wang, F., Ritsema, C. J. & Geissen, V. 2014 Domestic water consumption under intermittent and continuous modes of water supply. *Water Resources Management* **28**, 853–865. https://doi.org/10.1007/s11269-014-0520-7.

Freeman, M. C. & Caruso, B. A. 2020 Comment on 'Global access to handwashing: implications for COVID-19 control in low-income countries'. *Environmental Health Perspectives* **128**, 098001. https://doi.org/10.1289/EHP7852.

GoK 2021 *Government of Karnataka Department of Health & Family Welfare.* Available from: https://karunadu.karnataka.gov.in/hfw/pages/home.aspx.

Guragai, B., Takizawa, S., Hashimoto, T. & Oguma, K. 2017 Effects of inequality of supply hours on consumers' coping strategies and perceptions of intermittent water supply in Kathmandu Valley, Nepal. *Science of the Total Environment* **599–600**, 431–441. https://doi.org/10.1016/j.scitotenv.2017.04.182.

Gurung, Y., Zhao, J., Kumar KC, B., Wu, X., Suwal, B. & Whittington, D. 2017 The costs of delay in infrastructure investments: a comparison of 2001 and 2014 household water supply coping costs in the Kathmandu Valley, Nepal. *Water Resources Research* **53**, 7078–7102. https://doi.org/10.1002/2016WR019529.

Kumar, T., Post, A. E. & Ray, I. 2018 Flows, leaks and blockages in informational interventions: a field experimental study of Bangalore's water sector. *World Development* **106**, 149–160. https://doi.org/10.1016/j.worlddev.2018.01.022.

Kumpel, E. & Nelson, K. L. 2013 Comparing microbial water quality in an intermittent and continuous piped water supply. *Water Research* **47**, 5176–5188.

Kumpel, E. & Nelson, K. L. 2014 Mechanisms affecting water quality in an intermittent piped water supply. *Environmental Science & Technology* **48**, 2766–2775. https://doi.org/10.1021/es405054u.

Kumpel, E., Woelfle-Erskine, C., Ray, I. & Nelson, K. L. 2017 Measuring household consumption and waste in unmetered, intermittent piped water systems: water use in intermittent piped systems. *Water Resources Research* **53**. https://doi.org/10.1002/2016WR019702.

Lakhani, N. 2020 *Detroit Suspends Water Shutoffs over Covid-19 Fears*. The Guardian. Available from: http://www.theguardian.com/us-news/2020/mar/12/detroit-water-shutoffs-unpaid-bills-coronavirus (accessed 7 December 2021).

Menasinakai, S. 2020 One tests positive for Covid-19 in Karnataka's Dharwad | Hubballi News - Times of India. The Times of India.

NDTV 2020 *Delhi Extends Water Bill Waiver Scheme Till September 30 Amid Coronavirus Pandemic. NDTV.com.* Available from: https://www.ndtv.com/delhi-news/delhi-extends-water-bill-waiver-scheme-till-september-30-amid-coronavirus-pandemic-2254722 (accessed 7 December 2021).

Rajwi, T., 2021 *In Kerala, Electricity and Water Bill Arrears Pile Up During COVID-19 Pandemic.* The Hindu.

Ray, I. 2020 Viewpoint – handwashing and COVID-19: simple, right there…? *World Development* **135**, 105086. https://doi.org/10.1016/j.worlddev.2020.105086.

Ray, I., Billava, N., Burt, A., Colford Jr, J. M., Ercümen, A., Jayaramu, K. P., Kumpel, E., Nayak, N., Nelson, K. L. & Woelfle-Erskine, C. 2018 From intermittent to continuous water supply: A household-level evaluation of water system reforms in Hubli-Dharwad, Karnataka. *Economic and Political Weekly* **53** (49), 39–48.

Rohilla, S. K. 2020 *COVID-19 Outbreak: More Hand Washing Can Increase India's Water Woes. Down to Earth.* Available from: https://www.downtoearth.org.in/blog/water/covid-19-outbreak-more-hand-washing-canincrease-india-s-water-woes-69900 (accessed 7 December 2021).

Sayeed, A., Rahman, M. H., Bundschuh, J., Herath, I., Ahmed, F., Bhattacharya, P., Tariq, M. R., Rahman, F., Joy, M. T. I., Abid, M. T., Saha, N. & Hasan, M. T. 2021 Handwashing with soap: a concern for overuse of water amidst the COVID-19 pandemic in Bangladesh. *Groundwater for Sustainable Development* **13**, 100561. https://doi.org/10.1016/j.gsd.2021.100561.

Stoler, J., Brewis, A., Harris, L. M., Wutich, A., Pearson, A. L., Rosinger, A. Y., Schuster, R. C. & Young, S. L. 2019 Household water sharing: a missing link in international health. *International Health* **11**, 163–165. https://doi.org/10.1093/inthealth/ihy094.

Stoler, J., Jepson, W. E. & Wutich, A. 2020 Beyond handwashing: water insecurity undermines COVID-19 response in developing areas. *Journal of Global Health* **10**, 010355. https://doi.org/10.7189/jogh.10.010355.

Tortajada, C. & Biswas, A. K. 2020 COVID-19 heightens water problems around the world. *Water International* **45**, 441–442. https://doi.org/10.1080/02508060.2020.1790133.

Walton, B. 2020 Water Shutoffs Are Suspended, But the Bills Will Still Be Due. Circle of Blue. Available from: https://www.circleofblue.org/2020/world/water-shutoffs-are-suspended-but-the-bills-will-still-be-due/ (accessed 7 December 2021).

WHO/UNICEF 2021 *Progress on Household Drinking Water, Sanitation, and Hygiene 2015–2020: Five Years into the SDGs.* Joint Monitoring Programme, Geneva, Switzerland and New York, NY, USA.

Zvobgo, L. & Do, P. 2020 COVID-19 and the call for 'Safe hands': challenges facing the under-resourced municipalities that lack potable water access – a case study of Chitungwiza municipality, Zimbabwe. *Water Research X* **9**, 100074. https://doi.org/10.1016/j.wroa.2020.100074.

First received 30 July 2021; accepted in revised form 12 November 2021. Available online 24 November 2021

© 2022 The Authors

SARS-CoV-2 in Water doi: 10.2166/wh.2022.190

Effects of face masks and ventilation on the risk of SARS-CoV-2 respiratory transmission in public toilets: a quantitative microbial risk assessment

Thammanitchpol Denpetkul[a], Monchai Pumkaew[b], Oranoot Sittipunsakda[a], Pornsawan Leaungwutiwong[c], Skorn Mongkolsuk[d,e] and Kwanrawee Sirikanchana[d,e,*]

[a] Department of Social and Environmental Medicine, Faculty of Tropical Medicine, Mahidol University, Bangkok 10400, Thailand
[b] Environmental Engineering and Disaster Management Program, School of Multidisciplinary, Mahidol University, Kanchanaburi Campus, Sai Yok, Kanchanaburi 71150, Thailand
[c] Department of Microbiology and Immunology, Faculty of Tropical Medicine, Mahidol University, Bangkok, Thailand 10400
[d] Research Laboratory of Biotechnology, Chulabhorn Research Institute, Bangkok 10210, Thailand
[e] Center of Excellence on Environmental Health and Toxicology (EHT), OPS, MHESI, Thailand
*Corresponding author. E-mail: kwanrawee@cri.or.th

TD, 0000-0001-8427-6183; MP, 0000-0003-4286-6817; PL, 0000-0001-6215-8290; SM, 0000-0002-8951-4598; KS, 0000-0001-7273-4060

ABSTRACT

Public toilets may increase the risk of COVID-19 infection via airborne transmission; however, related research is limited. We aimed to estimate SARS-CoV-2 infection risk through respiratory transmission using a quantitative microbial risk assessment framework by retrieving SARS-CoV-2 concentrations from the swab tests of 251 Thai patients. Three virus-generating scenarios were investigated: an infector breathing, breathing with a cough, and breathing with a sneeze. The infection risk (95th percentile) was as high as 10^{-1} with breathing and increased to 1 with a cough or a sneeze. No significant gender differences for toilet users (receptors) were noted. The highest risk scenario, namely breathing with a sneeze, was further evaluated for risk mitigation measures. Mitigation to a lower risk under 10^{-3} succeeded only when the infector and the receptor both wore N95 respirators or surgical masks. Ventilation of up to 20 air changes per hour (ACH) did not decrease the risk. However, an extended waiting time of 10 min between an infector and a receptor resulted in approximately 1.0-$\log_{10}$ further risk reduction when both wore masks with the WHO-recommended 12 ACH. The volume of expelled droplets, virus concentrations, and receptor dwell time were identified as the main contributors to transmission risk.

Key words: aerosol, COVID-19, mask, restroom, risk management, toilet, ventilation

HIGHLIGHTS

- The use of public toilets poses a risk of SARS-CoV-2 respiratory transmission.
- No gender differences in risk by counteracting dwell times and inhalation rates.
- Ventilation alone did not reduce risk at 20 ACH with immediate receptor entrance.
- 10-min waiting time further mitigated risks beyond face masks at 12 ACH ventilation.
- N95 and surgical masks offer the most effective risk mitigation to toilet users.

GRAPHICAL ABSTRACT

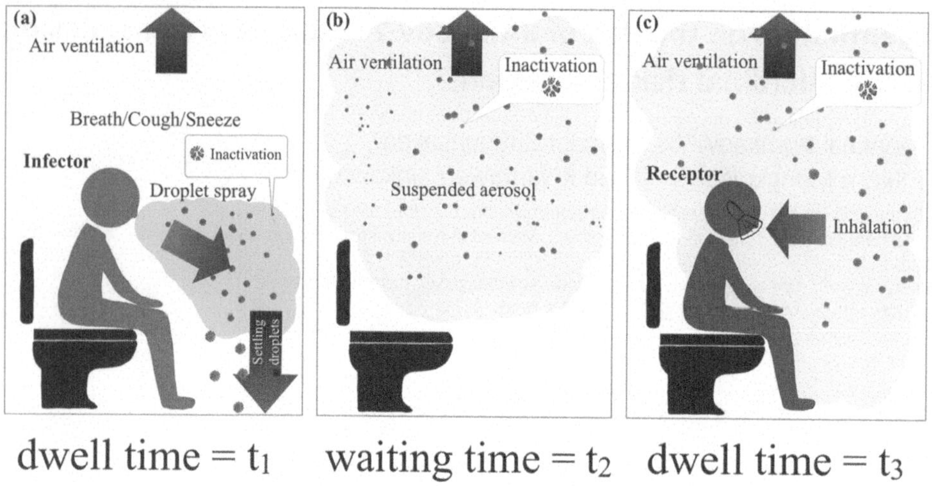

INTRODUCTION

The coronavirus disease 2019 (COVID-19) pandemic has affected the global population since its emergence in December 2019. The three main transmission routes of the severe acute respiratory syndrome coronavirus 2 (SARS-CoV-2), the etiological agent of COVID-19, have been identified as (1) the inhalation of respiratory fluids carrying infectious viruses, (2) direct splashes or sprays of infectious respiratory droplets and aerosol particles, and (3) the touching of contaminated surfaces (Centers for Disease Control & Prevention 2021). Communal confined spaces such as public toilets in shopping centers, schools, restaurants, airports, theaters, and hospitals may be significant areas for SARS-CoV-2 transmission (Dancer *et al.* 2021). Surface contamination with SARS-CoV-2 in toilets and bathrooms has been reported (Ding *et al.* 2021; Maestre *et al.* 2021); however, transmission risk from fomite exposure could be reduced significantly through the simple yet effective interventions of hand washing, hand sanitizing, and surface disinfection (Dancer *et al.* 2021; Pitol & Julian 2021; World Health Organization [WHO] 2020a). Airborne transmission, on the other hand, is deemed the main route of COVID-19 spread (Centers for Disease Control & Prevention 2021) and could be aggravated by the use of busy, confined public toilet spaces, especially if appropriate steps are not taken to mitigate the risk of virus transmission (Dancer *et al.* 2021).

The positive detection of SARS-CoV-2 RNA has been reported in 23.8% of air samples from hospital toilets, which have demonstrated higher viral loads than clinical areas (Birgand *et al.* 2020). However, risk assessments of respiratory exposure to SARS-CoV-2 in public toilets are limited. Potential sources of infectious respiratory droplets and aerosol particles in toilet settings include exhalation and expelling, such as sneezing, speaking, and coughing, by infected toilet users, and the aerosolization of infected feces and urine after toilet flushing (Dancer *et al.* 2021; Schijven *et al.* 2021). Although infectious SARS-CoV-2 was isolated from the feces of a severely infected patient (Xiao *et al.* 2020), studies confirming that feces and urine in wastewater remain infectious for SARS-CoV-2 are limited, with supporting evidence showing poor virus survival in gastrointestinal tracts due to the low pH of gastric fluids, bile, digestive enzymes, and bacterial byproducts (Jones *et al.* 2020; Zang *et al.* 2020; Albert *et al.* 2021). Consequently, even though flushing activities can produce airborne droplets and aerosols, the associated risks may be low because contamination by infectious virus particles is less likely (Shi *et al.* 2021). In this study, we, therefore, focused on characterizing the risk of SARS-CoV-2 respiratory transmission introduced by normal breathing and expelling (i.e., coughing and sneezing) in a public toilet setting.

Quantitative microbial risk assessment (QMRA) is a valuable tool used to quantitatively estimate human health risks associated with exposure to pathogens in different environmental matrices (Rose & Gerba 1991; Haas *et al.* 2014). The QMRA framework has been applied to estimate SARS-CoV-2 transmission risk among wastewater treatment plant workers (Dada & Gyawali 2021; Zaneti *et al.* 2021), Tokyo 2020 Olympic Games attendees (Murakami *et al.* 2021), and confined vehicle passengers and shared room users (Schijven *et al.* 2021). In the present study, we aimed to estimate the risk of infection associated with public toilet exposure to SARS-CoV-2 through airborne transmission using the QMRA approach. For convenience, we called a healthy person who is exposed to transmission risk a receptor, while a disease-carrying person, either symptomatic or asymptomatic, was termed an infector. We gathered the input parameters from a variety of sources. These included COVID-19 concentration data obtained by swab testing 251 Thai patients from a public hospital in Bangkok and the exposure factors related to three droplet- and aerosol-generating activities of infectors, namely breathing, breathing with a cough, and breathing with a sneeze, which were all identified from published sources (Duguid 1946; Loudon & Roberts 1967; Fabian *et al.* 2011; Han *et al.* 2013; Schijven *et al.* 2021). The risk of infection was calculated separately for the male and female receptors because of their different respiratory rates and periods of time spent in the toilet, the so-called dwell times. To include uncertainty and variability in the risk characterization, we applied the Monte Carlo simulation (MCs) technique to calculate the risks. The sensitivity of the model parameters was evaluated to determine which input parameters could help reduce the associated uncertainty. Finally, three risk mitigation measures, namely face mask wearing, increased ventilation, and extended waiting times, were assessed to ascertain their efficacy. The calculated risks and associated mitigation measures may be beneficial in the development of public health policies aimed at providing the effective control of SARS-CoV-2 transmission.

MATERIALS AND METHODS

Risk scenarios

We evaluated the infection risk in various scenarios with and without preventive measures. For the public toilet model, a Thailand standard cubicle size of $1.5\times0.8\times2.7$ m ($3.24\ m^3$) was set for the risk evaluation (Ministry of Public Health 2016). The three scenarios used in this study that can cause an infector to generate infectious droplets and aerosols included breathing (Br), breathing with a cough (Br+Co), and breathing with a sneeze (Br+Sn). The scenario that provided the highest risk was further investigated to determine the efficacy of the identified mitigation measures (i.e., face mask wearing, ventilation, and waiting times). To evaluate the effects of mask wearing, different types of masks (i.e., N95 respirator and surgical and denim fabric masks) were modeled when worn by either an infector or a receptor, or both. For the ventilation evaluation, the air changes per hour (ACH) were varied at 0 (no ventilation), 0.5 (poor ventilation), 10 (DIN 1946 ventilation standard for public toilets), 12 (WHO-recommended standard ventilation 2021), and 20 (extreme ventilation). In addition, the waiting times (i.e., the time the toilet was unoccupied) between an infector and a receptor were set at 0, 3, 5, and 10 min. An outline of the QMRA steps for all the scenarios is presented in Figure 1.

Virus levels generated by an infector

The SARS-CoV-2 concentrations used in this study were retrieved from the reverse-transcription quantitative polymerase chain reaction quantification cycle (C_t) values of 251 positive swab test results of an N2 gene from patients in a public hospital in Bangkok from March to May 2021. Due to the absence of a standard curve for clinical swab testing in Thailand, the viral concentrations were estimated using a published standard curve (Sherchan *et al.* 2020), which ranged from 4.4×10^{-1} to 6.4×10^{8} gene copies (gc)/μL. The SARS-CoV-2 concentrations (A) were fitted with a triangular distribution as shown in Supplementary Table S1. To convert the virus concentrations from gc to an infectious plaque-forming unit (PFU), the ratio of the PFU/gc (R) of SARS-CoV-2 was calculated as 1:10 because the ratio of the median tissue culture infectious dose (TCID50) to the viral RNA copies in the saliva of ferrets has been found to be 0.15 (Kim *et al.* 2020), and the ratio of PFU to TCID50 has been established as 0.7 (Covés-Datson *et al.* 2020). The product of 0.15 and 0.7 was rounded to 0.1 (Murakami *et al.* 2021).

The number of infectious viruses suspended in the ambient air of the toilet cubicle was calculated using the mass balance equations (Equations (1.1) and (1.2)) in which each term in the equation has units of mass per time. Under the completely mixed condition, the accumulation of virus particles as aerosols was obtained from the summation of the breathing of an infector, virus inactivation, and virus removal by mechanical ventilation and inhalation. Under typical conditions of 20–70% relative humidity, a 20 °C temperature, and no direct sunlight, an average SARS-CoV-2 inactivation rate of 0.008 (min^{-1}) was applied (Schuit *et al.* 2020). We assumed that the virus particles were released continuously during the time the infector was in the toilet cubicle (infector's dwell time=t_1 minutes). The dwell times for men and women, which were in line with those indicated in an airport study of 250 men and 237 women, were fitted with a log-normal distribution (Supplementary Table S1; Gwynne *et al.* 2019). The remaining infectious virus concentrations generated by the infector after leaving the toilet (C_{t_1}) were calculated according to Equation (2) by integrating Equation (1.2) with no initial virus particles (C=0). The inhalation rates (q_{in}) following the uniform distribution ranged between 8.40 and 14.93 L/min for men and between 7.71 and 11.32 L/min for women (Stifelman 2007). The additional concentrations of infectious SARS-CoV-2 expelled by coughing (C_{co}) and sneezing (C_{sn}) were calculated using Equations (3) and (4), respectively. The volumetric

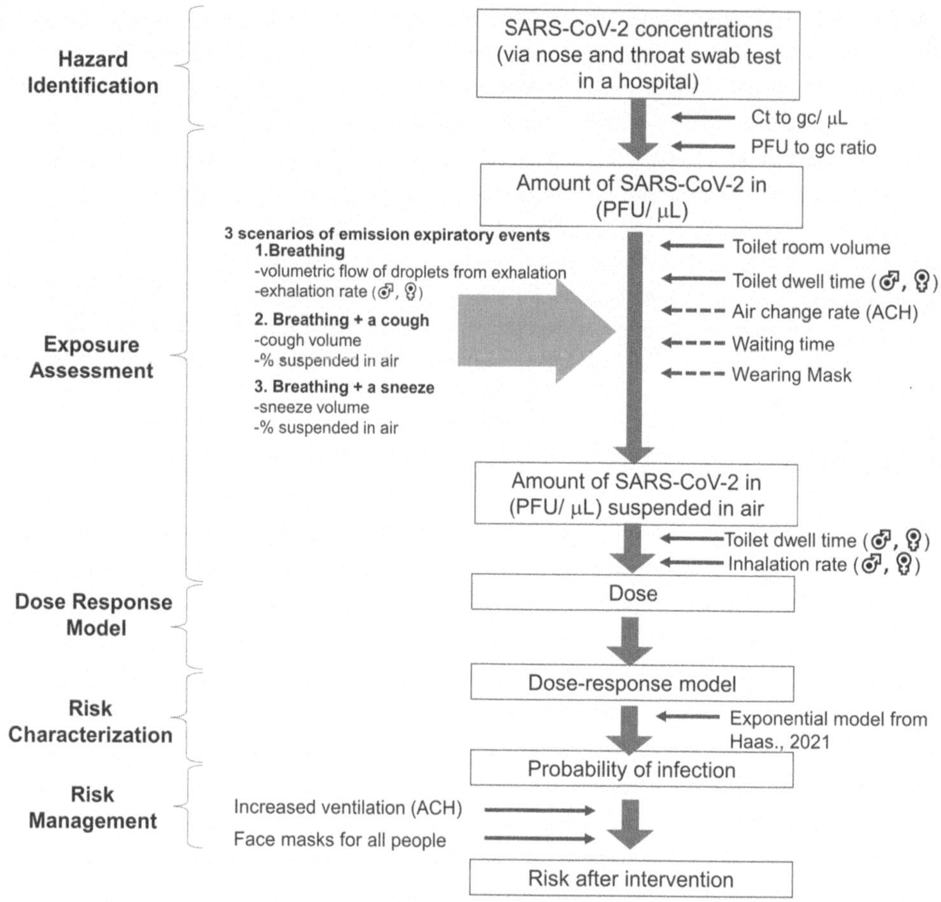

Figure 1 | Outline of the QMRA steps for the scenarios associated with SARS-CoV-2 respiratory transmission in a public toilet setting.

flow of droplets from an infector's exhalation (q_{br}) ranged from 5×10^{-9} to 6×10^{-6} μL/min (Schijven *et al.* 2021). In addition, the volume of aerosol droplets expelled per cough (V_{co}) and per sneeze (V_{sn}) was set in accordance with the results of the study by Schijven *et al.* (2021). However, the size of the droplets can play an important role in their activity. Larger droplets tend to deposit quickly, whereas smaller droplets (aerosols) can remain suspended in the air for a longer period. Thus, the volumetric ratios of aerosols to total droplets expelled (F) were considered using a droplet size ≤70 μm based on the size distribution of droplets in the literature for breathing (Fabian *et al.* 2011), coughing (Duguid 1946; Loudon & Roberts 1967; Han *et al.* 2013), and sneezing (Duguid 1946; Han *et al.* 2013; Supplementary Table S1).

$$\left\langle \begin{matrix}\text{Accumulation}\\ \text{within system}\end{matrix} \right\rangle = \left\langle \begin{matrix}\text{Flow in through}\\ \text{system boundary}\end{matrix} \right\rangle - \left\langle \begin{matrix}\text{Flow out through}\\ \text{system boundary}\end{matrix} \right\rangle - \left\langle \begin{matrix}\text{Reaction}\\ \text{within system}\end{matrix} \right\rangle \tag{1.1}$$

$$V\frac{dC}{dt} = q_{br}AR - \mu VC - q_{vent}C - q_{in}C \tag{1.2}$$

$$C_{t_1} = \frac{q_{br}AR}{(\mu V + q_{vent} + q_{in})}\left(1 - e^{-\frac{\mu V + q_{vent} + q_{in}}{V} t_1}\right) \tag{2}$$

$$C_{co} = \frac{A\, V_{co} R\, F}{V} \tag{3}$$

$$C_{sn} = \frac{A\, V_{sn}\, R\, F}{V} \tag{4}$$

where the parameters related to virus generation by an infector are: q_{br} is the volumetric flow rate of droplets from an infector's exhalation (μL-droplet/min); A is the virus concentrations in the genome copies per volume of droplets (gc/μL-droplet);

C_{co} is the additional infectious virus concentrations in the air caused by a cough (PFU/L); C_{sn} is the additional infectious virus concentrations in the air caused by a sneeze (PFU/L); R is the PFU/gc ratio; F is the fraction of aerosol volume per total volume of droplets expelled (dimensionless); V_{co} is the volume of droplets expelled per cough (μL/cough); V_{sn} is the volume of droplets expelled per sneeze (μL/sneeze); q_{in} is the inhalation rate (L/min); q_{vent} is the ventilation rate (L/min) that equals ACH×V/60; V is the volume of air in a cubicle (3,240 L-air); t_1 is infector's dwell time (min); μ is the inactivation rate in the air at 20–70% relative humidity levels (min^{-1}).

Virus levels during the waiting time

The toilet was unoccupied during the waiting time (t_2), and there was no virus input source (i.e., no infector). From Equations (1.1), (5), and (6), the infectious virus concentration decreased based on the ventilation and inactivation mechanism. The initial virus concentration (C_0) was applied in line with the three scenarios (i.e., breathing only, breathing with a cough, and breathing with a sneeze [Equations (7)–(9)]). In this study, practical waiting times of 3, 5, and 10 min were considered. If the receptor entered the cubicle immediately after the infector exited the cubicle, the waiting time was equal to 0.

$$V\frac{dC}{dt} = -\mu VC - q_{vent}C \tag{5}$$

$$C_{t_2} = C_0\left(e^{-\frac{\mu V + q_{vent}}{V}t_2}\right) \tag{6}$$

Given C_0 based on the following specific scenarios:

Breathing only $C_0 = C_{t_1}$ (7)

Breathing with a cough $C_0 = C_{t_1} + C_{co}$ (8)

Breathing with a sneeze $C_0 = C_{t_1} + C_{sn}$ (9)

Virus doses inhaled by the receptor

After the waiting time (t_2), the receptor entering the toilet was exposed to an infectious virus concentration of C_{t_2}. However, this infectious virus concentration was continuously removed by ventilation, inactivation, and inhalation. During the receptor's dwell time (t_3), the exposure to infectious virus concentrations (C_{t_3}) was calculated using Equation (10):

$$C_{t_3} = C_{t_2}\left(e^{-\frac{\mu V + q_{vent} + q_{in}}{V}t_3}\right) \tag{10}$$

The virus doses (d) that would be inhaled by a receptor were calculated by incorporating the inhalation rate (q_{in}) with a definite integral of the infectious virus concentration-time function (C_{t_3}). The limits of integration were set from the time of entering the cubicle to the receptor's dwell time (t_3) (Equations (11.1) and (11.2)). The initial exposure concentrations (C_{t_2}) also followed Equation (6) in line with the desired scenarios.

$$d = q_{in}\int_{t=0}^{t=t_3} C_{t_3}dt \tag{11.1}$$

$$d = \left(\frac{q_{in}VC_{t_2}}{\mu V + q_{vent} + q_{in}}\right)\left(1 - e^{-\frac{\mu V + q_{vent} + q_{in}}{V}t_3}\right) \tag{11.2}$$

where t_3 is receptor's dwell time (min) and d is SARS-CoV-2 infectious dose (PFU).

SARS-CoV-2 dose-response models

The risk assessment was conducted by following the QMRA framework. Given the lack of dose-response information for SARS-CoV-2, the SARS-CoV data sets (Watanabe *et al.* 2010) utilized in various SARS-CoV-2 QMRA studies (Cortellessa *et al.* 2021; Dada & Gyawali 2021; Murakami *et al.* 2021; Zaneti *et al.* 2021) were applied. The risk of infection followed the exponential model (Equation (12)):

$$P_{\text{event}} = 1 - e^{\left(-\frac{d}{k}\right)} \tag{12}$$

where P_{event} is the probability of infection per event (ppe) and k is the optimal dose-response function value of 18.54 obtained from Coronavirus 229E, which is equivalent to the chance that a single pathogen would initiate an infection response (Watanabe *et al.* 2010; Haas 2021).

Risk characterization and sensitivity analysis

To estimate the P_{event} for a receptor exposed to SARS-CoV-2, the data from the previous steps were integrated into the MCs with 10,000 iterations for each condition using Oracle Crystal Ball software version 11.1.2.4.850. The MCs is a randomization technique that uses repeated random sampling from distributions given to key input variables in a model, including corresponding sensitivity profiles. The risk of infection was displayed in the 5th percentile, mean, and 95th percentile using a forest plot in GraphPad Prism version 7.0. A sensitivity analysis was also conducted to determine the effects of the input variables on the risk calculation.

Risk management evaluation

Three risk mitigation interventions were investigated: face mask wearing, ventilation, and waiting times. The universal wearing of face masks has been recommended as a low-cost and efficient means of mitigating virus transmission (WHO 2020b). Among the different types of face masks, predominantly N95 respirators and surgical and fabric masks are used worldwide. Viral filtration efficiency (VFE) characterized using a bacteriophage MS2 following the ASTM F2101-14 standard testing method has revealed 99.8–100% VFE for N95 respirators, 99.3–99.8% VFE for surgical masks, and 54.8–92.1% VFE for denim fabric masks (Whiley *et al.* 2020; Supplementary Table S1). MS2 bacteriophages were selected as the model microbes because they are two to three times smaller in size than SARS-COV-2 (70–90 nm in diameter).

Ventilation is also an important element used to control indoor air quality in public toilets. Depending on the applicable regulatory building standard, either the installation of a mechanical ventilation system or the use of natural ventilation may be necessary. The effect of air change rates on SARS-CoV-2 transmission risk was considered in this study. For the pandemic, the WHO has suggested that ventilation in indoor spaces with aerosol-generating potential be greater than or equal to 12 ACH (WHO 2021). In this study, five air change rates were tested: 0 ACH (no ventilation), 0.5 ACH (poor ventilation), 10 ACH (DIN 1946 ventilation standard for public toilets), 12 ACH (WHO-recommended standard ventilation), and 20 ACH (extreme ventilation) (Supplementary Table S1). Furthermore, the waiting times, during which the infector exited the cubicle and the receptor entered the cubicle, of 3, 5, and 10 min were considered.

RESULTS AND DISCUSSION

Infection risk from respiratory transmission in public toilets

The risk of infection from SARS-CoV-2 transmission through three respiratory exposure scenarios, namely breathing, breathing with a cough, and breathing with a sneeze, were characterized in this study. The probability of infection per event was not found to be significantly different between men and women ($p>0.05$; Mann–Whitney U test) across all scenarios (Figure 2 and Supplementary Table S2). Although men usually have a higher breathing rate (Stifelman 2007), which results in greater exposure to viruses, they spend on average around 22% less time in toilets than women (Gwynne *et al.* 2019), leading to a reduced risk of virus transmission. When a receptor without a protective mask was in an unventilated public toilet, the mean infection risk was 2.5×10^{-2} ppe for men and 3.1×10^{-2} ppe for women in the infector breathing scenario (Figure 2 and Supplementary Table S2). Interestingly, the risk values increased sharply when additional viral loads were expelled into the air by an infector either sneezing or coughing. Coughing and sneezing can produce saliva droplets of various sizes (Duguid 1946; Loudon & Roberts 1967; Han *et al.* 2013) and thus generate infectious virus-containing aerosols in public toilet facilities. For breathing with a cough, the mean risk was higher at 4.5×10^{-1} ppe for men and 4.6×10^{-1} ppe for women. Similarly, sneezing increased the risk of infection to 8.5×10^{-1} ppe and 8.6×10^{-1} ppe for men and women,

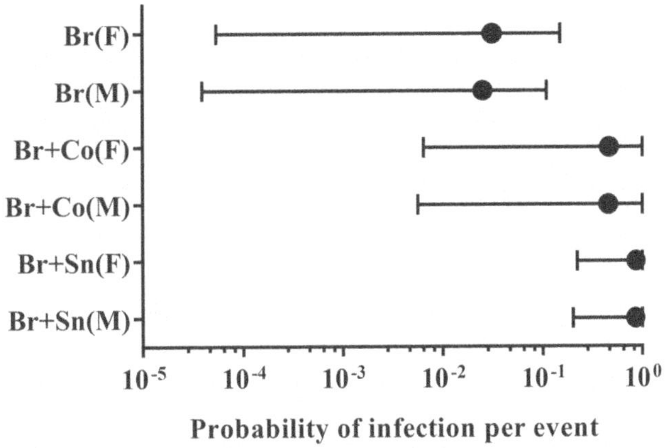

Figure 2 | The risk of infection per event for a male (M) receptor and a female (F) receptor in three virus-generating scenarios: an infector breathing (Br), breathing with a cough (Br+Co), and breathing with a sneeze (Br+Sn). The forest plots show the mean values in solid circles and 90% confidence intervals (ranging from the 5th [left whiskers] to the 95th [right whiskers] percentiles).

respectively. We, therefore, showed that receptors have a high risk of infection when using an unventilated public toilet without wearing a protective mask.

Risk mitigation

Face mask wearing by either an infector or a receptor

Because it delivered the highest risk, the scenario with an infector breathing with a sneeze was selected to further evaluate the effectiveness of face mask wearing to reduce infection risk in a receptor. The probabilities of infection per event for different mask types are shown in Figure 3 and Supplementary Table S3. The highest risk reduction at an approximately 2-log was observed when an N95 respirator was worn by either an infector or a receptor. The results indicated that a high risk of virus transmission in confined spaces like public toilets is, therefore, still possible even if an N95 respirator or surgical mask is worn. This could be because the risk of infection is associated with several factors, including a high concentration of virus aerosols in the ambient air due to insufficient ventilation, long dwell times by toilet users, the low inactivation rate of SAR-CoV-2, and the leakage of air around a mask's edges (Brooks *et al.* 2021; Gwynne *et al.* 2019; Schuit *et al.* 2020; Whiley *et al.* 2020; Stabile *et al.* 2021). We also observed a reduction in risk as low as 0.14-$\log_{10}$ when an infector or receptor wore a denim fabric mask.

Face mask wearing by both an infector and a receptor

In a scenario with an infector breathing with a sneeze, the infection risk could be further reduced if both the infector and receptor wear masks (Supplementary Table S3). The receptor's gender did not affect their risk significantly in any of the

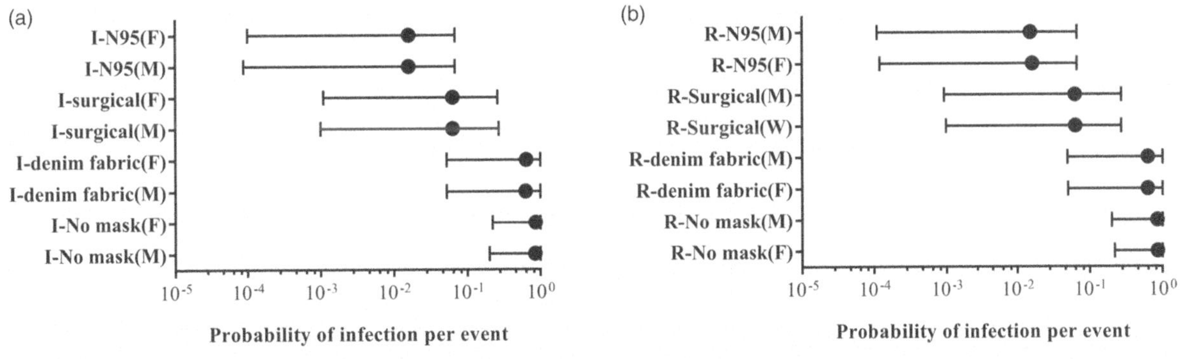

Figure 3 | The risk of infection per event among male (M) and female (F) receptors in the scenario with an infector (I) breathing with a sneeze (Br+Sn) when (a) only the infector wore different types of masks and (b) only the receptor (R) wore different types of masks. The forest plots show the mean values in solid circles and 90% confidence intervals (ranging from the 5th [left whiskers] to the 95th [right whiskers] percentiles).

conditions. The risks to a female receptor are illustrated in Figure 4. When an infector and a receptor both wore either a surgical mask or an N95 respirator, the mean infection risk decreased by 3.4–4.7 $\log_{10}$. With only a 0.4 $\log_{10}$ reduction, denim fabric masks may not provide sufficient protection, even when worn by both an infector and a receptor. Based on our study findings, we recommend that people, as receptors, wear either a surgical mask or an N95 respirator as personal protective equipment to minimize the risk of infection in unventilated public toilets and potentially in other confined communal spaces. In general, we found that wearing a mask was one of the most low-cost and simple yet effective intervention measures that can be used in daily life to minimize transmission risk, which is consistent with reports from other studies (Asadi *et al.* 2020b; Chu *et al.* 2020; WHO 2020b; Cheng *et al.* 2021; Goyal *et al.* 2021). Moreover, in confined communal spaces, other measures, including increasing ventilation, and extended waiting times, should be implemented to reduce the risk of infection.

Ventilation and extended waiting times

The effects of ventilation (0–20 ACH) were characterized for three virus-generating scenarios: infector breathing, breathing with a cough, and breathing with a sneeze (Supplementary Table S4). The results showed that increasing ACH did not significantly mitigate the risk of COVID-19 infection in a public toilet setting if there was a high turnover of people using a toilet. Although a high ventilation rate has been suggested as a way to reduce the number of virus-containing droplets and aerosols in the air (Morawska *et al.* 2020; Li *et al.* 2021; Stabile *et al.* 2021; WHO 2021), reducing infectious virus concentrations to a low-risk level is time-consuming under completely mixed conditions. Moreover, the continuous expelling of the SARS-CoV-2 virus from breathing and/or sneezing by an infector without a mask appeared to be a significant cause of virus aerosol accumulation in ambient air. To exchange the clean air in a toilet cubicle, a sufficient waiting time should be considered. For example, in a scenario with an infector breathing with a sneeze and at the 12 ACH suggested by the WHO (2021), the mean probabilities of infection per event for the female receptor were as high as 7.4×10^{-1}, 6.7×10^{-1} and 4.7×10^{-1} ppe for waiting times of 3, 5, and 10 min, respectively (Supplementary Figure S1). This study demonstrated that indoor ventilation alone cannot effectively reduce SARS-CoV-2 transmission risk in a public toilet setting with small cubicles and is less effective in risk reduction than face mask wearing. Another study similarly found that masks could reduce the infection risk caused by the Middle Eastern respiratory syndrome coronavirus in an indoor hospital setting better than ventilation (Adhikari *et al.* 2019). Multiple measures should, therefore, be implemented to mitigate infection risk to an acceptable level.

Combination of face mask wearing, ventilation, and waiting time

The addition of face mask wearing was investigated for its combined effectiveness in mitigating SARS-CoV-2 transmission risk when used simultaneously with increased ventilation and extended waiting times. In the virus-generating scenario with an infector breathing with a sneeze, mask wearing by both the infector and receptor was assessed using the 12 ACH

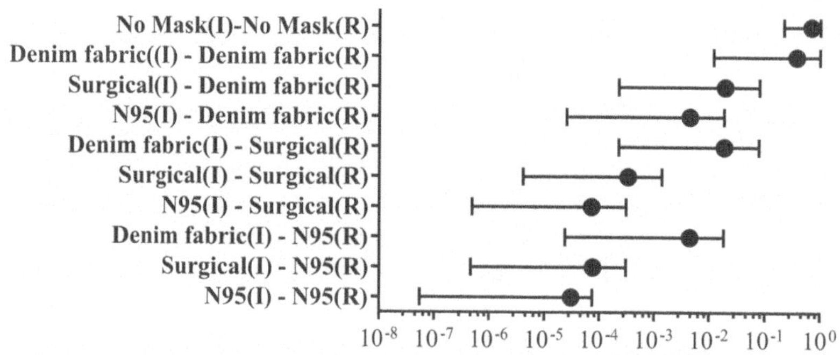

Figure 4 | The risk of infection per event among male (M) and female (F) receptors in the scenario of no ventilation (0 ACH) with an infector (I) breathing with a sneeze (Br+Sn) and the receptor (R) wearing different types of masks. The risk of infection per event in a scenario with an infector (I) breathing with a sneeze (Br+Sn) when both the infector and receptor (R) wore different types of masks. The risks to a female receptor are represented because no gender effect was evident. The forest plots show the mean values in solid circles and 90% confidence intervals (ranging from the 5th [left whiskers] to the 95th [right whiskers] percentiles).

WHO-recommended value and varied waiting times of 3, 5, and 10 min (Figure 5 and Supplementary Table S5). Compared with no waiting time (Figure 5(a)), increasing the waiting time to 10 min at 12 ACH showed a 1.0-$\log_{10}$ reduction in the mean and 95th percentile infection risks in most double-mask-wearing cases. Nevertheless, a waiting time of 10 min for public toilets may not be practical in certain circumstances. Consequently, we reiterate that proper ventilation did not impact the risk mitigation for SARS-CoV-2 transmission in a public toilet setting, especially in confined toilet cubicle conditions. Face mask wearing should, therefore, be promoted as a normal practice when entering public indoor spaces. Moreover, wearing a more tightly fitting mask is also critical for risk mitigation because it can reduce leakage at the mask's edges (Brooks *et al.* 2021).

Sensitivity analysis

A sensitivity analysis of the QMRA was conducted to identify the input variables that most contributed to the risk estimation. An infector's expelled volumes and concentrations of SARS-CoV-2 in droplets (saliva and mucus) play an important role in infection risk. For all three transmission scenarios, namely an infector breathing (Br), breathing with a cough (Br+Co), and breathing with a sneeze (Br+Sn), the infector's expelled volume (μL) of the virus (breathing, coughing, and sneezing volumes) was the most sensitive parameter, accounting for 71.7, 74.6, and 46.5% contribution in the probability of infection in women, respectively (Figure 6 and Supplementary Table S6). The second-most sensitive parameter was concentrations of the SARS-CoV-2 virus in gc/μL droplets (saliva and mucus), which accounted 16.2, 15.0, and 38.6% in women, respectively. The other parameters, namely a receptor's dwell time, an infector's dwell time, and the percentage of droplet suspension, were minor parameters in terms of sensitivity (Figure 6 and Supplementary Table S6). Since breathing with a sneeze was the highest virus-generating risk scenario, a sensitivity analysis was performed in which both the infector and receptor wore masks (Supplementary Table S7). Virus concentrations in gc/μL, sneeze volume, and the receptor's dwell time were the three parameters that most influenced infection risk. Since controlling for virus concentrations and an infector's expelled volumes are a challenge, particularly among asymptomatic patients, individuals should avoid spending prolonged time in closed indoor settings (Dancer *et al.* 2021; Stabile *et al.* 2021).

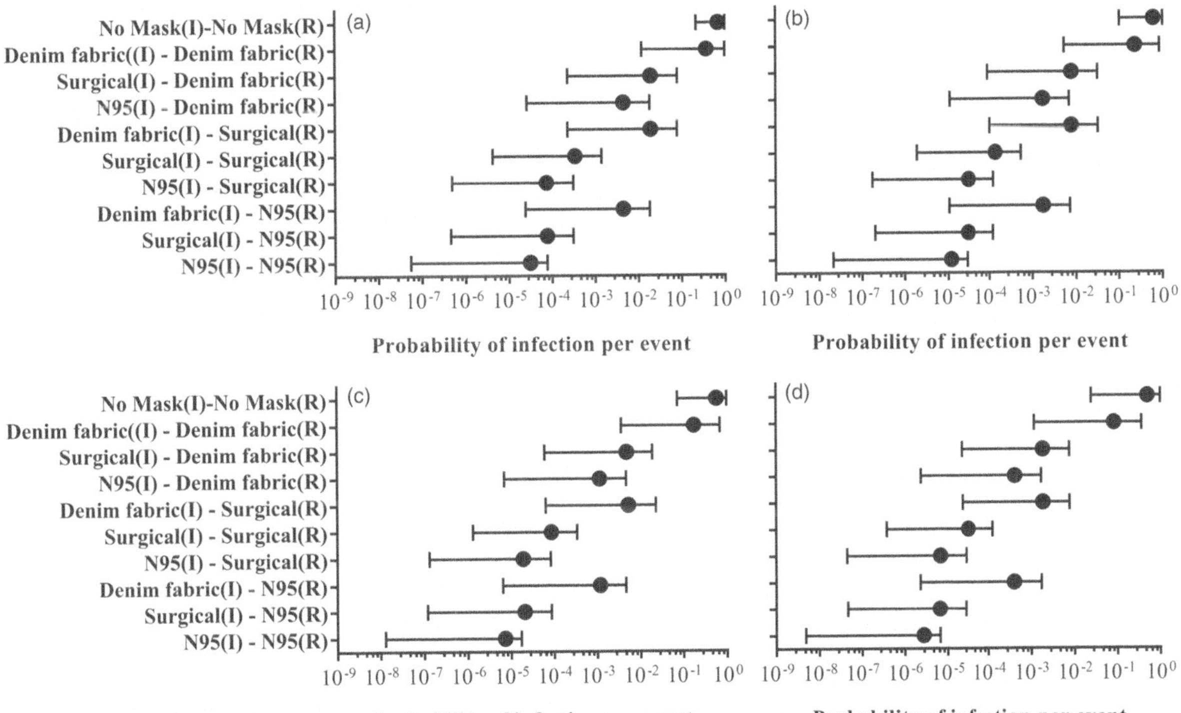

Figure 5 | The risk of infection per event in a scenario with an infector (I) breathing with a sneeze (Br+Sn) when both the infector and receptor (R) wore different types of masks. The risks to a female receptor are represented because no gender effect was evident. The influence of waiting times of (a) 0 min, (b) 3 min, (c) 5 min, and (d) 10 min at 12 ACH are displayed. The forest plots show the mean values in solid circles and 90% confidence intervals (ranging from the 5th [left whiskers] to the 95th [right whiskers] percentiles).

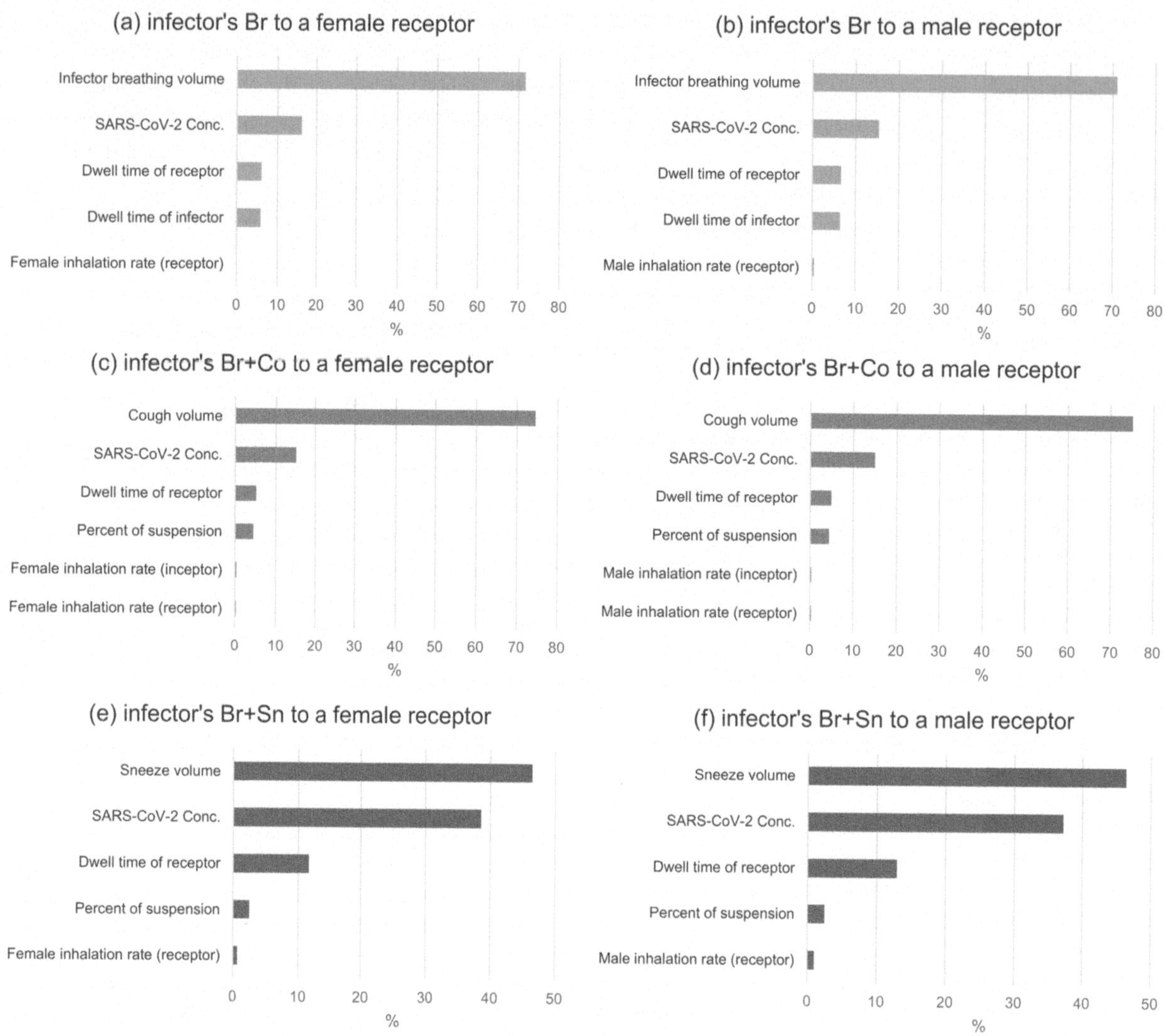

Figure 6 | The sensitivity analysis representing the contribution of input variables to the risk of infection per event for male and female receptors with no ventilation and no mask wearing in three transmission scenarios, namely with the infector breathing (Br), breathing with a cough (Br+Co), and breathing with a sneeze (Br+Sn): (a) infector's Br to a female receptor, (b) infector's Br to a male receptor, (c) infector's Br+Co to a female receptor, (d) infector's Br+Co to a male receptor, (e) infector's Br+Sn to a female receptor, and (f) infector's Br+Sn to a male receptor.

Limitations of this study and future perspectives

While this study evaluated the risk of SARS-CoV-2 transmission according to the QMRA framework, its limitations and uncertainties should be carefully acknowledged. SARS-CoV-2 concentrations in gc/μL, the most sensitive parameter affecting the calculation of risk, are subject to natural variations in the saliva and mucus of infected patients (Azzi *et al.* 2020; Wölfel *et al.* 2020). In this study, 251 swab test C_t values were used to represent the virus levels in Thai patients. Due to the lack of a standard curve from Thai hospital laboratories, we used a published standard curve of the N2 gene (Sherchan *et al.* 2020) to estimate the virus concentrations in this study. However, heterogeneity in published standard curves for SARS-CoV-2 has been observed (Bivins *et al.* 2021). Moreover, variations in technical and laboratory analyses (e.g., data analysis methods and control materials) could intensify bias, leading to variability in the calculated virus concentrations (Kongprajug *et al.* 2020; Bivins *et al.* 2021). Adhering to standards and quality control measures is, therefore, highlighted to support data sharing and referencing in future research, especially for emerging infectious diseases. Notwithstanding, even with consideration of the aforementioned uncertainties, the calculated virus concentrations in droplets used in this study, which ranged from 4.4×10^{-1} to 6.4×10^{8} gc/μL, were in agreement with those of Schijven *et al.* (2021).

We chose to evaluate three virus-generating scenarios: an infector breathing, breathing with a cough, and breathing with a sneeze. However, infectors may sneeze and/or cough more than once depending on the individuals' symptoms, which could increase the risk of infection. Coughing is the predominant symptom in COVID-19 (Wang *et al.* 2020), rendering coughing potentially more important than sneezing. Nevertheless, some studies have suggested that the airborne transmission of infectious diseases is possible without coughing or sneezing and simply through the exhaled breath of individuals with few symptoms (Asadi *et al.* 2020a). In addition, the lack of dose-response information and infection risk benchmarks for SARS-CoV-2 poses a challenge when evaluating the infection risk of this virus. We assumed that the dose-response of SARS-CoV-2 was similar to that indicated in the SARS-CoV data from Haas (2021) study, which refers to Coronavirus 229E. With the recent emergence of various SARS-CoV-2 variants, much remains unknown regarding the behavior and characteristics of this virus. In this study, we used the available inactivation coefficients of SARS-CoV-2 at 20 °C (Schuit *et al.* 2020), which could have overestimated the calculated risks in Thailand given its average daily temperature of 27 °C (Denpetkul & Phosri 2021).

Even though virus protection efficiencies using different mask types are important for risk protection, we excluded the factor of leakage from a mask's edges due to the different methods of mask wearing (i.e., double masks, knotted and tucked masks). Brooks *et al.* (2021) indicated that mask wearing methods play a role in mask protection efficiencies. In their study, wearing a surgical mask alone blocked 56.1% of the particles from a cough, while a cloth mask alone blocked 51.4% of such particles. Meanwhile, a knotted and tucked surgical mask blocked 77.0% of cough particles, and double masking blocked 85.4%. Leakage based on wearing methods should be included in future risk assessments.

The scope of this study excluded the risk of SARS-CoV-2 respiratory transmission potentially produced by toilet flushing, as well as other transmission risks (e.g., direct splashing and surface transmission) (Verani *et al.* 2014). As evident from the study by McKinney *et al.* (2006) on SARS-CoV-1 in-building transmission in Hong Kong, it must be emphasized that well-designed and standard maintenance procedures for the plumbing and ventilation systems in toilets could play an important role in controlling infectious droplets. By integrating all the known risk sources, comprehensive knowledge regarding risk estimation could be achieved to accurately inform public health policies and help further reduce transmission risk. It is apparent that research related to SARS-CoV-2 is continuing, and additional data will greatly benefit future studies that aim to better understand its characteristics. The QMRA-based risk models developed in this study could facilitate future risk assessments through modifications for particular risk scenarios and the updating of the input parameters based on newly available data. Such improved risk models will be crucial tools in assessing the impact of different risk mitigation strategies during the COVID-19 and future pandemics.

CONCLUSION

Indoor public toilet facilities could be hubs of virus transmission during the COVID-19 pandemic. This study investigated the risk of the airborne transmission of SARS-CoV-2 in public toilets for three virus-generating scenarios: an infector breathing, breathing with a cough, and breathing with a sneeze. The risk analysis, which followed the QMRA framework, revealed that the highest risk was when an asymptomatic or symptomatic infector sneezed. Both genders were found to be exposed to similar risks. A combination of measures that includes suitable ventilation, extended waiting times, and mask wearing should be applied as an effective intervention for public toilet users.

ACKNOWLEDGEMENTS

This work was financially supported by the Thailand Science Research and Innovation (TSRI), Chulabhorn Research Institute (grant no. 2536702/43410), the Mahidol University and the Kurita Overseas Research Grant (Grant No. 21Pth008-38U to T.D.) provided by Kurita Water and Environment Foundation.

CONFLICT OF INTEREST

The authors declare that they have no conflict of interest in this work. This study was approved by the Ethics Committee of the Faculty of Tropical Medicine, Mahidol University, Thailand (Ethical number MUTM 2021-060-01).

DATA AVAILABILITY STATEMENT

All relevant data are included in the paper or its Supplementary Information.

REFERENCES

Adhikari, U., Chabrelie, A., Weir, M., Boehnke, K., McKenzie, E., Ikner, L., Wang, M., Wang, Q., Young, K., Haas, C. N., Rose, J. & Mitchell, J. 2019 A case study evaluating the risk of infection from middle eastern respiratory syndrome coronavirus (MERS-CoV) in a hospital setting through bioaerosols. *Risk Analysis* **39** (12), 2608–2624.

Albert, S., Ruíz, A., Pemán, J., Salavert, M. & Domingo-Calap, P. 2021 Lack of evidence for infectious SARS-CoV-2 in feces and sewage. *European Journal of Clinical Microbiology & Infectious Diseases* **0123456789**, 2–4. https://doi.org/10.1007/s10096-021-04304-4.

Asadi, S., Bouvier, N., Wexler, A. S. & Ristenpart, W. D. 2020a The coronavirus pandemic and aerosols: does COVID-19 transmit via expiratory particles? *Aerosol Science and Technology* **54** (6), 635–638.

Asadi, S., Cappa, C. D., Barreda, S., Wexler, A. S., Bouvier, N. M. & Ristenpart, W. D. 2020b Efficacy of masks and face coverings in controlling outward aerosol particle emission from expiratory activities. *Scientific Reports* **10** (1), 15665.

Azzi, L., Carcano, G., Gianfagna, F., Grossi, P., Gasperina, D. D., Genoni, A., Fasano, M., Sessa, F., Tettamanti, L., Carinci, F., Maurino, V., Rossi, A., Tagliabue, A. & Baj, A. 2020 Saliva is a reliable tool to detect SARS-CoV-2. *Journal of Infection* **81** (1), e45–e50. https://doi.org/10.1016/j.jinf.2020.04.005.

Birgand, G., Peiffer-Smadja, N., Fournier, S., Kerneis, S., Lescure, F. X. & Lucet, J. C. 2020 Assessment of air contamination by SARS-CoV-2 in hospital settings. *JAMA Network Open* **3** (12), 1–14.

Bivins, A., Kaya, D., Bibby, K., Simpson, S. L., Bustin, S. A., Shanks, O. C. & Ahmed, W. 2021 Variability in RT-qPCR assay parameters indicates unreliable SARS-CoV-2 RNA quantification for wastewater surveillance. *Water Research* **117516**. https://doi.org/10.1016/j.watres.2021.117516.

Brooks, J. T., Beezhold, D. H., Noti, J. D., Coyle, J. P., Derk, R. C., Blachere, F. M. & Lindsley, W. G. 2021 Maximizing fit for cloth and medical procedure masks to improve performance and reduce SARS-CoV-2 transmission and exposure, 2021. *MMWR Surveillance Summaries* **70** (7), 254–257.

CDC 2021 COVID-19 Scientific Brief: SARS-CoV-2 Transmission. Updated May 7, 2021.

Cheng, Y., Cheng, Y., Ma, N., Witt, C., Rapp, S., Wild, P. S. & Andreae, M. O. 2021 Face masks effectively limit the probability of SARS-CoV-2 transmission. *Science* **6296**, 1–11.

Chu, D. K., Akl, E. A., Duda, S., Solo, K., Yaacoub, S., Schünemann, H. J., El-harakeh, A., Bognanni, A., Lotfi, T., Loeb, M., Hajizadeh, A., Bak, A., Izcovich, A., Cuello-Garcia, C. A., Chen, C., Harris, D. J., Borowiack, E., Chamseddine, F., Schünemann, F., Morgano, G. P., Muti Schünemann, G. E. U., Chen, G., Zhao, H., Neumann, I., Chan, J., Khabsa, J., Hneiny, L., Harrison, L., Smith, M., Rizk, N., Giorgi Rossi, P., AbiHanna, P., El-khoury, R., Stalteri, R., Baldeh, T., Piggott, T., Zhang, Y., Saad, Z., Khamis, A. & Reinap, M. 2020 Physical distancing, face masks, and eye protection to prevent person-to-person transmission of SARS-CoV-2 and COVID-19: a systematic review and meta-analysis. *The Lancet* **395** (10242), 1973–1987.

Cortellessa, G., Stabile, L., Arpino, F., Faleiros, D. E., van den Bos, W., Morawska, L. & Buonanno, G. 2021 Close proximity risk assessment for SARS-CoV-2 infection. *Science of the Total Environment* **794**, 148749. https://doi.org/10.1016/j.scitotenv.2021.148749.

Covés-Datson, E. M., King, S. R., Legendre, M., Gupta, A., Chan, S. M., Gitlin, E., Kulkarni, V. V., García, J. P., Smee, D. F., Lipka, E., Evans, S. E., Tarbet, E. B., Ono, A. & Markovitz, D. M. 2020 A molecularly engineered antiviral banana lectin inhibits fusion and is efficacious against influenza virus infection in vivo. *PNAS* **117** (4), 2122–2132.

Dada, A. C. & Gyawali, P. 2021 Quantitative microbial risk assessment (QMRA) of occupational exposure to SARS-CoV-2 in wastewater treatment plants. *Science of the Total Environment* **763**, 142989. https://doi.org/10.1016/j.scitotenv.2020.142989.

Dancer, S. J., Li, Y., Hart, A., Tang, J. W. & Jones, D. L. 2021 What is the risk of acquiring SARS-CoV-2 from the use of public toilets? *Science of the Total Environment* **148341**. https://doi.org/10.1016/j.phrs.2020.104743.

Denpetkul, T. & Phosri, A. 2021 Daily ambient temperature and mortality in Thailand: estimated effects, attributable risks, and effect modifications by greenness. *Science of the Total Environment* **791**, 148373. https://doi.org/10.1016/j.scitotenv.2021.148373.

Ding, Z., Qian, H., Xu, B., Huang, Y., Miao, T., Yen, H., Xiao, S., Cui, L., Wu, X., Shao, W., Song, Y., Sha, L., Zhou, L., Xu, Y., Zhu, B. & Li, Y. 2021 Toilets dominate environmental detection of severe acute respiratory syndrome coronavirus 2 in a hospital. *Science of the Total Environment* **753**, 141710. https://doi.org/10.1016/j.scitotenv.2020.141710.

Duguid, J. P. 1946 The size and the duration of air-carriage of respiratory droplets and droplet-nuclei. *Journal of Hygiene* **44** (6), 471–479.

Fabian, P., Brain, J., Houseman, E. A., Gern, J. & Milton, D. K. 2011 Origin of exhaled breath particles from healthy and human rhinovirus-infected subjects. *Journal of Aerosol Medicine and Pulmonary Drug Delivery* **24** (3), 137–147.

Goyal, A., Reeves, D. B., Thakkar, N., Famulare, M., Cardozo-Ojeda, E. F., Mayer, B. T. & Schiffer, J. T. 2021 Slight reduction in SARS-CoV-2 exposure viral load due to masking results in a significant reduction in transmission with widespread implementation. *Scientific Reports* **11** (1), 1–12. https://doi.org/10.1038/s41598-021-91338-5.

Gwynne, S. M. V., Hunt, A. L. E., Thomas, J. R., Thompson, A. J. L. & Séguin, L. 2019 The toilet paper: bathroom dwell time observations at an airport. *Journal of Building Engineering* **24**, 100751. https://doi.org/10.1016/j.jobe.2019.100751.

Haas, C. N. 2021 Action levels for SARS-CoV-2 in air: preliminary approach. *Risk Analysis: An Official Publication of the Society for Risk Analysis* **41** (5), 705–709.

Haas, C. N., Rose, J. B. & Gerba, C. P. 2014 *Quantitative Microbial Risk Assessment*. Wiley, Hoboken, NJ.

Han, Z. Y., Weng, W. G. & Huang, Q. Y. 2013 Characterizations of particle size distribution of the droplets exhaled by sneeze. *Journal of the Royal Society Interface* **10** (88), 24026469.

Jones, D. L., Baluja, M. Q., Graham, D. W., Corbishley, A., McDonald, J. E., Malham, S. K., Hillary, L. S., Connor, T. R., Gaze, W. H., Moura, I. B., Wilcox, M. H. & Farkas, K. 2020 Shedding of SARS-CoV-2 in feces and urine and its potential role in person-to-person transmission and the environment-based spread of COVID-19. *Science of the Total Environment* **749**, 141364. https://doi.org/10.1016/j.scitotenv.2020.141364.

Kim, Y. I., Kim, S. G., Kim, S. M., Kim, E. H., Park, S. J., Yu, K. M., Chang, J. H., Kim, E. J., Lee, S., Casel, M. A. B., Um, J., Song, M. S., Jeong, H. W., Lai, V. D., Kim, Y., Chin, B. S., Park, J. S., Chung, K. H., Foo, S. S., Poo, H., Mo, I. P., Lee, O. J., Webby, R. J., Jung, J. U. & Choi, Y. K. 2020 Infection and rapid transmission of SARS-CoV-2 in ferrets. *Cell Host and Microbe* **27** (5), 704–709. https://doi.org/10.1016/j.chom.2020.03.023.

Kongprajug, A., Chyerochana, N., Mongkolsuk, S. & Sirikanchana, K. 2020 Effect of quantitative polymerase chain reaction data analysis using sample amplification efficiency on microbial source tracking assay performance and source attribution. *Environmental Science & Technology* **54**, 8232–8244.

Li, Y., Qian, H., Hang, J., Chen, X., Cheng, P., Ling, H., Wang, S., Liang, P., Li, J., Xiao, S., Wei, J., Liu, L., Cowling, B. J. & Kang, M. 2021 Probable airborne transmission of SARS-CoV-2 in a poorly ventilated restaurant. *Building and Environment* **196**, 107788. https://doi.org/10.1016/j.buildenv.2021.107788.

Loudon, R. G. & Roberts, R. M. 1967 Droplet expulsion from the respiratory tract. *American Review of Respiratory Disease* **95** (3), 435–442.

Maestre, J. P., Jarma, D., Yu, C., Siegel, J., Horner, S. & Kinney, K. A. 2021 Distribution of SARS-CoV-2 RNA signal in a home with COVID-19 positive occupants. *Science of the Total Environment* **778**, 146201. https://doi.org/10.1016/j.scitotenv.2021.146201.

McKinney, K. R., Gong, Y. Y. & Lewis, T. G. 2006 Environmental transmission of SARS at Amoy Gardens. *Journal of Environmental Health* **68** (9), 26–30.

Ministry of Public Health. 2016 *Standard Construction Plan and Layout for Public Toilets*. Bureau of Environmental Health, Department of Health, Ministry of Public Health, Thailand.

Morawska, L., Tang, J. W., Bahnfleth, W., Bluyssen, P. M., Boerstra, A., Buonanno, G., Cao, J., Dancer, S., Floto, A., Franchimon, F., Haworth, C., Hogeling, J., Isaxon, C., Jimenez, J. L., Kurnitski, J., Li, Y., Loomans, M., Marks, G., Marr, L. C., Mazzarella, L., Melikov, A. K., Miller, S., Milton, D. K., Nazaroff, W., Nielsen, P. V., Noakes, C., Peccia, J., Querol, X., Sekhar, C., Seppänen, O., Tanabe, S. i., Tellier, R., Tham, K. W., Wargocki, P., Wierzbicka, A. & Yao, M. 2020 How can airborne transmission of COVID-19 indoors be minimised? *Environment International* **142**, 105832.

Murakami, M., Miura, F., Kitajima, M., Fujii, K., Yasutaka, T., Iwasaki, Y., Ono, K., Shimazu, Y., Sorano, S., Okuda, T., Ozaki, A., Katayama, K., Nishikawa, Y., Kobashi, Y., Sawano, T., Abe, T., Saito, M. M., Tsubokura, M., Naito, W. & Imoto, S. 2021 COVID-19 risk assessment at the opening ceremony of the Tokyo 2020 Olympic Games. *Microbial Risk Analysis* 100162. https://doi.org/10.1016/j.mran.2021.100162.

Pitol, A. K. & Julian, T. R. 2021 Community transmission of SARS-CoV-2 by surfaces: risks and risk reduction strategies. *Environmental Science and Technology Letters* **8** (3), 263–269.

Rose, J. B. & Gerba, C. P. 1991 Use of risk assessment for development of microbial standards. *Water Science and Technology* **24** (2), 29–34.

Schijven, J., Vermeulen, L. C., Swart, A., Meijer, A., Duizer, E. & de Roda Husman, A. M. 2021 Quantitative microbial risk assessment for airborne transmission of SARS-CoV-2 via breathing, speaking, singing, coughing, and sneezing. *Environmental Health Perspectives* **129** (4), 1–10.

Schuit, M., Ratnesar-Shumate, S., Yolitz, J., Williams, G., Weaver, W., Green, B., Miller, D., Krause, M., Beck, K., Wood, S., Holland, B., Bohannon, J., Freeburger, D., Hooper, I., Biryukov, J., Altamura, L. A., Wahl, V., Hevey, M. & Dabisch, P. 2020 Airborne SARS-CoV-2 is rapidly inactivated by simulated sunlight. *Journal of Infectious Diseases* **222** (4), 564–571.

Sherchan, S. P., Shahin, S., Ward, L. M., Tandukar, S., Aw, T. G., Schmitz, B., Ahmed, W. & Kitajima, M. 2020 First detection of SARS-CoV-2 RNA in wastewater in North America: a study in Louisiana, USA. *Science of the Total Environment* **743**, 140621. https://doi.org/10.1016/j.scitotenv.2020.140621.

Shi, K. W., Huang, Y. H., Quon, H., Ou-Yang, Z. L., Wang, C. & Jiang, S. C. 2021 Quantifying the risk of indoor drainage system in multi-unit apartment building as a transmission route of SARS-CoV-2. *Science of the Total Environment* **762**, 143056. https://doi.org/10.1016/j.scitotenv.2020.143056.

Stabile, L., Pacitto, A., Mikszewski, A., Morawska, L. & Buonanno, G. 2021 Ventilation procedures to minimize the airborne transmission of viruses in classrooms. *Building and Environment* **202**, 108042. https://doi.org/10.1016/j.buildenv.2021.108042.

Stifelman, M. 2007 Using doubly-labeled water measurements of human energy expenditure to estimate inhalation rates. *Science of the Total Environment* **373** (2–3), 585–590.

Verani, M., Bigazzi, R. & Carducci, A. 2014 Viral contamination of aerosol and surfaces through toilet use in health care and other settings. *American Journal of Infection Control* **42**, 758–762.

Wang, Z., Yang, B., Li, Q., Wen, L. & Zhang, R. 2020 Clinical features of 69 cases with coronavirus disease 2019 in Wuhan, China. *Clinical Infectious Diseases* **71** (15), 769–777.

Watanabe, T., Bartrand, T. A., Weir, M. H., Omura, T. & Haas, C. N. 2010 Development of a dose-response model for SARS coronavirus. *Risk Analysis: An International Journal* **30** (7), 1129–1138.

Whiley, H., Keerthirathne, T. P., Nisar, M. A., White, M. A. F. & Ross, K. E. 2020 Viral filtration efficiency of fabric masks compared with surgical and N95 masks. *Pathogens* **9** (9), 1–8.

WHO 2020a *Considerations for Public Health and Social Measures in the Workplace in the Context of COVID-19.* World Health Organisation, (May), 1. Available from: https://www.who.int/publications-detail/risk (accessed August 7, 2021).

WHO 2020b *Coronavirus Disease (COVID-19) Advice for the Public: When and How to Use Masks.* (December 2020). Available from: https://www.who.int/emergencies/diseases/novel-coronavirus-2019/advice-for-public/when-and-how-to-use-masks (accessed August 8, 2021).

WHO 2021 *Roadmap to Improve and Ensure Good Indoor Ventilation in the Context of COVID-19.* ISBN 978-92-4-002128-0 (electronic version). World Health Organization, Geneva.

Wölfel, R., Corman, V. M., Guggemos, W., Seilmaier, M., Zange, S., Müller, M. A., Niemeyer, D., Jones, T. C., Vollmar, P., Rothe, C., Hoelscher, M., Bleicker, T., Brünink, S., Schneider, J., Ehmann, R., Zwirglmaier, K., Drosten, C. & Wendtner, C. 2020 Virological assessment of hospitalized patients with COVID-2019. *Nature* **581** (7809), 465–469.

Xiao, F., Sun, J., Xu, Y., Li, F., Huang, X., Li, H., Zhao, J., Huang, J. & Zhao, J. 2020 Infectious SARS-CoV-2 in feces of patient with severe COVID-19. *Emerging Infectious Diseases* **26** (8), 1920–1922.

Zaneti, R. N., Girardi, V., Spilki, F. R., Mena, K., Westphalen, A. P. C., da Costa Colares, E. R., Pozzebon, A. G. & Etchepare, R. G. 2021 Quantitative microbial risk assessment of SARS-CoV-2 for workers in wastewater treatment plants. *Science of the Total Environment* **754**, 142163. https://doi.org/10.1016/j.scitotenv.2020.142163.

Zang, R., Castro, M. F. G., McCune, B. T., Zeng, Q., Rothlauf, P. W., Sonnek, N. M., Liu, Z., Brulois, K. F., Wang, X., Greenberg, H. B., Diamond, M. S., Ciorba, M. A., Whelan, S. P. J. & Ding, S. 2020 TMPRSS2 and TMPRSS4 promote SARS-CoV-2 infection of human small intestinal enterocytes. *Science Immunology* **5** (47), 1–14.

First received 13 August 2021; accepted in revised form 4 February 2022. Available online 11 February 2022

SARS-CoV-2 in Water doi: 10.2166/wh.2021.140

Determinants of hand-hygiene practices in India: reflections from the 76th round National Sample Survey, 2018

Sonali Smriti Biswas [a] and Ranjan Karmakar [b,*]

[a] Jawaharlal Nehru University, PhD Research Scholar, Centre for the Study of Regional Development, Jawaharlal Nehru University, New Delhi 110067, India
[b] Centre for the Study of Regional Development, Jawaharlal Nehru University, New Delhi 110067, India
*Corresponding author. E-mail: ranjan51_ssf@jnu.ac.in

SSB, 0000-0002-9869-6338; RK, 0000-0001-7446-9969

ABSTRACT

This paper studies the differences and determinants of handwashing practices in India and identifies sections of the population with poor handwashing practices who are relatively more vulnerable during the COVID-19 pandemic. We have used the data from the recent National Sample Survey (NSS, 76th round) for India (2018). Bivariate and logistic regression analyses have been performed to predict the determinants of handwashing practices across states and socio-economic groups. Levels of education of the household head, Usual Monthly Per Capita Expenditure (UMPCE) of the household, access to water (other than drinking water) resources and sanitation facilities, and the availability of water with soap in and around latrines are major socio-economic and demographic factors that impact handwashing practices. Higher access to principal sources of water for drinking and other purposes, access to bathrooms and latrines with soap, and the availability of water in or around latrines increase the likelihood of handwashing among the people. Universal handwashing across different sections of the population will be effective to prevent further infection. The available data help us to identify the vulnerable sections of the population which are towards the lower end of the handwashing compliance spectrum. The policymakers can outline specific planning and strategy implementation for them.

Key words: COVID-19, handwashing, hygiene practices, preventive care, sanitation

HIGHLIGHTS

- The National Sample Survey is a recent assessment of sanitation and hand-hygiene practices in India.
- It gives a regional- as well as community-level picture of handwashing practices.
- It provides a detailed account of the determinants of hand-hygiene practices.
- This survey is relevant in the current situation of the COVID-19 pandemic.
- It fulfils the gaps in the research of access to water resources and related hand-hygiene practices.

INTRODUCTION

Hands play a major role in the spread of infections (Teare 1999; Aiello & Larson 2002; Curtis & Cairncross 2003). Handwashing is a simple personal hygiene measure that has historically proved to be effective in reducing the transmission of infections (Larson 1988; Coignard *et al.* 1998). It is also a cost-effective strategy to reduce the pressure on the health system across the country and reduce the global burden of diseases (Hirai *et al.* 2016). Handwashing with soap at critical moments, such as prior to meals and post defaecation, can effectively prevent infectious diseases by interrupting the transmission of infectious agents (Hirai *et al.* 2016). An integrated approach, including improvements in personal hygiene with a simultaneous development of public health infrastructure, can effectively control infections (Esrey & Habicht 1986; Esrey *et al.* 1991; Curtis Cairncross & Yonli 2000; Aiello & Larson 2002).

Inadequate handwashing post defaecation and anal cleaning practices are common in the Indian subcontinent and are a major source of faeco-oral transmission of enteric diseases (Hoque *et al.* 1995; Hoque 2003). Improvement in water and sanitation facilities has proved to have reduced infections (Esrey & Habicht 1986; Esrey *et al.* 1991). The World Health Organization (WHO) has advised frequent handwashing with water and soap (among several others) as an effective preventive measure against COVID-19. However, like other hygiene practices, handwashing is highly influenced by individual behaviour and usually has biological and social origins (Deodhar 2003). Even during crucial moments such as post defaecation and before meals, handwashing behaviour varies widely across different social and economic groups of the population. Therefore, if handwashing compliance is to be ensured to prevent infections, it is important to delve into the complex set of factors that influence it – the most important being accessibility to basic facilities required for handwashing compliance. Hand-sanitizing gels and alcohol-based hand rubs are effective alternatives to handwashing with water and soap except when hands are visibly soiled (Widmer 2000) but are not accessible universally. In this paper, an attempt has been made to study the differential handwashing behaviour across social and economic groups of the population in two instances – prior to meals and post defaecation. Since these two instances are crucial ones, it can give a picture of the general anticipated handwashing behaviour and help to identify the vulnerable sections of the population.

Objectives

This paper attempts to study the differential handwashing practices and their determinants in India and identify the sections of the population that are less compliant with handwashing practices. This will be helpful to outline some public health planning and strategies that can be implemented to facilitate handwashing practices among these sections of the population to prevent the spread of infections.

MATERIALS AND METHODS

Data source

Data from the 76th round of National Sample Survey (NSS) (Schedule 1.2) in India have been used, which include information on the issues of drinking water, sanitation, hygiene and housing conditions. It is a nationally representative multistage stratified sample survey of households, wherein the total number of households surveyed was 106,838 (63,736 in rural areas and 43,102 in urban areas).

Outcome variables

Two questions related to handwashing practices were asked during the survey: '(i) whether household members regularly wash their hands before meals?' and '(ii) whether household members regularly wash their hands after defaecation?' A set of four options were available to choose from: if household members wash their hands 'with water and soap/detergent – 1; with water and ash/mud/sand, etc. – 2; with water only – 3; no – 4'. All these responses have been recoded into a binary variable, i.e. 'with water and soap/detergent – 1 and all others – 0', as only handwashing with water and soap/detergent is effective against COVID-19 among the four categories of responses.

Explanatory variables

Various regional, demographic and socio-economic determinants of handwashing have been considered. In addition, access to water resources, the availability of sanitation facilities and institutional factors have also been considered. The place of residence and Empowered Action Group (EAG)[1] status of the states have been considered to examine the regional effects. Main demographic and socio-economic variables are the size of the family, highest education level of household head, Usual Monthly Per Capita Expenditure (UMPCE) of the household, caste and religion. Variables of access to water resources and sanitation include the availability of water in or around latrines used, access to the principal sources of drinking water and other water and access to bathroom and latrine at the household or community level. Benefits obtained from the government scheme on drinking water and sanitation are considered as an institutional factor. The detailed categories of independent variables and sample characteristics are given in Table 1.

Statistical analysis

Chi-square test and bivariate and multivariate statistical tools have been applied in this study using the STATA-14 statistical software. The relationship between the dependent variables (hand washing before meals and after defaecation) and

[1] In connection with this low achievement of some states in controlling population growth, health and educational progress, the Government of India had constituted EAG states in 2001 constituting eight states to provide them with better support for faster growth and sustainable development through self-empowered action. The states under the EAG Group include Bihar, Jharkhand, Uttar Pradesh, Uttaranchal, Rajasthan, Orissa, Madhya Pradesh and Chhattisgarh.

Table 1 | Sample characteristics

Background variables	Categories	*N*	Percent
Place of residence	Rural	63,736	59.66
	Urban	43,101	40.34
EAG status	EAG states	39,909	37.36
	Non-EAG states	66,928	62.64
Family size	Small (up to three)	35,514	33.24
	Medium (four to six)	58,021	54.31
	Large (more than six)	13,302	12.45
Levels of education of the household head (HH)	Illiterate	26,512	24.82
	Literate without formal schooling	998	0.93
	Below primary and primary	22,769	21.31
	Upper primary	17,272	16.17
	Secondary	16,223	15.18
	Higher secondary and above	23,063	21.59
Social group	Scheduled tribe (ST)	14,767	13.82
	Scheduled caste (SC)	18,157	17
	Other backward class (OBC)	43,640	40.85
	Others	30,273	28.34
Religious group	Muslim	13,789	12.91
	Hindu	81,825	76.59
	Christians	6,338	5.93
	Others	4,885	4.57
Usual monthly per capita expenditure	Poorest	15,803	14.79
	Poor	17,549	16.43
	Middle	20,528	19.21
	Rich	23,868	22.34
	Richest	29,089	27.23
Access to the principal source of drinking water	Community	24,762	23.18
	Neighbours source	3,508	3.28
	Common use of HHs in building	11,823	11.07
	HHs exclusive	58,001	54.29
	Others	8,743	8.18
Principal source of drinking water for all household activities	Bottle/piped in HH	32,298	30.23
	Piped in plot/neighbour	14,772	13.83
	Public tap/stand pipe	9,593	8.98
	Hand pump/tube well	39,361	36.84
	Well	7,002	6.55
	Others	3,811	3.57
Principal source of water for all household activities excluding drinking	Bottle/piped in HH	30,659	28.7
	Piped in plot/neighbour	12,774	11.96
	Public tap/pipe	7,717	7.22
	Hand pump/tube well	39,565	37.03
	Well	7,471	6.99
	Others	8,651	8.1
Access of the household to bathroom	No bathroom	30,324	28.38
	Own	66,354	62.11
	Common	9,707	9.09
	Public/community with or without payment	452	0.42
Access of the household to latrine	No latrine	17,771	16.63
	Own	77,159	72.22
	Common	10,108	9.46
	Others	1,799	1.68

(*Continued.*)

Table 1 | Continued

Background variables	Categories	*N*	Percent
Availability of water in or around the latrine used	Not available	3,073	2.88
	Water/water with mud, etc.	18,221	17.05
	Water with soap/detergent	67,030	62.74
Benefits received by the household from government schemes for drinking water during last 3 years	Received	1,759	1.65
	Not received	1,556	1.46
	Don't know	1,03,522	96.90
Benefits received by the household from government schemes for sanitation during last 3 years	Received	12,436	11.64
	Not received	2,161	2.02
	Don't know	92,240	86.34

Estimated from NSS 76th Round, Schedule 1.2.

independent socio-economic and demographic variables has been examined using the Chi-square test. Two separate logistic regression models are executed to examine the impacts of various determinants of handwashing before meals and after defaecation.

RESULTS

Regional variation in handwashing practices

Handwashing practices in two critical moments – before meals and after defaecation – are shown in Table 2 across rural and urban areas in each state. It is observed that on an average, about 35% of people wash their hands with soap/detergent and water before meals. While only 25% of people in rural areas wash hands with soap/detergent and water before meals, the same figure in urban areas is 56%. Union territories record the highest (70%), followed by non-EAG states (43%) and north eastern states (42%), and the lowest is recorded in the EAG states (23%). Variation in handwashing practices before meals across states is shown in Figure 1.

On the other hand, it is seen that about 74% of people wash their hands with soap/detergent and water post defaecation. While only 67% of people in rural areas wash hands with soap/detergent and water post defaecation, the same figure in urban areas is 88%. The highest percentage is observed in union territories (96%), followed by EAG states (74%) and non-EAG states (73%), and the lowest is recorded in north eastern states (62%). Rural areas in Chandigarh and urban areas in Goa record 100% of people washing hands post defaecation. Variation in handwashing practices after defecation across states is shown in Figure 2.

Relation between handwashing and socio-economic and demographic characteristics

Table 3 reveals the association of handwashing practices with socio-economic and demographic characteristics of the households. People who wash hands prior to meals are comparatively high in the urban areas (55.28%, $P<0.001$) compared to rural areas (24.96%, $P<0.001$). Handwashing prior to meals is highest in small families (38.70%, $P<0.001$), and post defaecation is highest in large families (76.25%, $P<0.001$). People who wash hands prior to meals and post defaecation are comparatively high among those with education at higher secondary and above (55.90 and 89.36%, respectively, $P<0.001$). The rate of handwashing prior to meals and post defaecation is satisfactorily high (49.04 and 86.71%, respectively, $P<0.001$) among the non-scheduled and backward social group. Handwashing before meals is lowest among Muslims (32.35%, $P<0.001$) and after defaecation among Christians (67%, $P<0.001$). There is a significantly progressive association observed between both outcome variables and the increasing wealth status of the household ($P<0.001$). Non-EAG states have a higher percentage of population washing hands prior to meals (44.11%, $P<0.001$). The highest tendency of handwashing in both cases is observed among households who used bottled/piped water as principal sources of household activities excluding drinking (57.55 and 88.64%, respectively, $P<0.001$). A similar pattern was observed in the case of principal source of water for all household activities (56.65 and 87.02%, respectively, $P<0.001$). It must be noted that bottled water is packaged drinking water that meets certain safety standards and is therefore considered safe for drinking; piped water is provided by corporation, municipality, panchayat or other local authorities and can also be considered safe (Government of India 2018). The highest percentage of handwashing before meals is observed among households that possess their own bathrooms (45.67,

Table 2 | Regional variation in handwashing across India, 2018

State/UTs		Whether household members regularly wash their hands with water and soap/detergent before meal? Rural	Urban	Total	Whether household members regularly wash their hands with water and soap/detergent after defaecation? Rural	Urban	Total	No. of sample households
Non-EAG others	Jammu and Kashmir	46.21	60.06	49.63	72.58	90.74	77.06	1,714
	Himachal Pradesh	87.6	77.56	86.24	97.61	99.23	97.83	947
	Punjab	74.36	81.85	77.53	96.33	99.36	97.61	2,361
	Haryana	56.07	76.77	64.02	87.22	95.82	90.53	2,145
	Assam	28.2	44.72	30.7	69.41	94.3	73.18	3,600
	West Bengal	18.57	47.72	28.3	70.28	93.24	77.95	7,789
	Gujarat	32.62	60.71	46.35	73.42	89.51	81.28	4,840
	Maharashtra	42.92	73.2	56.51	75.59	96.59	85.02	9,298
	Andhra Pradesh	25.84	51.79	34.61	44.31	72.95	53.99	3,863
	Karnataka	35.11	67.28	49.53	57.4	83.82	69.25	4,895
	Goa	30.69	69.88	55.09	98.06	100	99.27	239
	Kerala	51.57	56.04	53.76	82.76	93.11	87.83	3,383
	Tamil Nadu	13.19	41.27	27.35	30.74	61.51	46.26	6,108
	Telangana	28.9	52.47	39.89	61.31	86.77	73.18	2,949
	Total	33.04	58.34	43.46	65.09	85.54	73.51	54,131
EAG	Uttarakhand	32.91	71.4	44.01	89.03	99.26	91.98	984
	Rajasthan	19.66	53.37	28.67	64.92	93.27	72.49	5,240
	Uttar Pradesh	16.87	44.78	23.79	75.6	96.15	80.69	12,423
	Bihar	12.22	30.81	14.29	66.65	90.99	69.36	6,993
	Jharkhand	3.28	32.84	10.56	50.56	86.53	59.41	2,565
	Odisha	9.22	42.16	15.11	46.34	83.42	52.96	3,671
	Chhattisgarh	33.27	56.87	38.14	84.07	98.23	86.99	2,125
	Madhya Pradesh	26.64	57.41	35.24	74.95	95.81	80.78	5,908
	Total	17.04	47.14	23.73	68.8	93.79	74.35	39,909
Non-EAG North Eastern States	Sikkim	85.38	90.56	87.05	100	98.53	99.53	816
	Arunachal Pradesh	50.38	69.05	54.82	75.2	92.7	79.37	1,143
	Nagaland	32.17	55.08	39.25	45.28	68.85	52.56	912
	Manipur	40.69	54.72	45.29	44.86	61.37	50.28	2,242
	Mizoram	36.22	59.48	47.53	43	75.36	58.74	1,200
	Tripura	28.06	44.04	31.78	62.18	82.09	66.82	2,256
	Meghalaya	31.94	60.74	37.53	48.57	75.01	53.71	1,292
	Total	36.72	56.82	42.23	56.38	75.75	61.69	9,861
Union Territories	Chandigarh	100	79.96	80.97	100	97.27	97.41	192
	Delhi	55.97	73.92	73.5	99.67	97.51	97.56	1,616
	Daman and Diu	23.39	5.85	7.97	95.44	97.24	97.03	192
	Dadra and Nagar Haveli	12.98	36.22	27.98	69.51	90.62	83.14	192
	Ladakh	73.9	34.24	38.42	95.16	97.03	96.84	144
	Puducherry	59.61	78.58	70.78	59.61	87.06	75.78	360
	Andaman and Nicobar Islands	14.55	82.17	53.1	95.96	99.04	97.72	240
	Total	47.24	72.32	70.72	80.37	97.02	95.95	2,936
India Total		25.31	56.05	35.82	66.81	88.26	74.15	1,06,837

Estimated by the author from NSSO 76th Round.

$P<0.001$) and latrines (41.2, $P<0.001$). The percentage of handwashing in both cases is higher where water is available in or around the latrine used (48.91 and 94.04%, respectively, $P<0.001$). Households that received benefits from government schemes for drinking water have a higher percentage of people washing hands in both situations (36.12 and 77.57%, respectively, $P<0.001$). On the other hand, households that do not receive benefits from government schemes for sanitation have a higher percentage of people washing hands before meals (27.29%, $P<0.001$), while the percentage of people washing hands

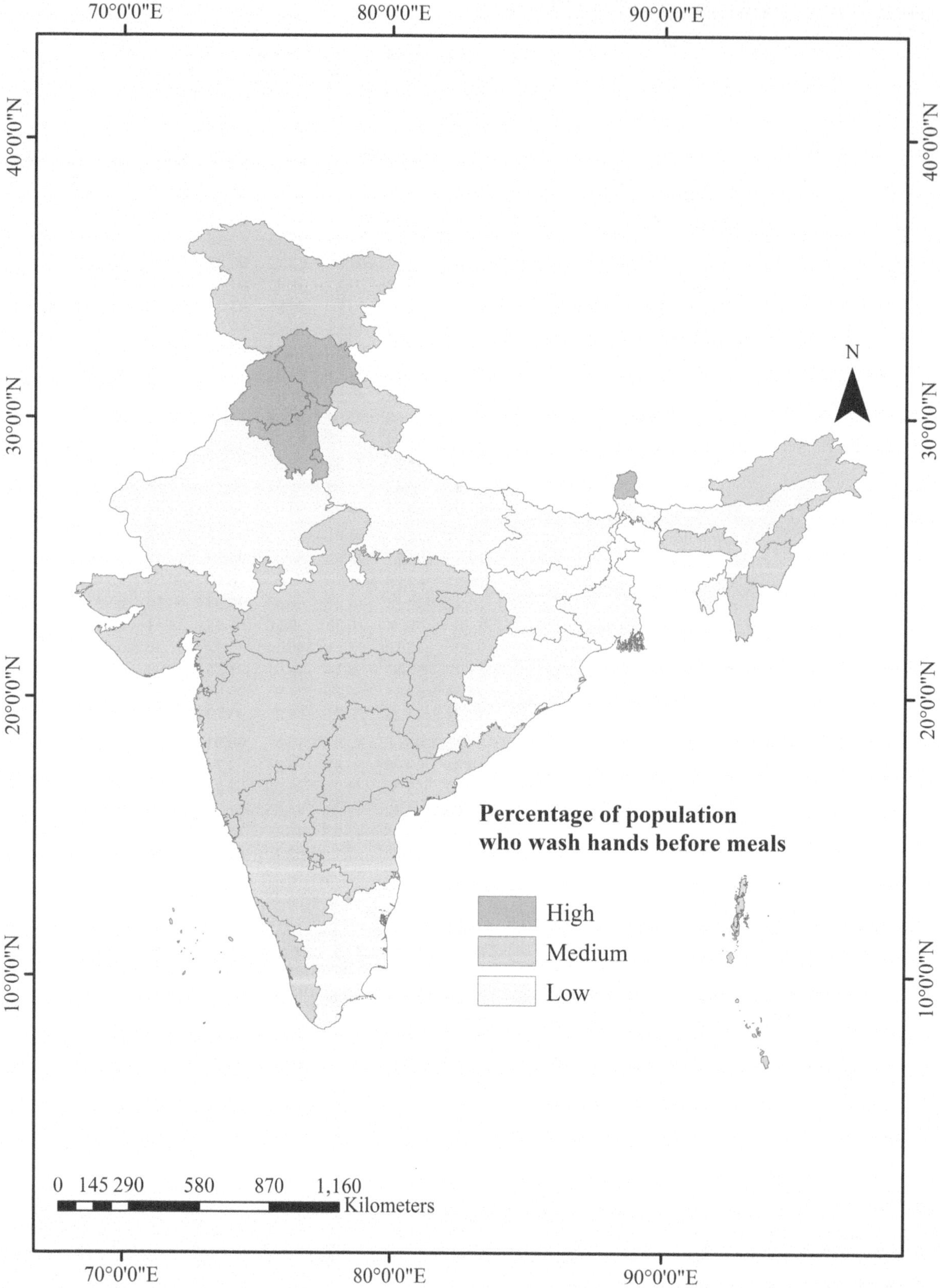

Figure 1 | Percentage of population who wash hands prior to meals in India, 2018 (data source: NSS 76th round, Schedule 1.2).

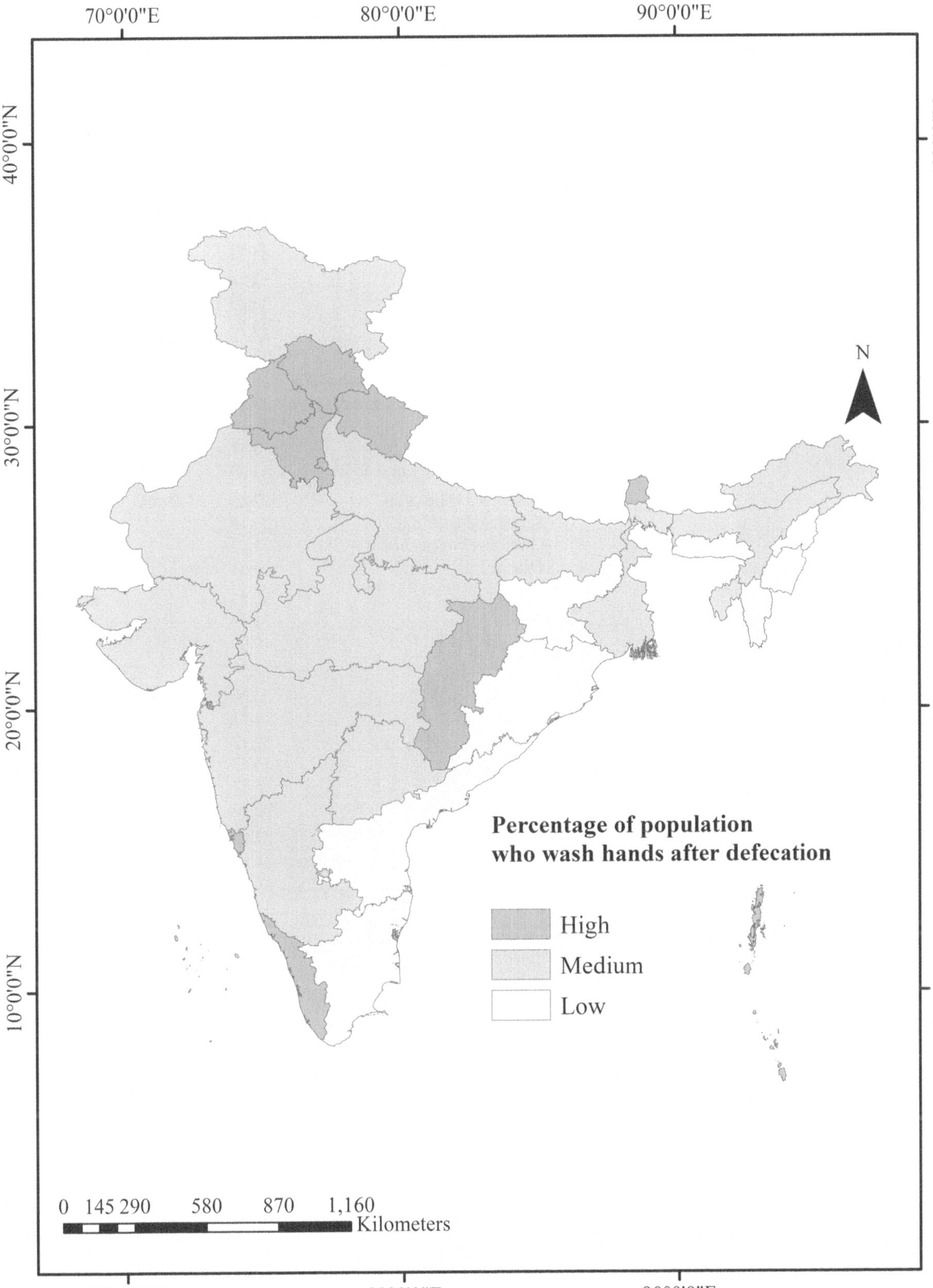

Figure 2 | Percentage of population who wash hands post defaecation in India, 2018 (data source: NSS 76th round, Schedule 1.2).

Table 3 | Bivariate statistics for the association between handwashing before meals and after defaecation with different background variables

Background variables	Categories	Washing hands with water and soap/detergent			
		Before meal		After defaecation	
		Percentage	*P*-value	Percentage	*P*-value
Place of residence	Rural	24.96	0.001	68.35	0.001
	Urban	55.28		89.19	
Empowered Action Group	EAG states	22.75	0.001	74.66	0.001
	Non-EAG states	44.11		74.82	
Family size	Small (up to three)	38.70	0.001	73.25	0.001
	Medium (four to six)	34.69		74.6	
	Large (more than six)	29.77		76.25	
Levels of education of HH	Illiterate	22.20	0.001	64.87	0.001
	Literate without formal schooling	29.70		73.29	
	Below primary and primary	29.27		71.02	
	Upper primary	31.82		76.15	
	Secondary	43.11		81.47	
	Higher secondary and above	55.90		89.36	
Social group	Scheduled tribe (ST)	22.8	0.001	57.8	0.001
	Scheduled caste (SC)	26.18		67.12	
	Other backward class (OBC)	31.42		74.55	
	Others	49.04		86.71	
Religious group	Muslim	32.35		80.71	0.001
	Hindu	33.49		73.4	
	Christians	40.00		67	
	Others	56.41		86.66	
Usual Monthly Per Capita Expenditure	Poorest	22.08	0.001	63.56	0.001
	Poor	25.71		68.19	
	Middle	30.12		72.49	
	Rich	38.48		79.46	
	Richest	55.37		90.41	
Access to the principal source of drinking water	Community	18.65	0.001	59.19	0.001
	Neighbours source	21.52		63.26	
	Common use of HHs in building	34.19		77.95	
	HH exclusive	40.85		82.05	
	Others	44.35		73.63	
Principal source of drinking water for all households	Bottle/piped in HH	56.65	0.001	87.02	0.001
	Piped in plot/neighbour	38.59		74.11	
	Public tap/stand pipe	23.49		56.51	
	Hand pump/tube well	21.91		71.94	
	Well	33.69		70.76	
	Other sources	29.67		65.37	
Principal source of drinking water for all household activities excluding drinking	Bottle/piped in HH	57.55	0.001	88.64	0.001
	Piped in plot/neighbour	35.09		72.75	
	Public tap/stand pipe	24.07		53.08	
	Hand pump/tube well	24.81		73.75	
	Well	34.09		71.85	
	Other sources	18.2		56.47	
Access of the household to bathroom	No bathroom	12.45	0.001	58.65	0.001
	Own	45.67		82.74	
	Common	39.74		82.76	
	Public/community with or without payment	23.12		66.14	

(Continued.)

Table 3 | Continued

Background variables	Categories	Washing hands with water and soap/detergent			
		Before meal		After defaecation	
		Percentage	*P*-value	Percentage	*P*-value
Access of the household to latrine	No latrine	11.18	0.001	47.02	0.001
	Own	41.2		82.21	
	Common	33.19		82.1	
	Others	39.85		81.06	
Availability of water in or around the latrine used	Not available	19.78	0.001	72.12	0.001
	Water/water with mud, etc.	12.85		40.22	
	Water with soap/detergent	48.91		94.04	
Benefits received by the household from government schemes for sanitation	Not received	27.29	0.001	73.86	0.001
	Received	20.45		73.21	
Benefits received by the household from government schemes for drinking water	Not received	31.68	0.001	74.66	0.001
	Received	36.12		77.57	

Estimated by the author from NSSO 76th Round, Schedule 1.20.
Note: *P*-values at a 95% confidence interval, *P*-values $<=0.05$ is considered as the significance level.

after defaecation is the same for those who have received (73.21%, $P<0.001$) benefits from government schemes and those who have not (73.86%, $P<0.001$).

Factors associated with handwashing with soap/detergent and water prior to meals

Table 4 presents the regression models predicting handwashing with soap/detergent and water prior to meals. Several factors determine whether a person uses soap/detergent and water for washing his or her hands before having meals. Significant predictors include the place of residence, levels of education of a household head, state-/region-specific characteristics, social group, religion, monthly per capita expenditure, family size, benefits received by the household from the government on schemes for drinking water, access to the principal source of drinking water, the principal source of water for all household activities excluding drinking water, access of the household to bathroom, the availability of water in or around the latrine used and whether the household has benefitted from government schemes related to drinking water.

People residing in the urban areas are 1.40 (adjusted odds ratio or AOR=1.40, 95% confidence interval or CI: 1.35–1.45) times more likely to wash hands before meals than those in rural areas. The non-EAG states are 2.04 (AOR=2.04, 95% CI: 1.97–2.11) times more likely to wash hands than EAG states. It is observed that medium (four to six members) and small (up to three members) families are 0.93 (AOR=0.93, 95% CI: 0.89–0.97) and 0.86 (AOR=0.86, 95% CI: 0.81–0.90) times less likely to wash hands than large families (above six members). The odds of washing hands with soap/detergent and water prior to meals increase with an increase in levels of education of a household head and those with higher secondary and above education are 1.70 (AOR=1.70, 95% CI: 1.62–1.78) times more likely than those who are illiterate. The odds of washing hands with soap/detergent and water prior to meals were 1.16 (AOR=1.16, 95% CI: 1.09–1.23), 1.07 (AOR=1.07, 95% CI: 1.02–1.12) and 1.31 (AOR=1.31, 95% CI: 1.25–1.37) times more likely among people belonging to scheduled tribes, other backward classes and others, respectively, than those who belong to scheduled castes. Hindus, Christians and people belonging to other religious groups are 1.28 (AOR=1.28, 95% CI: 1.22–1.34), 1.39 (AOR=1.39, 95% CI: 1.28–1.50) and 2.17 (AOR=2.17, 95% CI: 2.01–2.35) times, respectively, more likely to wash hands than Muslims. People belonging to poor, middle, rich and richest quintiles are 1.07 (AOR=1.07, 95% CI: 1.01–1.13), 1.13 (AOR=1.13, 95% 1.08–1.20), 1.18 (AOR=1.18, 95% CI: 1.12–1.25) and 1.4 (AOR=1.4, 95% CI: 1.32–1.48) times more likely, respectively, to wash hands than the poorest quintile. Households that have received benefits from government schemes for drinking water are 1.57 (AOR=1.57, 95% CI: 1.34–1.83) times more likely to wash hands with soap/detergent and water prior to meals.

Households for which the access to the principal source of drinking water is from the common use of households in building, exclusively for the households, are 1.14 (AOR=1.14, 95% CI: 1.06–1.18) and 1.13 (AOR=1.13, 95% CI: 1.07–1.19) times more likely, and other sources are 0.90 (AOR=0.90, 95%, CI: 0.84–0.97) times less likely to wash hands with soap/detergent and water, respectively, than those using water from community sources. Households that depend on bottled and piped water

Table 4 | Factors associated with handwashing with soap/detergent and water prior to meals

		Handwashing before meal			
Background characteristics	**Categories**	**Unadjusted odds ratio (UOR) Odds ratio (95% CI)**	***P*-values *P*>*z***	**AOR Odds ratio (95% CI)**	***P*-values *P*>*z***
Place of residence	Rural	Reference		Reference	
	Urban	3.14 (3.06–3.22)	0.001	1.40 (1.35–1.45)	0.001
Empowered Action Group	EAG states	Reference		Reference	
	Non-EAG states	2.54 (2.47–2.61)	0.001	2.04 (1.97–2.11)	0.001
Family size	Large (above six)	Reference		Reference	
	Small (up to three)	1.36 (1.3–1.42)	0.001	0.86 (0.81–0.9)	0.001
	Medium (four to six)	1.22 (1.17–1.27)	0.001	0.93 (0.89–0.97)	0.002
Levels of education of the HH head	Illiterate	Reference		Reference	
	Literate without formal schooling	1.47 (1.28–1.69)	0.001	1.09 (0.94–1.27)	0.268
	Below primary and primary	1.46 (1.4–1.52)	0.001	1.10 (1.05–1.15)	0.001
	Upper primary	1.76 (1.68–1.83)	0.001	1.20 (1.14–1.26)	0.001
	Secondary	2.74 (2.63–2.86)	0.001	1.38 (1.32–1.45)	0.001
	Higher secondary and above	4.45 (4.28–4.62)	0.001	1.70 (1.62–1.78)	0.001
Social group	Scheduled caste (SC)	Reference		Reference	
	Scheduled tribe (ST)	1.21 (1.15–1.27)	0.001	1.16 (1.09–1.23)	0.001
	Other backward class (OBC)	1.31 (1.26–1.36)	0.001	1.07 (1.02–1.12)	0.003
	Others	2.55 (2.45–2.65)	0.001	1.31 (1.25–1.37)	0.001
Religious group	Muslim	Reference		Reference	
	Hindu	1.14 (1.1–1.18)	0.001	1.28 (1.22–1.34)	0.001
	Christians	1.49 (1.4–1.58)	0.001	1.39 (1.28–1.5)	0.001
	Others	2.68 (2.51–2.87)	0.001	2.17 (2.01–2.35)	0.001
Usual Monthly Per Capita Expenditure	Poorest	Reference		Reference	
	Poor	1.22 (1.17–1.28)	0.001	1.07 (1.01–1.13)	0.013
	Middle	1.46 (1.39–1.53)	0.001	1.13 (1.08–1.2)	0.001
	Rich	2.01 (1.92–2.1)	0.001	1.18 (1.12–1.25)	0.001
	Richest	3.57 (3.42–3.72)	0.001	1.4 (1.32–1.48)	0.001
Access to the principal source of drinking water	Community	Reference		Reference	
	Neighbours source	1.2 (1.1–1.3)	0.001	1.07 (0.97–1.18)	0.169
	Common use of HHs in building	2.32 (2.21–2.44)	0.001	1.14 (1.06–1.22)	0.001
	Exclusive use of HH	3.25 (3.14–3.36)	0.001	1.13 (1.07–1.19)	0.001
	Others	3.21 (3.04–3.38)	0.001	0.90 (0.84–0.97)	0.008
Principal source of drinking water for household	Public tap/stand pipe	Reference		Reference	
	Bottle and piped water into dwelling	4.44 (4.22–4.68)	0.001	1.50 (1.36–1.65)	0.001
	Piped water in plot and from neighbour	2.18 (2.06–2.31)	0.001	1.30 (1.17–1.44)	0.001
	Tube well/hand pump	1.01 (0.95–1.06)	0.842	1.10 (1.01–1.2)	0.028
	Well: protected and unprotected	1.85 (1.73–1.98)	0.001	1.40 (1.24–1.59)	0.001
	Other sources[a]	1.89 (1.75–2.05)	0.001	2.01 (1.78–2.28)	0.001
Principal source of water for all household activities excluding drinking	Public tap/stand pipe	Reference		Reference	
	Bottle and piped water into dwelling	4.42 (4.18–4.68)	0.001	1.34 (1.23–1.47)	0.001
	Piped water in plot and from neighbour	1.93 (1.82–2.06)	0.001	1.02 (0.92–1.12)	0.736
	Tube well/hand pump	1.13 (1.07–1.19)	0.001	1.02 (0.94–1.12)	0.608
	Well: protected and unprotected	1.79 (1.67–1.91)	0.001	1.00 (0.89–1.12)	0.997
	Other sources[a]	0.99 (0.92–1.06)	0.715	0.88 (0.8–0.97)	0.011
Access of the household to bathroom	No bathroom	Reference		Reference	
	Exclusive use of HH	5.74 (5.54–5.95)	0.001	1.83 (1.75–1.91)	0.001
	Common use of HHs in building	4.51 (4.28–4.75)	0.001	1.69 (1.59–1.8)	0.001
	Public/community use with or without payment and others	2.34 (1.9–2.88)	0.001	1.48 (1.18–1.86)	0.001

(*Continued.*)

Table 4 | Continued

Background characteristics	Categories	Handwashing before meal			
		Unadjusted odds ratio (UOR) Odds ratio (95% CI)	*P*-values *P*>*z*	AOR Odds ratio (95% CI)	*P*-values *P*>*z*
Availability of water in or around the latrine used	Not available	Reference		Reference	
	Water/water with mud, etc.	0.75 (0.68–0.83)	0.001	0.49 (0.44–0.54)	0.001
	Water with soap/detergent	4.55 (4.16–4.98)	0.001	2.41 (2.18–2.65)	0.001
Whether received any benefits from government scheme during last 3 years (related to drinking water)	Not received	Reference		Reference	
	Received	1.46 (1.27–1.67)	0.001	1.57 (1.34–1.83)	0.001

Estimated by the author from NSSO 76th Round, Schedule 1.20.

Note: *P*-values at a 95% confidence interval, *P*-values <= 0.05 is considered as the significance level.

[a]Other sources include – tanker-truck: public and private; spring: protected and unprotected;rainwater collection, surface water: tank/pond and other surface water (river, dam, stream,canal, lake, etc.) and others (cart with small tank or drum, etc.).

into dwelling, piped water in plot and from neighbour, protected and unprotected wells and other sources for the principal source of drinking water for household are 1.50 (AOR=1.50, 95% CI: 1.36–1.65), 1.30 (AOR=1.30, 95% CI: 1.17–1.44), 1.40 (AOR=1.40, 95% CI: 1.24–1.59) and 2.01 (AOR=2.01, 95% CI: 1.78–2.28) times, respectively, more likely to wash hands than who use water from public tap/stand pipe. Households that depend on bottled and piped water into dwellings for the principal source of water for all household activities excluding drinking are 1.34 (AOR=1.34, 95% CI: 1.23–1.47) times, respectively, more likely to wash hands before meals than those who use water from public taps/stand pipes.

Households that have their own bathrooms, use common bathrooms for all households in the building and use public or community bathrooms are 1.83 (AOR=1.83, 95% CI: 1.75–1.91), 1.69 (AOR=1.69, 95% CI: 1.59–1.80) and 1.48 (AOR=1.48, 95% CI: 1.18–1.86) times, respectively, more likely to wash their hands before meals than those with no bathrooms. Households that have water available along with mud/sand around the latrines are 0.49 (AOR=0.49, 95% CI: 0.44–0.54) times less likely and those households that have soap/detergent available with water around latrines are 2.41 (AOR=2.41, 95% CI: 2.18–2.65) times more likely to wash hands before meals than those who do not have water available around latrines.

Factors associated with handwashing with soap/detergent and water after defaecation

Table 5 presents the regression models predicting handwashing with soap/detergent and water post defaecation. Several factors influence whether a person uses soap/detergent and water for washing their hands after defaecation. Significant predictors include the place of residence, levels of education of a household head, EAG characteristics, social group, religion, monthly per capita expenditure, family size, access to the principal source of drinking water, the principal source of water for all household activities excluding drinking, the principal source of drinking water for the household, access of the household to bathroom, the availability of latrine facilities and the availability of water in or around the latrine used and whether the household has benefitted from government schemes related to sanitation.

People residing in the urban areas are 1.50 (AOR=1.50, 95% CI: 1.42–1.57) times more likely to wash hands after defaecation than people in rural areas. The non-EAG states are 0.54 (AOR=0.54, 95% CI: 0.51–0.56) times less likely to wash hands than EAG states. Medium (four to six members) and small (upto three members) families are 0.85 (AOR=0.85, 95% CI: 0.80–0.90) and 0.73 (AOR=0.73, 95% CI: 0.69–0.78) times less likely to wash hands than large families (more than six members). The odds of washing hands with soap/detergent and water after defaecation were 1.68 (AOR=1.68, 95% CI: 1.58–1.80) times more likely among households with the education level of household head as higher secondary and above than those who are illiterate. The odds of washing hands with soap/detergent and water post defaecation were 0.95 (AOR=0.95, 95% CI: 0.89–1.01) times less and 1.02 (AOR=1.02, 95% CI: 0.97–1.07) and 1.68 (AOR=1.68, 95% CI: 1.58–1.80) times more likely among people belonging to scheduled tribes, other backward classes and others, respectively, than those who belong to scheduled castes. Hindus and Christians are 0.92 (AOR=0.92, 95% CI: 0.87–0.97) and 0.64 (AOR=0.64, 95% CI: 0.58–0.70) times less likely and people belonging to other religions are 1.29 (AOR=1.29, 95%, CI: 1.15–1.44) times, respectively, more likely to wash hands than Muslims. People belonging to rich and richest quintiles are 1.10 (AOR=1.10, 95% CI: 1.03–1.16) and 1.35 (AOR=1.35, 95% 1.26–1.45) times, respectively, more likely to wash hands than the poorest quintile. Households for

Table 5 | Factors associated with handwashing with soap/detergent and water post defaecation

		Handwashing after defaecation			
		UOR Odds ratio (95% CI)	*P*-values *P*>*z*	AOR Odds ratio (95% CI)	*P*-values *P*>*z*
Place of residence	Rural	Reference		Reference	
	Urban	3.50 (3.38–3.62)	0.001	1.5 (1.42–1.57)	0.001
Empowered Action Group	EAG states	Reference		Reference	
	Non-EAG states	0.92 (0.89–0.95)	0.001	0.54 (0.51–0.56)	0.001
Family size	Large (above six)	Reference		Reference	
	Small (up to three)	0.93 (0.89–0.98)	0.006	0.73 (0.69–0.78)	0.001
	Medium (four to six)	0.96 (0.92–1.00)	0.058	0.85 (0.8–0.9)	0.001
Levels of education of the HH head	Illiterate	Reference		Reference	
	Literate without formal schooling	1.45 (1.26–1.66)	0.001	1.27 (1.07–1.52)	0.008
	Below primary and primary	1.36 (1.31–1.42)	0.001	1.1 (1.05–1.16)	0.001
	Upper primary	1.81 (1.73–1.89)	0.001	1.27 (1.21–1.34)	0.001
	Secondary	2.55 (2.43–2.67)	0.001	1.34 (1.26–1.42)	0.001
	Higher sec. and above	4.88 (4.65–5.13)	0.001	1.68 (1.58–1.8)	0.001
Social group	Scheduled caste (SC)	Reference		Reference	
	Scheduled tribe (ST)	0.73 (0.69–0.76)	0.001	0.95 (0.89–1.01)	0.001
	Other backward class (OBC)	1.39 (1.33–1.44)	0.001	1.02 (0.97–1.07)	0.003
	Others	3.22 (2.10–2.24)	0.001	1.68 (1.58–1.78)	0.001
Religious group	Muslim	Reference		Reference	
	Hindu	0.75 (0.71–0.78)	0.001	0.92 (0.87–0.97)	0.001
	Christians	0.39 (0.37–0.42)	0.001	0.64 (0.58–0.7)	0.001
	Others	1.27 (1.16–1.38)	0.001	1.29 (1.15–1.44)	0.001
Usual Monthly Per Capita Expenditure	Poorest	Reference		Reference	
	Poor	1.22 (1.16–1.27)	0.001	1.04 (0.98–1.1)	0.205
	Middle	1.50 (1.43–1.57)	0.001	1.03 (0.98–1.09)	0.257
	Rich	2.10 (2.01–2.20)	0.001	1.1 (1.03–1.16)	0.002
	Richest	5.14 (4.89–5.40)	0.001	1.35 (1.26–1.45)	0.001
Access to the principal source of drinking water	Community	Reference		Reference	
	Neighbours source	1.16 (1.08–1.24)	0.001	1.09 (0.99–1.2)	0.075
	Common use of HHs in building	2.92 (2.78–3.08)	0.001	1.21 (1.12–1.31)	0.001
	Exclusive use of HH	3.40 (3.29–3.52)	0.001	1.34 (1.27–1.42)	0.001
	Others	1.82 (1.72–1.92)	0.001	0.89 (0.81–0.98)	0.021
Principal source of drinking water for household	Public tap/stand pipe	Reference		Reference	
	Bottle and piped water into dwelling	5.63 (5.34–5.93)	0.001	0.94 (0.84–1.05)	0.282
	Piped water in plot and from neighbour	2.36 (2.24–2.50)	0.001	0.82 (0.73–0.92)	0.001
	Tube well/hand pump	2.08 (1.98–2.17)	0.842	1.05 (0.95–1.15)	0.337
	Well: protected and unprotected	1.87 (1.75–1.99)	0.001	0.93 (0.81–1.06)	0.250
	Other sources	1.15 (1.06–1.24)	0.001	1.07 (0.94–1.23)	0.313
Principal source of water for all household activities excluding drinking	Public tap/stand pipe	Reference		Reference	
	Bottle and piped water into dwelling	7.62 (7.19–8.07)	0.001	1.97 (1.76–2.2)	0.001
	Piped water in plot and from neighbour	2.68 (2.53–2.85)	0.001	1.76 (1.57–1.97)	0.001
	Tube well/hand pump	2.71 (2.58–2.85)	0.001	1.47 (1.33–1.61)	0.001
	Well: protected and unprotected	2.37 (2.21–2.53)	0.001	1.44 (1.26–1.64)	0.001
	Other sources	1.16 (1.08–1.23)	0.715	1.08 (0.98–1.19)	0.133
Access of the household to bathroom	No bathroom	Reference		Reference	
	Exclusive use of HH	3.41 (3.31–3.51)	0.001	1.17 (1.12–1.23)	0.001
	Common use of HHs in building	3.65 (3.44–3.87)	0.001	1.12 (1.02–1.22)	0.019

(*Continued.*)

Table 5 | Continued

		Handwashing after defaecation			
		UOR Odds ratio (95% CI)	***P*-values *P*>*z***	**AOR Odds ratio (95% CI)**	***P*-values *P*>*z***
	Public/community use with or without payment and others	1.29 (1.07–1.57)	0.001	0.99 (0.78–1.27)	0.952
Access of the household to latrine	No latrine	Reference		Reference	
	Exclusive use of HH	5.26 (5.08–5.44)	0.001	0.80 (0.68–0.95)	0.009
	Common use of HHs in building	5.78 (5.45–6.14)	0.001	1.21 (1.01–1.44)	0.043
	Public/community latrine with or without payment and others	4.60 (4.09–5.18)	0.001	2.79 (2.29–3.41)	0.001
Availability of water in or around the latrine used	Not available	Reference		Reference	
	Water/water with mud, etc.	0.28 (0.26–0.30)	0.001	0.24 (0.22–0.27)	0.001
	Water with soap/detergent	6.88 (6.33–7.48)	0.001	4.95 (4.52–5.42)	0.001
Whether received any benefits from government scheme during last 3 years for sanitation	Not received	Reference		Reference	
	Received	1.01 (0.91–1.12)	0.001	0.88 (0.77–1)	0.058

Estimated by the author from NSSO 76th Round, Schedule 1.20.

Note: *P*-values at a 95% confidence interval, *P*-values <= 0.05 is considered as the significance level.

whom the access to the principal source of drinking water is from common use of households in the building and is exclusively for the households are 1.21 (AOR=1.21, 95% CI: 1.12–1.31) and 1.34 (AOR=1.34, 95% CI: 1.27–1.42) times more likely to wash hands with soap/detergent and water, respectively, than those using water from community sources. Households that depend on bottled and piped water into dwelling, piped water in plot from neighbour, tube well/hand pump, protected and unprotected wells for the principal source of water for all household activities excluding drinking are 1.97 (AOR=1.97, 95% CI: 1.76–2.20), 1.76 (AOR=1.76, 95% CI: 1.57–1.97), 1.47 (AOR=1.47, 95% CI: 1.33–1.61) and 1.44 (AOR=1.44, 95% CI: 1.26–1.64) times more likely to wash hands than those who use water from public taps/stand pipes. Households that depend on piped water in plot and from neighbours for the principal source of drinking water are 0.82 (AOR=0.82, 95% CI: 0.73–0.92) times less likely to wash hands after defaecation than those who use water from public taps/stand pipes. Households that have their own bathrooms are 1.17 (AOR=1.17, 95% CI: 1.12–1.23) times more likely to wash their hands after defaecation than those with no bathrooms. Households that use public/community latrines with or without payment are 2.79 (AOR=2.79, 95% CI: 2.29–3.41) times more likely to wash hands after defaecation than those who have no access to latrines. Households that have water available along with mud/sand around the latrines and those households that have soap/detergent available with water around latrines are 0.24 (AOR=0.24, 95% CI: 0.22–0.27) and 4.95 (AOR=4.95, 95% CI: 4.52–5.42) times, respectively, more likely to wash hands after defaecation than those who do not have water available around latrines.

DISCUSSION

The WHO's preventive advisory guidelines related to hand hygiene to contain COVID-19 are not easy to follow by all sections of the population. Although water and sanitation occupy a place in Millennium Development Goals and Sustainable Development Goals, hygiene promotion and monitoring has received limited attention (Hirai *et al.* 2016). The differential handwashing practices across caste, class, consumption expenditure groups, regions and various other background characteristics reveal that although handwashing is a vital hygiene practice, it is far from being universal. The lack of adequate handwashing practices makes the population vulnerable to infections. The differential access to essential services across caste, ethnicity and classes is evident, and inclusive policies can help in the reduction of disparities (Desai & Dubey 2011; Kumar 2015). The EAG states are mostly in the lower phase of epidemiological transition and face health threats due to unsafe water, sanitation and handwashing (India State-Level Disease Burden Initiative Collaborators 2017). Handwashing practices are also towards the lower end of the spectrum in these states.

While it is understood that the dataset captures the picture before the onset of the COVID-19 pandemic, it must be noted that this paper aims to identify the vulnerable sections of the society which are in the lower end of the spectrum of handwashing compliance. The dataset is large enough to provide a macro picture of the situation prior to the pandemic and gives a fair idea of the handwashing behaviour of the people captured through two crucial moments of the day. Information on handwashing during two daily crucial instances helps us to understand the behaviour of the people regarding handwashing practices. Although minor changes may have been implemented, large-scale changes in behaviour are unlikely to have occurred. If compliance is not available during these vital moments, how much compliance can be anticipated during the newly emerging disease, is a question. The awareness creation strategies by the government shall definitely have an impact on the handwashing behaviour. But the general characteristics might act as an impediment towards its implementation. Identifying these vulnerable sections will help policymakers decide which sections of the population might need more attention regarding planning and implementation of strategies to combat COVID-19.

CDC (Centers for Disease Control and Prevention) recommends handwashing with soap and water whenever possible, but alcohol-based hand sanitizers may also be used (Kumar 2020). In both cases, awareness among people is important which can be effectively done through mass media. People are often aware of several health measures through mass media campaigns but are unable to practice them due to the lack of availability of proper facilities. Besides creating the awareness, the government also has to provide the facilities of water and soap through various avenues since not all people have access. Our study provides an insight that the beneficiaries of government schemes (related to drinking water) are more likely to wash hands regularly than others in the same category. Due to the localized nature of problems with the availability and accessibility of water and soap, steps can be taken at the disaggregated level for the creation of awareness as well as the distribution of soaps and sanitizers. Local volunteers may be directed to encourage people to practice frequent handwashing and distribute soaps, liquid soap and hand-sanitizing gel. These may also be provided in the PDS (Public Distribution System) shops along with basic food items, as mentioned in the Revamped PDS (1992). Regular supply of water in the localities has to be ensured to encourage people to practice handwashing.

CONCLUSION

Handwashing as an important preventive measure against infectious diseases needs more attention and monitoring. Handwashing practices are not universal across all sections of the population, making them vulnerable to infectious diseases. During the COVID-19 pandemic, it is crucial that every person, irrespective of their background, has access to basic facilities required to practice handwashing and maintain other hygiene practices. This paper helps to identify the vulnerable sections of the population who would require more attention in planning and implementing strategies to combat COVID-19 infections. Hand hygiene forms a part of the comprehensive package, which should be followed to prevent COVID-19 infections including the use of masks, physical distancing and following respiratory etiquette.

DATA AVAILABILITY STATEMENT

All relevant data are available from http://mospi.nic.in/unit-level-data-report-nss-76th-round-schedule-12-july-december-2018-drinking-water-sanitation.

REFERENCES

Aiello, A. E. & Larson, E. L. 2002 What is the evidence for a causal link between hygiene and infections? *The Lancet Infectious Diseases* **2**, 103–110. https://doi.org/10.1016/S1473-3099(02)00268-2.

Coignard, B., Grandbastien, B., Berrouane, Y., Krembel, C., Source Infection Control, Hospital Epidemiology, No JulCoignard, B. & Grandbastien, B. 1998 Handwashing quality: impact of a special program. *Infection Control and Hospital Epidemiology* **19** (7), 510–513.

Curtis, V. & Cairncross, S. 2003 Effect of washing hands with soap on diarrhoea risk in the community: a systematic review. *The Lancet Infectious Diseases* **3**, 275–281. https://doi.org/10.1016/S1473-3099(03)00606-6.

Curtis, V., Cairncross, S. & Yonli, R. 2000 Review: domestic hygiene and diarrhoea – pinpointing the problem. *Tropical Medicine and International Health* **5** (1), 22–32. https://doi.org/10.1046/j.1365-3156.2000.00512.x.

Deodhar, N. S. 2003 Epidemiological perspective of domestic and personal hygiene in India. *International Journal of Environmental Health Research* **13**, 47–56. https://doi.org/10.1080/0960312031000102796.

Desai, S. & Dubey, A. 2011 Caste in 21st century India: competing narratives. *Economic and Political Weekly* **46** (11), 40–49.

Esrey, S. A. & Habicht, J. P. 1986 Epidemiologic evidence for health benefits. *Epidemiologic Reviews* **8**, 117–128.

Esrey, S. A., Potash, J. B., Roberts, L. & Shiff, C. 1991 Effects of improved water supply and sanitation on ascariasis, diarrhoea, dracunculiasis, hookworm infection, schistosomiasis, and trachoma. *Bulletin of the World Health Organization* **69** (5), 609–621.

Government of India 2018 *NSS 76th Round: Drinking Water, Sanitation, Hygiene and Housing Condition in India*. Ministry of Statistics and Programme Implementation, National Statistical Office, New Delhi, India.
Hirai, M., Graham, J. P., Mattson, K. D., Kelsey, A., Mukherji, S. & Cronin, A. A. 2016 Exploring determinants of handwashing with soap in Indonesia: a quantitative analysis. *International Journal of Environmental Research and Public Health* **13**. https://doi.org/10.3390/ijerph13090868.
Hoque, B. A. 2003 Handwashing practices and challenges in Bangladesh. *International Journal of Environmental Health Research* **13**, S81–S87. https://doi.org/10.1080/0960312031000102831.
Hoque, B. A., Mahalanabis, D., Alam, M. J. & Islam, M. S. 1995 Post-defecation handwashing in Bangladesh: practice and efficiency perspectives. *Public Health* **109** (1), 15–24. https://doi.org/10.1016/S0033-3506(95)80071-9.
India State-Level Disease Burden Initiative Collaborators 2017 Nations within a nation: variations in epidemiological transition across the states of India, 1990–2016 in the Global Burden of Disease Study. *The Lancet* **390** (10111), 2437–2460. https://doi.org/10.1016/S0140-6736(17)32804-0.
Jumaa, P. A. 2005 Hand hygiene: simple and complex. *International Journal of Infectious Diseases* **9**, 3–14. https://doi.org/10.1016/j.ijid.2004.05.005.
Kumar, A. 2015 Disparities in access to basic amenities across caste, ethnicity and classes in rural and urban India. *The Indian Economic Journal* **62** (4), 1226–1250. https://doi.org/10.1177/0019466220150404.
Kumar, R. P. 2020 You might be using hand sanitizers incorrectly. Here's how to use it effectively. *Live Mint*. Available from: https://www.livemint.com/news/india/you-might-be-using-hand-sanitizers-incorrectly-here-s-how-to-use-it-effectively-11592281037936.html.
Larson, E. 1988 A causal link between handwashing and risk of infection? Examination of the evidence. *Infection Control and Hospital Epidemiology* **9** (1), 28–36. https://doi.org/10.2307/30144131.
Teare, L. 1999 Hand washing: a modest measure – with big effects. *British Medical Journal* **318**, 686. https://doi.org/10.1007/bf02737760.
Widmer, A. F. 2000 Replace hand washing with use of a waterless alcohol hand rub? *Clinical Infectious Diseases* **31**, 136–143.

First received 13 June 2021; accepted in revised form 29 September 2021. Available online 14 October 2021

SARS-CoV-2 in Water doi: 10.2166/wh.2021.186

The devil is in the details: emerging insights on the relevance of wastewater surveillance for SARS-CoV-2 to public health

Steve E. Hrudey [a,*] and Bernadette Conant [b]

[a] Analytical & Environmental Toxicology, Faculty of Medicine & Dentistry, University of Alberta, Edmonton, AB T6G 2G3, Canada
[b] Canadian Water Network, University of Waterloo, 200 University Avenue W, Waterloo ON N2L 3G1, Canada
*Corresponding author. E-mail: steve.hrudey@ualberta.ca

SEH, 0000-0002-6865-3855; BC, 0000-0002-2651-7927

ABSTRACT

The severe health consequences and global spread of the COVID-19 pandemic have necessitated the rapid development of surveillance programs to inform public health responses. Efforts to support surveillance capacity have included an unprecedented global research response into the use of genetic signals of SARS-CoV-2 in wastewater following the initial demonstration of the virus' detectability in wastewater in early 2020. The confirmation of fecal shedding of SARS-CoV-2 from asymptomatic, infected and recovering individuals further supports the potential for wastewater analysis to augment public health conventional surveillance techniques based on clinical testing of symptomatic individuals. We have reviewed possible capabilities projected for wastewater surveillance to support pandemic management, including independent, objective and cost-effective data generation that complements and addresses attendant limitations of clinical surveillance, early detection (i.e., prior to clinical reporting) of infection, estimation of disease prevalence, tracking of trends as possible indicators of success or failure of public health measures (mask mandates, lockdowns, vaccination, etc.), informing and engaging the public about pandemic trends, an application within sewer networks to identify infection hotspots, monitoring for presence or changes in infections from institutions (e.g., long-term care facilities, prisons, educational institutions and vulnerable industrial plants) and tracking of appearance/progression of viral variants of concern.

Key words: COVID-19, critical assessment, pandemic surveillance, public health, sewage

HIGHLIGHTS

- Wastewater-based surveillance (WBS) for the pathogen SARS-CoV-2 has experienced phenomenal growth in application across much of the world.
- Widespread adoption of WBS is relevant and useful where it can provide evidence that can reliably inform public health responses.
- The potential for WBS to provide early warning of the emergence of COVID-19 infection in relation to clinical testing is attractive but capability depends on specific details of WBS logistics.
- WBS data offers the advantage of being independent of extraneous factors such as policies governing clinical testing and data can be regarded as objective provided adequate QA/QC programs are provided to ensure validity of genetic signals of SARS-CoV-2 that are measured in wastewater.
- The potential for WBS to track the appearance and progression of viral variants of concern (VOC) in a sewered population has assumed increasing importance as the emergence of VOCs has proven challenging for managing the COVID-19 pandemic particularly as rapidly rising case counts pose a challenge for clinical testing.

GRAPHICAL ABSTRACT

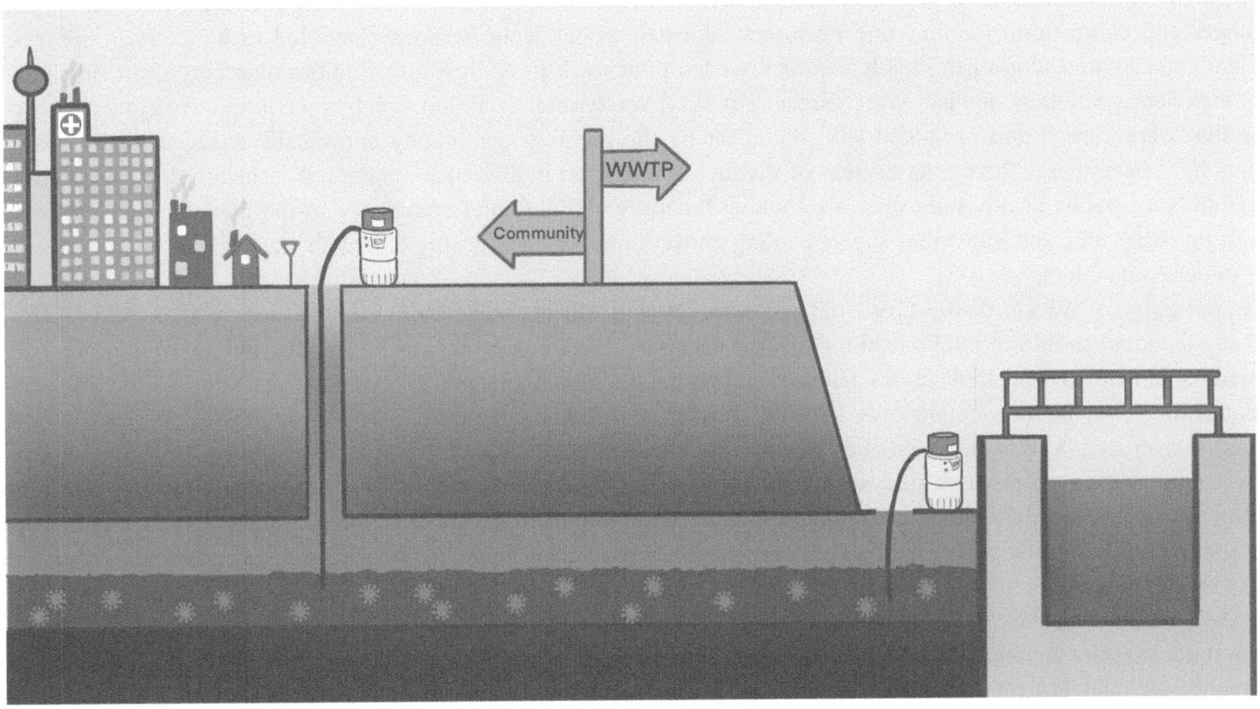

INTRODUCTION

The COVID-19 pandemic has been the most serious, acute public health epidemic of the past century in terms of infections, illness, deaths, economic and social impacts. On January 13, 2020, the World Health Organization (WHO; WHO 2021a) distributed the reverse transcription-polymerase chain reaction (RT-PCR) laboratory method for detecting genetic fragments of SARS-CoV-2, the virus that causes COVID-19, which has formed the backbone of clinical testing to identify if individuals were infected.

Based on experience with facilitating wastewater SARS-CoV-2 surveillance initiatives, collaboration and methods assessment through Canadian Water Network, combined with careful examination of international experience, we undertook a high-level review of the demonstrated strengths, limitations and future promise of wastewater surveillance, including the particular and practical challenges of ensuring surveillance results are relevant to informing meaningful public health measures.

A few experienced and insightful water researchers recognized early in the pandemic the opportunity to apply this highly sensitive, very specific analytical method to monitor wastewater for SARS-CoV-2. Several had submitted proof-of-concept results for refereed publication by April to early June 2020 (Ahmed *et al.* 2020; Gonzalez *et al.* 2020; LaRosa *et al.* 2020; Lodder & deRoda Huisman 2020; Medema *et al.* 2020a; Peccia *et al.* 2020; Randazzo *et al.* 2020a, 2020b; Sherchan *et al.* 2020; Wurtzer *et al.* 2020). This remarkably rapid dissemination of water and public health research led to rapid research applications evaluating this approach for tracking the pandemic around the world. The resulting research activity that arose has been unprecedented, but there needs to be a fair assessment of what wastewater surveillance for SARS-CoV-2 (WBS[1]) can and cannot be relied upon to achieve.

[1] Prior practice in this field has been described as wastewater-based epidemiology but we find this term to be inaccurate in its reference to epidemiology. According to the Centres for Disease Control and Prevention: '*Epidemiology is the method used to find the causes of health outcomes and diseases in populations. In epidemiology, the patient is the community and individuals are viewed collectively. By definition, epidemiology is the study (scientific, systematic, and data-driven) of the distribution (frequency, pattern) and determinants (causes, risk factors) of health-related states and events (not just diseases) in specified populations (neighborhood, school, city, state, country, global).*' https://www.cdc.gov/careerpaths/k12teacherroadmap/epidemiology.html. Surveillance is an accurate description of what we are reviewing. Surveillance is an element of epidemiology, so we have chosen to refer to wastewater-based surveillance (WBS): of the pathogen SARS-CoV-2.

However, the rapid development of the techniques applied to WBS for SARS-CoV-2 has led to inevitable uncertainties in dependability and meaning of the results that determine its value for use in public health decisions. In addition, the challenges and competition for time and resources allocated to pandemic response have led to difficulty in securing sufficient capacity and support to rapidly conduct needed pilot work to address uncertainties over very short timelines. In our experience with the Canadian Water Network (CWN) Wastewater Coalition, much of the initial work and advancements that were viewed and promoted with optimism by the research community and locally engaged public health partners have been met with varying degrees of skepticism and even indifference among public health decision-makers. Understandably, public health authorities are looking for more certainty and confidence in the meaning of WBS data, prior to investing time and effort into supplementing conventional clinical testing data with wastewater-based data as a guide to decision-making.

The application of WBS for SARS-CoV-2 has expanded dramatically over the past 21 months, along with increased interest in the potential to inform public health concerns generally, including by the media and the public. In addition to a multitude of individual local applications, national and regional wastewater surveillance programs are now either in development or underway in many countries as a result (Gawlik *et al.* 2021). The rapid expansion of knowledge, research and application over such a short timeframe has significantly challenged the ability to vet processes using conventional peer-reviewed publications and research study and evaluation approaches. However, WBS for SARS-CoV-2 now has an initially established body of peer-reviewed literature supporting this young, but rapidly evolving area. This development is occurring in parallel with the ongoing and changing needs of the pandemic management, as the very demographics and characteristics of the disease itself continue to change. There is a strong need to view this research and experience in the context of the significance of the key findings and uncertainties to ongoing public health decision needs. The research community will continue to refine understanding and identify gaps and uncertainties in response to an expressed ongoing need. But to serve pandemic management and further engage the public health community, *the devil will always be in the details* in terms of the relevance of the different knowns and unknowns to informing public health practitioners. By use of this idiom here and in our title, we assert that although the meaning of evidence gathered by this approach may appear straightforward, there are many subtle but important details that may well justify conclusions that differ from those apparent upon first consideration.

METHODS

The CWN COVID-19 Wastewater Coalition (CWN 2021a) was established to pursue WBS for SARS-CoV-2 among capable Canadian investigators, wastewater utilities and public health authorities with the express intent of ensuring from the outset of the work that such an undertaking was driven by a goal of determining if or how the technique could serve the needs of public health authorities. In addition, CWN participated as a member of the Global Water Research Coalition (GWRC 2021), providing an important window on relevant initiatives among research agencies in the nine-member countries and expanded inputs from invited participants from other countries, together with interactions involving the International Water Association (IWA) and the WHO. These collaborations provided us with an early window on which groups were pursuing WBS and the literature that was emerging and how and where the approach was being supported and adopted in Canada. We have followed this targeted head start with a focused literature review to provide a more comprehensive perspective on the documented experience with WBS. We obtained a potentially relevant list of references by searching Web of Science (all databases) covering the period from March 2020 to mid-July 2021 with a topic search strategy of (SARS-CoV-2 OR Covid-19) AND (wastewater) AND (monitoring OR surveillance). From this list of 251 references, we considered those that were directly relevant to the WBS applications being pursued by the CWN coalition (surveillance at influent to community wastewater treatment plants (WWTPs) or at targeted locations in sewer systems). We supplemented references from this approach with those that were cited or otherwise brought to our attention by mid-2021 but we did not attempt to perform a formal systematic review. Because of the rapidly emerging nature of this topic, we chose to include a consideration of 23 preprints that had not been refereed. Readers need to recognize the caution included in these preprints about relying on them because they have not been refereed. We have found that four of these were subsequently published so that we have cited the refereed published versions.

Through the foregoing approach, we have collected publications that describe the implementation of WBS, and we have reviewed that published experience in the context of our own experience in seeking to enable accelerated collective action and progress for WBS in Canada. As such, we reviewed this evolving literature for what it suggests to us about the strengths and limitations of WBS for SARS-CoV-2 toward the goal of ensuring findings that are relevant to public health needs.

RESULTS

The literature has experienced an overwhelming proliferation of publications over the past year on a variety of research studies that are relevant to WBS. This paper does not attempt to provide an exhaustive review of the practice of WBS, nor a comprehensive review of all the publications relevant to WBS. Rather we seek to provide a high-level critical perspective about the strengths and limitations of WBS from our extensive involvement and assessment over more than a year, using a summary of what we deemed, based on our experience with the CWN Wastewater Coalition, to be a significant and relevant selection of some of the most applicable literature for that assessment.

Geographic distribution of WBS activity

Applications of WBS have not only been rapidly implemented, but they have also been global in scope. WBS papers had already been published[2] by investigators from 36 countries, including Argentina, Australia, Bangladesh, Brazil, Canada, Chile, China, Czech Republic, Ecuador, Finland, France, Germany, Greece, Hungary, India, Iran, Israel, Italy, Japan, Mexico, Nepal, the Netherlands, Pakistan, Portugal, Qatar, Saudi Arabia, Singapore, South Africa, Spain, Sweden, Switzerland, Thailand, Turkey, United Arab Emirates, United Kingdom and the United States (U.S.). The global coverage of these publication locations is illustrated and cited in Figure 1.

These publications must be understood as representing only a very small fraction of actual WBS activity that has and continues to occur. Given the speed of development of this research over such a short time (less than 1.5 years), the ability of publication systems to keep up and advance knowledge in a timely fashion has been a major challenge. In our experience with the CWN Wastewater Coalition, this challenge is highlighted in the resulting disparity between those anxious that the public health community embraces WBS and those reluctant to invest time and resources in such a challenging and rapidly evolving approach until there is a proven record of success.

To give a sense of the much greater activity currently ongoing, relative to published results, the University of California Merced website lists over 3,145 sites (as of November 30, 2021; Naughton 2021; Naughton *et al.* 2021). The CWN is aware of WBS activity at over 80 locations in Canada (CWN 2021b). EU countries are widely adopting WBS (Gawlik *et al.* 2021), including the Netherlands, which began monitoring WWTPs in February 2020 and has now been monitoring all 352 WWTPs in the country daily since June 2021 as well as 250 WWTPs in Spain since mid-2020 (Randazzo *et al.* 2020b).

Early assessment of WBS

The rapid advancement of WBS has been driven by the demands of the severe public health challenges posed by the pandemic and has taken place, while procedures for WBS have been developed and assessed in progress. Medema *et al.* (2020b) provided an early, very detailed review and summary of the potential benefits and constraints for WBS being able to generate evidence that can be useful for informing public health decision-making. This overview was founded on understanding the successful history of wastewater surveillance for polio, which provides some similar characteristics to COVID-19 for detection in wastewater, and a review of 21 WBS publications (11 peer-reviewed and 10 preprints). Several concerns were elaborated, including how well wastewater surveillance can be understood in terms of how it represents the cumulative human sources of virus, how quantitative results are reported to be meaningful and how wastewater SARS-CoV-2 results can be understood in relation to community COVID-19 prevalence and data limitations that restrict the ability of wastewater results to accurately estimate community COVID-19 prevalence.

Occupational and environmental risk of COVID-19 transmission

Although much of the research and applications have been focused on WBS and its potential and benefits, once evidence was released that SARS-CoV-2 was found in clinical samples of faces and urine from hospitalized cases (Lo *et al.* 2020), a primary question arose whether there was a substantial occupational risk to sewer and wastewater treatment plant

[2] As of July 31, 2021.

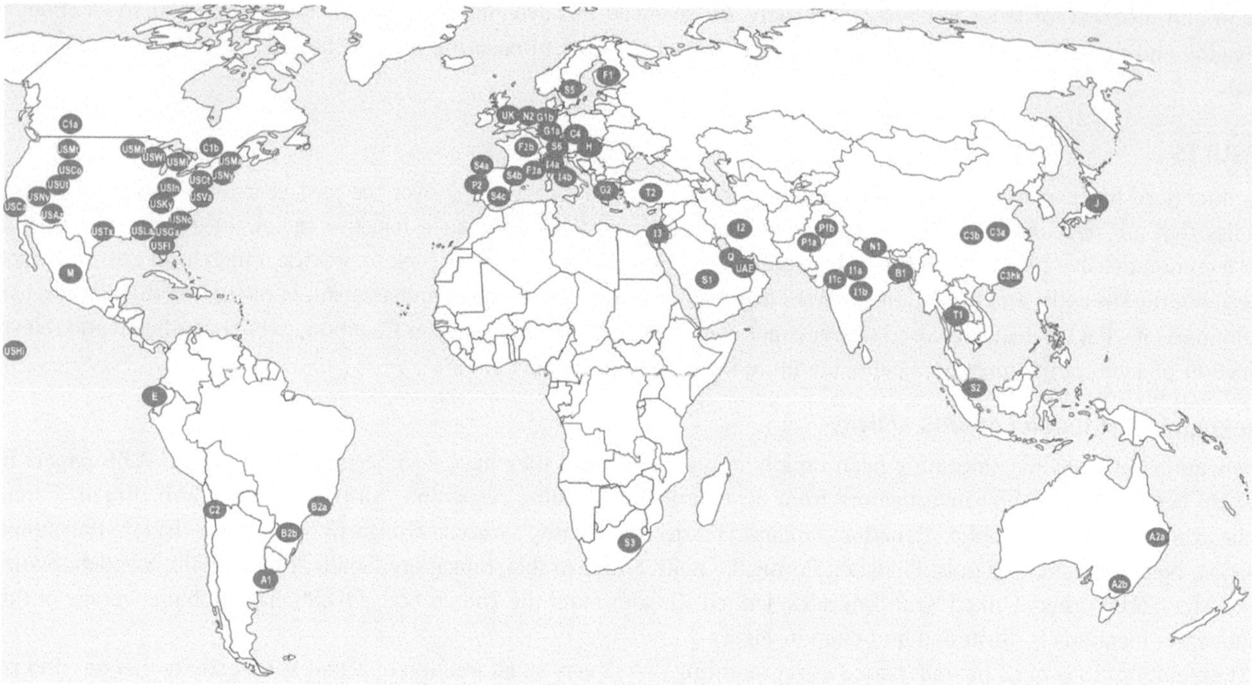

Figure 1 | Geographic distribution of available publications on WBS (**A1**: Iglesias *et al.* 2021; **A2**: Ahmed *et al.* 2020; Black *et al.* 2021; **B1**: Ahmed *et al.* 2021a; **B2a**: Fongaro *et al.* 2021; **B2b**: Prado *et al.* 2021; **C1a**: Acosta *et al.* 2021; C1b: D'Aoust *et al.* 2021a, 2021b; **C2**: Ampuero *et al.* 2020; **C3a**: Zhang *et al.* 2020a; **C3b**: Zhao *et al.* 2021; **C3hk**: Xu *et al.* 2021; **C4**: Mlejnkova *et al.* 2020; **E**: Guerrero-Latorre *et al.* 2020; **F1**: Hokajarvi *et al.* 2021; **F2a**: Trottier *et al.* 2020; **F2b**: Wurtzer *et al.* 2020; **G1a**: Agrawal *et al.* 2021; **G1b**: Westhaus *et al.* 2021; **G2**: Pechlivanis *et al.* 2021; **H**: Roka *et al.* 2021; **I1a**: Arora *et al.* 2020; **I1b**: Hemalatha *et al.* 2021; **I1c**: Kumar *et al.* 2021; **I2**: Nasseri *et al.* 2021; **I3**: Bar Or *et al.* 2020, 2021; **I4a**: Rimoldi *et al.* 2020; **I4b**: LaRosa *et al.* 2020, 2021; **J**: Haramoto *et al.* 2020; Hata *et al.* 2021; Kitamura *et al.* 2021; Miura *et al.* 2021; **M**: Carillo-Reyes *et al.* 2021; **N1**: Napit *et al.* 2021; **N2**: Medema *et al.* 2020a; **P1a**: Sharif *et al.* 2021; **P1b**: Yaqub *et al.* 2021; **P2**: Tomasino *et al.* 2021; **Q**: Saththasivam *et al.* 2021; **S1**: Hong *et al.* 2021; **S2**: Wong *et al.* 2021; **S3**: Johnson *et al.* 2021; Street *et al.* 2020; **S4a**: Balboa *et al.* 2021; **S4b**: Chavarria-Miró *et al.* 2021; Fernandez-Cassi *et al.* 2021; Rusinol *et al.* 2021; **S4c**: Randazzo *et al.* 2020a, 2020b; **S5**: Saguti *et al.* 2021; **S6**: Goncalves Cabecinhas *et al.* 2021; **T1**: Wannigama *et al.* 2021; **T2**: Alpaslan Kocamemi *et al.* 2020; Kocamemia *et al.* 2020; **UAE**: Albastaki *et al.* 2021; Hamouda *et al.* 2021; **UK**: Hillary *et al.* 2021; **USAz**: Betancourt *et al.* 2021; **USCa**: Graham *et al.* 2021; **USCo**: Reeves *et al.* 2021; **USCt**: Peccia *et al.* 2020; **USFl**: Gering *et al.* 2021; **USGa**: Liu *et al.* 2020; **USHi**: Li *et al.* 2021; **USIn**: Bivins *et al.* 2021; **USKy**: Fuqua *et al.* 2021; **USLa**: Sherchan *et al.* 2020; **USMa**: Wu *et al.* 2020; **USMi**: Miyani *et al.* 2020; **USMn**: Melvin *et al.* 2021; **USMt**: Nemudryi *et al.* 2020; **USNv**: Gerrity *et al.* 2021; **USNy**: Green *et al.* 2020; Wilder *et al.* 2021; **USNc**: Gibas *et al.* 2021; **USTx**: Stadler *et al.* 2020; **USUt**: Weidhaas *et al.* 2021; **USVa**: Curtis *et al.* 2020; Gonzalez *et al.* 2020; **USWi**: Feng *et al.* 2021).

workers and others to environmental exposures via the water route (Lodder & de Rosa Huisman 2020). Jones *et al.* (2020) performed a detailed review of available evidence about those possibilities. They reported the abundance of detected gene copies[3] (gc) signaling the presence of SARS-CoV-2 in urine ($\sim10^2$–10^5 gc/mL) and feces ($\sim10^2$–10^7 gc/mL), which are both lower than the virus count in nasopharyngeal fluids ($\sim10^5$–10^{11} gc/mL). Domestic sewage will contain all three of these excretions because it is not limited to toilet drainage, but also includes bath and laundry wastewaters. Jones *et al.* (2020) concluded that other than direct person-to-person contact, sewage is very unlikely to transmit COVID-19 in the environment, although Kang *et al.* (2020) provided probable evidence for sewage aerosol transmission in an apartment building. Giacobbo *et al.* (2021) agreed that extensive aquatic environmental transmission of COVID-19 was not likely and noted that as of the submission date of their review (February 1, 2021), there had been no case of COVID-19 attributed to SARS-CoV-2-contaminated water or wastewater among the more than 100 million cases worldwide (262 million cases as of November 30, 2021). Ahmed *et al.* (2021b) reviewed the critical distinction between possible and probable wastewater transmission of COVID-19. Buonerba *et al.* (2021), in a wide-ranging review about the hazards associated

[3] Gene copies that can be detected by PCR need not and most likely are not from intact, viable virus particles, meaning that those detections likely do not reflect infectious risk.

with SARS-CoV-2, reported that conventional water disinfection inactivated this virus and this finding was also verified by Greaves *et al.* (2021).

Dada & Gyawali (2021) performed a quantitative microbial risk assessment (QMRA) of occupational exposures to wastewater containing SARS-CoV-2 under three levels of assumptions of community COVID-19 infection and judged the occupational risks to be low. Zaneti *et al.* (2021) also performed QMRA for moderate, aggressive and extreme exposure scenarios, causing worker ingestion of wastewater, and estimated that the two higher scenarios exceeded the WHO population infection risk criteria based on Disability-Adjusted Life Years. Regardless, both QMRAs had to rely on limited relevant knowledge about SARS-CoV-2, notably for estimating the critical dose–response relationship using an estimate for SAR-CoV-1 and no direct evidence to estimate worker exposure to SARS-CoV-2. There has been, as yet (as of November 30, 2021), no epidemiological evidence of occupational risk of COVID-19 among wastewater personnel. In any case, domestic wastewater is known to pose an infection risk from a wide variety of enteric pathogens, making it essential that anyone occupationally exposed to untreated wastewater be provided with and effectively use appropriate personal protective equipment. Independent of these quantitative risk predictions, it should be clear that sampling for WBS must be conducted under established occupational health and safety practices for any exposure to wastewater.

Variability of methods used for detecting SARS-CoV-2 in wastewater

The majority of WBS has relied on detecting and quantifying the presence of genetic fragments of SARS-CoV-2 in wastewater using reverse transcription-quantitative polymerase chain reaction (RT-qPCR, described in detail along with other analytical options by Feng *et al.* 2020), as the foundational method for the detection of genetic signals of SARS-CoV-2 in wastewater. However, the dominant focus of much of the research and discussions on the strengths and limitations of the PCR analysis methods, while clearly critical to the validity of WBS, can serve to oversimplify the reality that there are multiple steps involved in sample collection and preparation required to process and concentrate wastewater samples. Each of these steps can introduce variability, errors or uncertainty to the utility and ultimate interpretation at many points in the process, prior to running the ultimate PCR analysis (Kumblathan *et al.* 2021).

Analytical techniques for the detection of SARS-CoV-2 to support WBS rapidly evolved over 2020 and early 2021 in parallel across many research laboratories. Rather than a single standard technique or protocol, the result has been the development of a variety of different, but apparently effective approaches. Hamouda *et al.* (2021) and Zhou *et al.* (2021) provided overviews of the different methods involved in surveillance of SARS-CoV-2 in wastewater considering: sample preliminary treatment, virus concentration, virus extraction, virus analysis and data analysis/interpretation. Philo *et al.* (2021) reviewed the wastewater concentration processes and evaluated a few of them by comparing the recovery of a spiked surrogate human betacoronavirus (OC43). They found that the best-performing concentration method evaluated had an average recovery of 24.3% (95% CI: 23.8–24.9%). Gerrity *et al.* (2021) found greater variability in recovery of bovine coronavirus for four concentration methods with recoveries ranging from only $2.1 \pm 0.87\%$ up to $55 \pm 38\%$. Such studies face the challenge that the behavior of spike additions necessary to perform recovery evaluations will be influenced to an unknown degree by the partitioning of the spike between the solid and liquid phase, compared with the partitioning behavior of native SARS-CoV-2 in the authentic wastewater.

Methods seeking to detect and quantify genetic markers for SARS-CoV-2 from wastewater samples, which present a complex matrix for analysis compared to either clinical or clean water samples, face recognized challenges from inhibition of the PCR step due to complexities of the sample matrix that occur to an unpredictable degree from one wastewater sample to another. Yet, in a survey of published results for SARS-CoV-2 in wastewater or sludge, Buonerba *et al.* (2021) found only 3 out of 43 publications specifically reported treating samples for inhibitor removal.

Taken together, the results from these and other studies about the range of different methods and protocols currently in active use suggest that, with WBS for SARS-CoV-2, there is unlikely to be a single standardized methodology established in the near future. In lieu of the development of a single standardized method for WBS for SARS-CoV-2, there is a clear need for the development of standardized quality assurance and quality control (QA/QC) requirements that can streamline the approach to the evaluation of the various methods, improving confidence in the reliability of the analytical results from the different methods and making more useful and reliable any comparisons between different sites or laboratories.

Gawlik *et al.* (2021) have called for developing a QA/QC framework, noted that the EU is committed to implementing a proficiency test and called for more interlaboratory testing such as reported by Chik *et al.* (2021). The establishment of standardized QA/QC requirements should significantly improve the basis for greater confidence in the use of the analytical data for public health applications going forward. Ultimately, there is a need to seek an approximation of proficiency testing that is normally done by public health laboratories for diagnostic analyses. Currently, given the variety of actual techniques in use and the variability of site conditions, although confidence in the consistency of results over time for a given location and lab is likely to be improved, these various sources of uncertainty are likely to make it a major challenge to compare results across different laboratories using different methods.

Representativeness of samples and analytical results

In addition to potential sources of variability in quantifying the genetic signal present in collected samples, the use of WBS data to inform public health analyses must also consider the representativeness of the samples themselves with respect to actual SARS-CoV-2 inputs to the system, including the variability in samples relative to sewer system conditions. For example, the interpretation of concentration results for SARS-CoV-2 in wastewater is subject to variable dilution of sanitary sewage (household wastewater) by groundwater infiltration into sewers and mixing with stormwater. The variable nature of wastewater, the fecal strength of wastewater and the proportion of SARS-CoV-2 contribution to ultimate samples will also differ between and within North America and Europe/Asia/Africa, e.g., because of different per capita domestic water usages and different system designs (Jones *et al.* 2021). The use of fecal indicator substances as part of the sample analysis has become a common means to attempt to 'normalize' wastewater results for variation in sewage fecal strength, with common choices being pepper mild mottle virus (PMMoV) and crAssphage (Wilder *et al.* 2021; Zhu *et al.* 2021). Such normalization techniques are continuing to be assessed to determine whether they will be more useful for informing public health decisions than normalizing by flow.

Under the conditions of rapid uptake and implementation of emerging sampling and analytical techniques, there have been relatively limited opportunities to standardize and validate techniques while seeking to respond to urgent needs for rapid results. Although there is general agreement about the desirability of having standardized sampling and analytical protocols, it has not been possible to implement such standardization while simultaneously performing WBS during the course of the pandemic. Chik *et al.* (2021) reported on an interlaboratory study among eight laboratories organized by the CWN. Wastewater from the City of Winnipeg, at a time when it had only 85 known cases in a total population of 700,000 (1.2 cases per 10,000), was provided as a blank. This was spiked at low and high levels with Gamma-irradiated inactivated SARS-CoV-2 and human coronavirus strain 229E at low (18 ± 2 gc/mL and 10 infectious units per mL, respectively) and high ($1{,}800 \pm 200$ gc/mL and 1,000 infectious units per mL, respectively) by Canada's National Microbiology Laboratory (Public Health Agency of Canada). The samples were then distributed to the participating labs which were blinded to sample content. All laboratories distinguished the spikes within a total range of 1.34 $\log_{10}$ on the high spike.

The Water Research Foundation funded an interlaboratory study involving 32 U.S. laboratories using a total of 36 specific standard operating procedures representing eight groups of methods (Pecson *et al.* 2021). Each participating laboratory received two raw wastewater samples believed to contain native SARS-CoV-2. Recoveries of a spike of OC43 showed a 7 $\log_{10}$ range, indicating that reported results needed to be corrected for recovery, thereby showing 80% of recovery-corrected results were within a band of $\pm 1.15 \log_{10}$.

Detection of low levels of SARS-CoV-2

The lowest concentration of SARS-CoV-2 that can be reliably detected in wastewater using an analytical method has been a key consideration in the usefulness of WBS ever since several early studies established that it could be realistically detected at relatively low levels (Ahmed *et al.* 2020; Gonzalez *et al.* 2020; LaRosa *et al.* 2020; Lodder & deRoda Huisman 2020; Medema *et al.* 2020a; Randazzo *et al.* 2020a, 2020b; Sherchan *et al.* 2020; Wurtzer *et al.* 2020). The application of the technique in a surveillance capacity to provide early warning of the presence of the pathogen, prior to its recognition through clinical testing, requires confidence in the ability to reliably detect the presence of the pathogen at very low levels when there are few cases contributing to wastewater.

The lowest concentration of gc that is consistently detectable by a method is commonly referred to in the WBS literature as 'sensitivity' because that use of the term is common in the environmental and analytical sciences. However, medical sciences and epidemiology ascribe a different meaning to the term sensitivity: the conditional probability that, given that the analyte is present, how likely it is that it will be detected (Rizak & Hrudey 2006). Therefore, we will use the expression minimum detectability level to avoid that potential confusion. Regardless of the term used, a consideration of how well WBS can detect the presence of genetic indicators of SARS-CoV-2 at very low levels in wastewater is an important assessment to be made when considering the findings of different labs and studies. The minimum detectability levels that are too high will allow for false negatives with WBS for SARS-CoV-2. In contrast, other than the possibility of cross-contamination during sample handling in the field or the lab, contamination of reagents or poor specificity of target genes (Ahmed *et al.* 2022), false positives with WBS are generally a less likely concern. False positives would be a concern in low COVID-19 prevalence areas where detection in wastewater may not be expected, thereby making any detection a matter of interest.

A summary of reported detection levels was provided by Medema *et al.* (2020b), reporting that for inlet wastewater to WWTPs the lowest value was 19–120 gc/L (Ahmed *et al.* 2020). However, it is difficult to assess levels of detection across studies with confidence based on the comparison of published results to date because of the differences in the way that they are interpreted or reported. Ahmed *et al.* (2021b) seek to better clarify this point on the relevant elements of minimum detectability level in their paper by providing a distinction between assay and method sensitivity/precision.

Monitoring for SARS-CoV-2 in wastewater and capability for early warning

Monitoring at WWTPs and early warning

From the outset, a common reason for promoting public health use of WBS as a contributing data source for COVID-19 management has been the potential for it to provide an 'early warning' of the presence of COVID-19 in the population contributing to that sewer system. As discussed below, assessing this potential for practical application to public health decisions based on reported results must consider the actual timelines. The ability to detect the genetic signature of SARS-CoV-2 in wastewater samples that were collected, but not reported, prior to dates of reported COVID-19 detection from clinical samples must be distinguished with sufficient confidence from real-time detection and WBS reporting prior to clinical reporting being able to support decisions about public health measures.

The possibility of the quantitative WBS signals providing early warning of the appearance of the virus in community wastewater, prior to detection in clinical samples, or the possibility of showing trends in community prevalence either to predict or to confirm changes in community prevalence of COVID-19 has been a common theme of published studies.

A summary of 17 studies that provide suggestive evidence that WBS might be able to deliver these predictive capabilities is provided in Table 1, but the devil is truly in the details. Fongaro *et al.* (2021) reported advanced wastewater SARS-CoV-2 detections: 107, 93, 22 and 9 days before the first clinically reported clinical COVID-19 case on March 13, 2020 in Florianopolis, Brazil. Chavarria-Miró *et al.* (2021) reported the same 41 days before the first reported COVID-19 case on January 15, 2020 in Barcelona, Spain. LaRosa *et al.* (2021) reported SARS-CoV-2 detections in Milan and Turin 65 days and Bologna 23 days before the first reported COVID-19 case on February 21, 2020. All these reports were clearly based on the analysis of archived wastewater samples dating back to 2019 before there was any awareness of COVID-19 and clinical testing at those locations. As reported earlier and widely cited, Medema *et al.* 2020a analyzed contemporary wastewater samples that were reported to provide a SARS-CoV-2 signal 6 days before the first COVID case was reported in Amersfoort, The Netherlands. Peccia *et al.* (2020) reported a SARS-CoV-2 signal in primary sludge 0–2 days ahead of clinical test results, according to date of collection, and 6–8 days ahead by date of clinical reporting for New Haven, CT, USA. Nemudryi *et al.* (2020) in Bozeman, MT, USA, D'Aoust *et al.* (2021a) in Ottawa, Canada, Roka *et al.* (2021) in Budapest, Hungary and Kumar *et al.* (2021) in Gandhinagar, India all inferred advanced wastewater signals from 2 days to up to 2 weeks compared with clinical COVID-19 case reports.

All of these reports are subject to the variability in clinical testing programs that were being implemented to varying degrees as the pandemic emerged in early 2020. Multiple factors were involved, such as the availability of testing, timing of reporting relative to test date and policies about who could be tested for whatever reasons (e.g., persons who are asymptomatic, symptomatic, under active treatment or exposed to active cases). Even without these challenges, most of these reports about early warning signals are based on retrospective analysis. This distinction is critical because an expectation, based on retrospective, not 'real-time', analysis, that WBS can provide 'real-time' advance warning, is subject to important details of timelines even given the potential advantages of WBS (can respond to asymptomatic cases, is not subject to inevitable clinical testing limitations about patient choice, policies about who can get tested, availability of testing and time delays in reporting clinical testing). As an illustration of the variability of the clinical testing numbers, Gonzalez *et al.* (2020) reported that for a WBS

Table 1 | Publications reporting possible early warning from sampling WWTPs

Location	Sampling	Monitoring period 2019/2020	WW detection vs. relevant clinical case	Comments	References
Brazil, *Santa Catarina, Florianópolis*	Sewer manhole 24 h composites	Frozen samples Oct 30, 2019 to Mar 4, 2020	WW detections Nov 27, Dec 11 archived, Feb 20, Mar 4	First confirmed case on Mar 13 in region, WW detection in frozen −0 °C, archived samples	Fongaro *et al.* (2021)
Spain, *Barcelona*	Two WWTPs, 24 h composites	Archived samples back to Dec 2019 until May June 2020	First detection in archived WW sample, 41 days before the first case reported 15 Jan	Clearly, a retroactive determination based on an archived WW sample. Later WW samples tracked reported cases	Chavarria-Miró *et al.* (2021)
USA, LA, *southern Louisiana*	Influent WW, 2 WWTPs, 9–24 h composites, 6 grabs	Jan 13 to Apr 29 4 sample rounds	Detected in last 2 rounds Apr 8 and Apr 29 when cumulative cases were already ~4,000 and ~6,200	No WW signal when no cases were known, but WW signal was clear when cases were documented. No insight was provided on early warning	Sherchan *et al.* (2020)
Italy, *Milan, Turin, Bologna*	Influent WW, 5 WWTPs, 24 h composites	Oct 9, 2019 to Feb 28, 2020	First detections in archived WW were Dec 18, 2019 in Milan & Turin and Jan 29, 2020 in Bologna	First clinical case reported on Feb 21, 2020, so Milan & Turin results were 65 days earlier and Bologna was 23 days earlier	LaRosa *et al.* (2021)
The Netherlands, *Amsterdam, Amersfoort Apeldoorn, Den Haag, Schipol, Tillburg, Utrecht*	Influent WW, 7 WWTPs, 24 h composites	4 sample rounds over 48 days Feb 5 to Mar 25	Gene copies per mL of wastewater correlated (log-log) with cumulative clinical cases	First WW signal detected in Amersfoort on Mar 5, 6 days before the first cases were reported, the WW lead time was established retroactively	Medema *et al.* (2020a)
Sweden, *Gothenburg*	Influent WW, 1 WWTP 24 h composites	Feb 10 to Jul 5 weekly samples over 20 weeks	WW showed 4 peaks over the 20 weeks. No data on cases, only hospitalizations	Limited granularity of data, but the claim was that WW data provide 3–4 week warning of hospitalization peaks	Saguti *et al.* (2021)
Australia, QLD, *Brisbane (SEQ)*	Influent WW, 2 WWTPs, composites	Feb 24 to Apr 4; 8 sample days over 39 days	1st WW 'weak' detection on Mar 26, when total cases in prior 28 days were ~120 cases, 2nd WW 'weak detection on Apr 4 with ~250 cases in prior 28 days	WW signals presented as proof of concept for detecting SARS-CoV-2 signals in WW, but this paper does not provide evidence for meaningful WW early warning of cases	Ahmed *et al.* (2020)
USA, NV, *Las Vegas, South Nevada*	Primary effluent WW, 2 WWTPs, 1 large, 1 small grab at daily low flow, some 24 h composites	Mar 2 to May 27, 29 daily samples over 61 days	First WW detection occurred on Mar 09 when clinical cases had first been reported to health. Now, archival data show Mar 9 symptom onset date had a 7-day running avg=12.4 cases/days for ~2 million population	WW detection was sensitive to detect a signal of SARS-CoV-2 for a WWTP serving a population of ~1 million (not entire ~2 million), but WW sampling started after case onsets and the first WW sample on Mar 2 was not positive. Does not support early warning for a large WW system	Gerrity *et al.* (2021)
France, *Paris region*	Influent WW 3 WWTPs, composites	Mar 5 to Apr 23	Mar 05 WW positive vs. Mar 4 consultation about COVID symptoms	The WW results do not clearly support advanced early detection in WW of clinical cases	Wurtzer *et al.* (2020)
Japan, *Ishikawa preft., Toyama preft.*	Influent WW, 5 WWTPs, morning grabs	Mar 5 to May 29	Mar 19 detection in Ishikawa WW when there were 4 clinical cases (0.3/100,000), Apr 3 detection in Toyama WW when there were 8 cases (0.8/100,000)	The first WW detections happened when the reported clinical cases were very low, but the WW signal cannot be claimed to be found in advance of clinical cases being reported and do not account for under-reporting of cases	Hata *et al.* (2021)
USA, VA Southeast Virginia	Influent WW 9 major WWTPs, composites	Mar 9 to Jul 28 weekly	Mar 9 first WW positive, Mar 9 first clinical cases confirmed		Gonzalez *et al.* 2020

				The clinical case numbers were limited by testing which covered only 69 patients on Mar 9, rose to 1.18 M by end of 21 weeks	
Spain, *Murcia region Murcia, Cartagena, Molina de Segura, Lorca, Cieza, Totana*	Influent WW 6 WWTPs, morning grabs	Mar 12 to Apr 14 approximately every 5th day	WW detection Mar 12 in Lorca, Mar 16 in Murcia and Cieza, first clinical case, Mar 8. Cases prevalent in Murcia Mar 20, all 3 by Mar 25	The data for detected cases vs. detection in wastewater provide no firm basis for claiming early warning because data are not readily comparable	Randazzo *et al.* (2020a)
USA, CT *New Haven metro Hamden, East Haven, Woodbridge*	Primary sludge, 1 WWTP morning grabs	Mar 19 to Jun 1 daily	RNA in sludge were 0–2 days ahead of clinical test results by date of collection, 6–8 days ahead of positive test results by the reporting date	Sludge results were not a leading indicator vs. positive test results by date of specimen collection. WW led lagging indicator – hospitalizations by 1–4 days and test results by clinical result reporting date by ~1 week.	Peccia *et al.* (2020)
USA, MT, *Bozeman*	Influent WW, 1 WWTP	Mar 30 to Jun 12, 17 daily samples over 74 day period	WW detections near detection limit and time in a decline of the first wave and a rise of the second wave. Limited data, paper inferred WW could precede clinical reports by 2–4 days	Based on retrospective interviews, clinical testing for COVID-19 typically occurs 3–9 days after symptom onset and may vary, depending on multiple policy factors. WW testing can detect signals during this window before reporting clinical testing	Nemudryi *et al.* (2020)
Canada, ON, *Ottawa*	Primary sludge 1 WWTP, 24 h manual (4×6 h) composites	Every 2nd day, Jun 20 to Aug 4	WW signals correlated with case signals: 1. new daily cases, 2. % positivity, 3. hospitatlization	Extensive WW testing program provided credible evidence of WW signal preceding case increases by 48 h, clinical testing at 81/100 k population	D'Aoust *et al.* (2021a)
Hungary, *Budapest*	Influent WW, morning grabs at 2 WWTPs, composites at 1	Weekly for 44 weeks, Jun 5 to Nov 3	WW signal r^2=0.72 (p<0.001) with daily new cases during 17 week rising case period, 6–333 new daily cases	Clear, significant correlations of WW signal with daily new cases and active cases. Possible ~1 week advance WW signal rise apparent.	Roka *et al.* (2021)
India, *Gandhinagar*	Influent WW, 4 WWTPs, grab samples	Bi-weekly, then weekly Aug 7 to Sep 30	Based on inference from patterns for infrequent WW sampling claim 1–2 week WW advance warning	No information provided on the timeliness of COVID case reporting and infrequent data make the claims of 1–2 week advance warning unpersuasive	Kumar *et al.* (2021)

sampling program that ran from early March to the end of July, 2020, the number of individuals who had been clinically tested over that period increased from 69 to 1,180,000.

Bibby *et al.* (2021) provided an insightful analysis of the capability of WBS to provide early warning by considering estimates for the time lag for fecal shedding of SARS-CoV-2 before or after the appearance of symptoms (a major driver for individual clinical testing) and the time delays in reporting clinical results before WBS results. They concluded that WBS could, at best, normally be expected to provide about 4 days of early warning for circumstances where clinical testing is readily available and reported rapidly.

Authentic early warning with WBS is inevitably subject to the frequency of wastewater sampling, i.e., sampling only once a week could not guarantee advance warning of less than a week regardless of the inherent meaning of the WBS signal. Likewise, sample shipping, processing, analysis and result reporting time lags will determine how much 'real-time' early warning could actually be achieved for any particular WBS program.

Sampling wastewater in sewers to identify community hot spots or institutions

The potential for sampling wastewater directly from sewers serving known populations offers a number of potential advantages, including less dilution of viral load and greater potential to detect relative increases in viral load, e.g., from non-detectable to readily detectable and an ability to target wastewater results to possible remedial actions. Unavoidable or potential disadvantages include a smaller contributing population to provide a viral signal, samples from sewers that are not routinely available, thereby requiring additional targeted sampling, sewer sampling being inherently more dangerous than obtaining samples from a WWTP and requiring professionally trained personnel. There will be greater uncertainty about how representative a sample can be that is taken from a sewer with varying depth of flow. Some success has been reported with the use of Moore swabs (Rafiee *et al.* 2021) and purpose-designed passive samplers (Schang *et al.* 2021) for obtaining what are effectively composite samples from sewers.

Table 2 summarizes results from nine selected studies that made use of in-sewer sampling for the detection of SARS-CoV-2 in wastewater. Chavarria-Miró *et al.* (2021) were monitoring both wastewater influents to WWTPs and grab samples from urban sewers in Barcelona, Spain and when the WWTP samples declined to non-detectable levels, sewer monitoring identified COVID-19 hot spots that facilitated the adoption of local mitigation measures. Wong *et al.* (2021) and Xu *et al.* (2021) used sewer sampling in Singapore and Hong Kong, respectively, to identify hot spots in apartment buildings that guided clinical testing and quarantining requirements. Acosta *et al.* (2021) sampled sewers that served acute care hospitals in Calgary, Canada, finding that the viral signals indicated the occurrence of hospital-acquired infections. Betancourt *et al.* (2021) and Gibas *et al.* (2021) used on-campus sewer sampling of student residences to identify the need for targeted clinical testing of students that allowed identification and quarantining of infected students before outbreaks occurred. Gibas *et al.* (2021) discussed the practical realities and logistics of operating sewer sampling such that it would provide results in a manner timely enough to allow for effective public health interventions. Harris-Lovett *et al.* (2021) reported on a consortium of 25 U.S. colleges that applied WBS in campus sewer systems. Black *et al.* (2021) identified the presence of COVID-19 cases residing in regions of the sewer network. This study was facilitated by a comprehensive sewer sampling program combined with an effective case identification and quarantine program that documented the geographic locations of cases. They found that it was possible, though not likely, that a single infected person in a sewer catchment could be detected, and that virus detection in sewer samples justified further investigation in an area of low or no COVID-19 prevalence.

The sewer monitoring applications mentioned above provide clear illustrations of the need for appropriate ethics guidance over WBS that was suggested for CWN by Hrudey *et al.* (2021). Gawlik *et al.* (2021) noted that the European Commission has adopted this ethics guidance for its proposed sentinel sewage surveillance system for Europe.

Estimating COVID-19 prevalence, incidence and measures of epidemic spread

The possibility of being able to estimate the prevalence of COVID-19 in a community by means of WBS has been a particularly attractive aspiration. The logistical and practical limits of clinical testing to establish disease prevalence are clearly problematic in a pandemic when cases undergo waves of exponential growth, including providing effective access to clinical testing for a large and rapidly expanding population of interest, initial periods of no or mild symptoms among infected individuals, individual reluctance to be tested until medical care becomes necessary, uncertainty about quantitative levels of virus

Table 2 | Publications reporting sampling from sewer networks to identify specific locations

Location	Sampling	Monitoring time frame – 2020	WW detection with respect to relevant clinical case	Comments	References
United Arab Emirates, *Dubai*	9 pump stations grabs, weekly 49 sewer regions grabs, bi-weekly	Apr 22 to May 4 then May 7 to Jul 7	Monitoring period did not allow for the assessment of possible early detection	Overall decline of sewage signal with overall decline of clinical cases. Did show geographic distribution of sewage signal, but not related to known cases	Albastaki *et al.* (2021)
Spain, *Barcelona*	Grab, sewer sampling after prior WWTP sampling	May 18, May 25	WWTPs had declined to not detectable, but sewer samples showed detections	WWTP detections recurred June 2–8, followed by an outbreak (300 cases) declared in early July	Chavarria-Miró *et al.* (2021)
Singapore	Hourly composites, 2 diurnal periods from building sewer	Jul 4 to Jul 20, post: Jun 23 cluster of cases in building	WW was positive with spikes from Jul 4 to Jul 9 led to identification of 2 cases	1 case identified by WW was asymptomatic, 1 case had diarrhea explaining the observed WW spikes	Wong *et al.* (2021)
Canada, AB *Calgary*	Sewers – 3 acute care hospitals, 24 h composite, 2 per week	Aug to Dec	Spikes in cases were attributed to hospital-acquired infections	Patients being treated for severe COVID-19 are not using toilet facilities therefore not appearing in WW was recognized	Acosta *et al.* (2021)
USA, AZ *Tuscon*	Sewer manholes focussed on student residences, morning grabs, initially daily, then 2 per week,	Pilot program Aug 18 to Aug 31 then throughout fall semester Aug 24 to Nov 20	During 2 week pilot, positive WW samples led to clinical testing that detected positives the next day for 3 cases that were removed	WW testing provided early warning of 2 asymptomatic and 1 symptomatic students who were tested because of positive WW results for the residence, likely preventing an outbreak	Betancourt *et al.* (2021)
Australia, VIC *Melbourne & southern Victoria*	Sewer manhole sampling at 46 catchments, mostly weekly, composites and grabs	Aug 25 to Oct 27	Determined odds ratios – change in odds of detecting virus in WW if one or more COVID cases are in the catchment.	486,193 person-location days were recorded – 72.8% were within −2 to +55 days of illness onset. WW is the best at detecting cases within the first 2–3 weeks of illness. Positive WW justifies a search for cases	Black *et al.* (2021)
USA, NC *Charlotte*	19 sewer sites covering 17 student residences grabs 3 per week	Sep 28 to Nov 23 for fall semester	WW results were reported 26–30 h after sample collection, 40 positives for 332 samples analyzed	Sep 30 sample, analyzed Oct 1 positive, lockdown of 151 students Oct 2, testing revealed 1 positive isolated Oct 3. Over period, 8 positive WW identified 13 cases	Gibas *et al.* (2021)
USA, 17 states	25 campuses surveyed, 1–7 per week, 18/25 were composite samples	Aug–Sep (one in May) throughout fall academic session	WW program widely variable. Used experience to develop a framework for surveillance program design	Some programs definitely prevented outbreaks by WW detection. Some programs with extensive clinical testing were skeptical of the value of WW	Harris-Lovett *et al.* (2021)
China, Hong Kong	3 manholes, 3 h composites	Jun 8 to Sep 29	MH samples positive Jul 26, start clinical testing Jul 27	Cases identified in buildings Jul 29, giving 2 day lead time in WW detection	Xu *et al.* (2021)

shedding of individuals and challenges of timely data collection, reporting and analysis across the population of interest. In contrast, wastewater can represent viral signal contributions from an entire sewered population with a single daily composite sample. As attractive as this prospect for WBS appears, there are major challenges to delivering meaningful estimates of COVID-19 prevalence in the community based on WBS.

Hart & Halden (2020) provided an early proposal for modeling COVID-19 prevalence from WBS in promoting the merits of widespread adoption of WBS. The latter has clearly happened, but the analysis of WBS results to demonstrate the validity of resulting estimates of community COVID-19 prevalence remains to be shown. Farkas *et al.* (2020) noted that: '*translating the viral titres from wastewater into the actual number of cases within a community is highly challenging, if not impossible*' because of the number and difficulty of verifying assumptions such as the quantitative and temporal dynamics of viral shedding in feces at various stages of infection and the persistence of the viral signal in wastewater as a function of sewer travel time and environmental conditions. Cao & Francis (2021) retrospectively analyzed 9.5 months of WBS and clinical case data from a WWTP in Pennsylvania, applying a multivariate, time-series model to conclude, retroactively, that WBS data could predict the clinical case data. Fernandez-Cassi *et al.* (2021) provided a retrospective analysis of wastewater monitoring in Switzerland during the first wave of the pandemic in March 2020 and determined that WBS and confirmed clinical case numbers are useful and independent measures of the incidence[4] of COVID-19 infection. They concluded that WBS provided a better estimate of the rise of incidence while confirmed case numbers provided a better estimate of the decline of incidence following the peak. Zhu *et al.* (2021) reported that understanding the viral shedding dynamics of infected individuals was the main limitation of translating WBS data into estimates of COVID-19 prevalence. Li *et al.* (2021) considered viral shedding dynamics among what they believed were five major elements contributing to uncertainty (virus excretion rates, virus decay in sewers, sampling location, preservation of samples and analytical detection) for estimating prevalence. They concluded that the analytical uncertainty of PCR detection of SARS-CoV-2 was the largest factor contributing to uncertainty in estimating prevalence. In making that judgment, Li *et al.* (2021) may not have sufficiently weighed that analytical accuracy is much more amenable to control by investigators than the highly variable and largely unknown virus excretion rates among individual cases over the course of infection during the rise and fall of the epidemiological curve.

The reproductive rate, R_0, is an estimate of the number of new cases created for each incident case and, as such, R_0 provides critical insight for public health decision-making about the need for interventions to reduce the spread of the disease. Zhang *et al.* (2020b) estimated the reproductive rate R_0 for the COVID-19 outbreak on the Diamond Princess cruise ship as a median of 2.28 (95% CI: 2.06–2.52) using the clinical case data. This high R_0 demonstrates the COVID-19 transmission risk posed by the cruise ship environment. Huisman *et al.* (2021) estimated the *effective* reproductive rate, R_e, using longitudinal WBS data in Zurich, Switzerland and San Jose, CA, USA finding their R_e estimates to be as similar to those from case report data as were the R_0 estimates based on clinical data (observed cases, hospitalizations and deaths) to each other.

The proportion of clinical tests found to be positive (positivity rate) is another public health indicator relied upon to judge the status of an epidemic. Stadler *et al.* (2020) analyzed 22 weeks of WBS data for Houston, USA to determine that they could achieve a strong predictive indicator of clinical testing positivity rates 2 weeks in advance, providing evidence that proved useful for making real-time public health responses such as deploying clinical testing resources.

Detecting genetic variants of SARS-CoV-2

As the COVID-19 pandemic has persisted internationally since being declared in March 2020, the emergence of mutations of the original native strain of SARS-CoV-2, so-called variants of concern (VoCs), has frustrated efforts to control the pandemic. Even with the benefit of having remarkably effective vaccines, VoCs that are proving more transmissible than the native strain are causing new waves of infection primarily among unvaccinated populations. Consequently, the literature concerning the detection of VoCs has been growing extremely rapidly in recent months. As of the end of November 2021, WHO (2021b) has

---

[4] '***Incidence** refers to the occurrence of new cases of disease or injury in a population over a specified period of time. Although some epidemiologists use incidence to mean the number of new cases in a community, others use incidence to mean the number of new cases per unit of population.... Prevalence differs from incidence in that prevalence includes all cases, both new and pre-existing, in the population at the specified time, whereas incidence is limited to new cases only.*' https://www.cdc.gov/csels/dsepd/ss1978/lesson3/section2.html.

designated five mutated strains to be VoC: alpha, beta, gamma, delta and omicron (added on November 26, 2021). Delta cases have dominated and rose rapidly in many countries during 2021. WHO has also declared two variants of interest, lambda and mu, and seven variants being monitored for evidence of behavior that may justify concern. These variants reflect the reality that viruses will inevitably mutate during a pandemic.

WBS is proving to be a valuable approach to aiding in the tracking of VoCs. Jahn *et al.* (2021) used WBS in Switzerland to monitor for the alpha and beta variants between July and December 2020, finding the alpha variant in wastewater from a Swiss ski resort 2 weeks before it was first verified in a clinical sample. Bar-Or *et al.* (2021), Goncalves Cabecinhas *et al.* (2021), Hillary *et al.* (2021), LaRosa *et al.* (2021), Martin *et al.* (2021), Pechlivanis *et al.* (2021), Prado *et al.* (2021) and Yaniv *et al.* (2021) have all reported a variety of promising applications for detecting and tracking the first appearance of VoCs in wastewater which creates a particularly valuable prospect for WBS to enhance our knowledge about the evolutionary dynamics and behavior of a virus during the course of a pandemic. These insights will inevitably have to come from retrospective analyses.

Cost-effectiveness of WBS

Determining costs for implementing WBS is a challenge because costs will be determined by local and regional circumstances as well as by whether the investigators have existing capabilities and capacity. An informative attempt to estimate costs for WBS has been provided by Gawlik *et al.* (2021) who surveyed a group of European laboratories engaged in WBS; 13 groups from 7 countries responded to a questionnaire with 17 practical questions concerning costs. They concluded that a WBS monitoring program involving twice weekly samples at a European WWTP would cost approximately €25,000 per year. Weidhaas *et al.* (2021) estimated WBS programs performed in Utah, USA, which include weekly sampling of WWTPs treating greater than 4 mL/day, would cost approximately US$220 per sample and cover 79% of the Utah population (US$0.005 per person/week).

Supporting public health decision-making

In making the case that WBS should be ethically governed as public health surveillance, Hrudey *et al.* (2021) note that the first guideline offered by the WHO (2017) guidance for public health surveillance included the statement that: '*Surveillance systems should have a clear purpose and a plan for data collection, analysis, use and dissemination based on relevant public health priorities.*' Although this implied connection between WBS and public health decision-making may be self-evident to many, it is not clear to us from the WBS literature that all investigators do recognize that WBS should be first and foremost about informing public health decision-making.

Medema *et al.* (2020b) make the public health priority explicit and describe the factors bearing on the ability of WBS to provide the requisite support to public health decision-making, including the design of adequate sampling strategies, considering the spatial and temporal resolution of sampling, reliable sample storage, replicate sampling and analysis, and quality controls for the molecular methods used for the reliable quantification of SARS-CoV-2 in wastewater. Prado *et al.* (2021) describe how a combination of WWTP and sewer sampling in the state of Rio de Janeiro, Brazil was able to use geoprocessing of data to support weekly updates to a heat map to promote public awareness of the local epidemic situation and need for public health interventions. Hillary *et al.* (2021) demonstrated with a longitudinal analysis of wastewater at WWTPs for six major urban centers in northwest England and Wales that tracking variants in wastewater reflected findings in clinical samples and WBS were able to confirm the success of lockdown measures (restricted movements and social distancing). However, they also note the need for a better understanding of analytical quantification of the genetic signals for SARS-CoV-2 to improve confidence in applying WBS data to support public health decision-making. The latter observation is consistent with the conclusions of Li *et al.* (2021) concerning analytical variability being the largest factor affecting the estimation of COVID-19 prevalence from WBS data.

DISCUSSION

Our analysis of possible capabilities projected for wastewater surveillance of SARS-CoV-2 focused on what information can be provided by WBS results that would be useful for informing public health decision-making. Perhaps one of the enduring benefits from all of the WBS initiatives has been that it required and achieved extensive interdisciplinary collaboration between environmental scientists/engineers and public health professionals. The recognition of the engagement of these groups led to the initiation of our ethics guidance for this activity (CWN 2020; Hrudey *et al.* 2021). That guidance reported that the first ethical obligation for any public health surveillance activity is to ensure having a clear and meaningful purpose that is explained through a plan that outlines how data will be collected, analyzed, used for decision-making and, ultimately, made publicly available.

Our review of what has been published or otherwise disseminated about programs of WBS for SARS-CoV-2 has identified 10 factors that are potential advantages for this approach as a means of providing supplementary evidence that can inform public health decision-makers in a meaningful way. The relative importance of some of these factors may change as effective vaccination becomes more common, new VoCs emerge and the dynamics of the pandemic change.

Independence of WBS data from extraneous factors

Strengths

The measures of SARS-CoV-2 in wastewater are independent of policies and practices that inevitably govern the clinical testing of individual patients in the population. Policies for clinical testing have been variable between and within jurisdictions based on factors such as testing capacity, and economic, human and technical resource limitations. Wastewater is collected from individual residences, commercial and industrial buildings as well as community healthcare facilities or other institutions. Wastewater collection captures the fecal and household discharges from all persons using the sanitary facilities of those buildings. The nature of these characteristics is entirely independent of all clinical testing policies applied to individuals.

Limitations

The composition of the contributing population is a function of the sewer network design and in some areas, substantial populations are served by septic tank systems, not connected to the local sewer system, and septic tank pumpouts are only intermittently discharged to a WWTP. Knowledge of the population served by a sewer network can only be known in overall potential terms, with the actual contributing resident populations subject to unknown variation.

Objectivity of data

Strengths

The measures of signals of SARS-CoV-2, provided they are subject to effective QA/QC programs to ensure they provide an accurate representation of what is actually present in the wastewater, can be largely free from biases arising from individual behaviors in seeking clinical testing. All residents of sewered communities, except those who are incontinent, will contribute fecal and other bodily wastes to the wastewater system.

Limitations

The understanding of how SARS-CoV-2 in wastewater relates to important factors like incidence and prevalence of COVID-19 infection is likely to be site-specific and will be subject to effective data collection and interpretation of SARS-CoV-2 fecal shedding rates.

Cost-effective data generation

Strengths

Wastewater sampling whether at a WWTP or in a sewer provides a sample over a knowable or estimable time period that is the composite contribution of a knowable or estimable population that goes beyond individuals, meaning that analytical costs per individual should be very cost-effective in comparison with individual clinical testing.

Limitations

The costs of an individual wastewater analysis are higher than an individual clinical test because there is more sample preparation for wastewater analysis. Costs for sampling from sewers will be higher than sampling from WWTPs where composite sampling is often already available and only entails splitting of an existing sample.

Early detection of infection

Strengths

The potential for early detection is provided by the reality that asymptomatic and low symptom individuals who are less likely to be given clinical tests will still contribute SARS-CoV-2 to wastewater. Likewise, WBS is not dependent on the logistics of reporting individual clinical test data that may contribute to delays.

Limitations

Real-time early detection is governed by wastewater sampling frequency, sample delivery time, analytical processing time and reporting time, all of which can substantially reduce the capability of WBS to provide an early warning.

Tracking of trends

Strengths

Quantitative WBS data so far have been shown to provide possible indicators of success or failure of public health measures (mask mandates, lockdowns and vaccination) as well as apparent changes in the prevalence of COVID-19 in the community.

Limitations

Real-time WBS trend data would be the most useful for informing public health decision-making, but how closely WBS data can approximate real-time data is subject to many practical limitations such as frequency of sampling, sample delivery time, analytical time and reporting time.

Estimation of disease prevalence

Strengths

WBS data are not subject to the vagaries of virus shedding by asymptomatic individuals and arbitrary constraints that govern clinical testing such as policies (e.g., testing only individuals under medical treatment), personal choices to be tested and clinical testing capacity.

Limitations

WBS data cannot be translated into COVID-19 prevalence without site-specific knowledge about the quantitative and temporal dynamics of viral shedding in feces at various stages of infection, the persistence of the viral signal in wastewater as a function of sewer travel time and relevant environmental conditions.

Potential for informing and engaging the public

Strengths

WBS data have the capability to provide, subject to the limitations noted in the discussion above and adequate confidence in the validity of the data, geographically specific and timely information for the public about the status of the pandemic in their community and the need for public health interventions.

Limitations

The utility of WBS for informing the public will be subject to the quality and extent of communication and explanation of the meaning of the WBS data and the resource requirements noted above for providing timely information.

Application within sewer networks to detect hotspots

Strengths

There have been a number of reports of successful detection of locations with infected individuals who were then confirmed by clinical testing and who could be quarantined to reduce the risk of further spread of COVID-19.

Limitations

WBS within sewer networks is more resource-intensive than WWTP monitoring, requires knowledge about the nature and extent of the catchment served by any sewer location and, inevitably, invokes greater ethics concerns about handling the data collected.

Monitor infections within institutions

Strengths

There have been reports of successful detection of infected university students in on-campus residences who were confirmed by clinical testing and who could be quarantined to reduce the risk of further spread of COVID-19. Other institutions such as nursing homes and prisons are prospective candidates for this approach.

Limitations

WBS within sewer networks is resource-intensive, and sewer locations to cover the target institution may be difficult to identify and to access for sampling. This approach may also entail ethical considerations that will be affected by the nature of the institution involved.

Tracking of appearance/progression of viral VoCs

Strengths

As the pandemic has progressed, the emergence of variants of SARS-CoV-2 has become a matter of increasing concern as some have proven to be much more transmissible and VoCs are perpetuating the pandemic causing infections even while extremely effective vaccines have been widely applied. WBS is adaptable and capable of tracking the emergence of such VoCs in a sewered population, but also has the capability, with the use of appropriate technology, of tracking the evolution of viral infection over time through tracking of less notable mutations.

Limitations

The capability of WBS to track VoCs requires adaptation of analytical methods, and the more general tracking of minor mutations requires more sophisticated analytical technology that is not as widely available as the basic RT-qPCR most commonly used in WBS.

CONCLUSIONS

Since the initial identifications of SARS-CoV-2 viral fragments in wastewater in March 2020, there has been an explosive growth of research and development related to WBS applications worldwide. The result has been a proliferation of different methods and approaches, unprecedented levels of peer exchange and the emergence of a significant and expanding research literature base.

WBS should first and foremost be aimed at providing meaningful contributions to informing public health decision-making. To achieve the intended contribution to pandemic management, WBS initiatives should be designed with a clear up-front recognition of how WBS results can best address the questions and needs of the public health community. The devil will inevitably be in the details for assessing the significance of new findings, unknowns and data limitations in this rapidly developing field. As the pandemic continues to evolve, the reality will be that assumptions must be continuously re-evaluated and valid generalizations will be few. A particularly strong potential for WBS is to complement clinical data by providing evidence that is independent of the biases introduced by policies, practices and performance of population clinical testing, thereby providing more objective indicators of community COVID-19 infections.

Reliance on WBS data by the public health community depends upon assurance that such data are valid. Such assurance is necessary to be confident in using WBS data for public communications. Given the rapidly evolving WBS activity, there is not likely to be convergence on a 'single' or 'standard' method in the immediate future. Consequently, there is a clear need to develop standard protocols and approaches to providing QA/QC of the different approaches to substitute for authentic proficiency testing that is normally done by public health laboratories.

The potential for WBS to provide early warning should not be oversold because that potential can currently only be realized by a substantial commitment to frequent sampling and rapid sample delivery, processing, analysis and reporting. WBS can provide particular potential value in surveillance for the emergence of COVID-19 in areas of low disease prevalence and for disease appearance or progression within institutions (e.g., schools and long-term care homes).

WBS can provide complementary evidence about disease trends within a population, which may become increasingly important if clinical testing rates decline. However, obtaining meaningful estimates of disease prevalence (total case numbers) by WBS alone would involve many unknowns at a local level that are not likely to be obtained without substantial, resource-intensive local research that undermines the cost-effective advantages otherwise offered by WBS.

Since early 2021, the significance of dominant genetic VoC of the SARS-CoV-2 virus to pandemic conditions and responses has become a major focus. Some WBS programs are already providing a valuable approach for tracking the spread of VoCs. On a longer-term basis, WBS applications involving genetic sequencing will have the potential to retroactively reveal the evolution of the pandemic within a specific community.

For the future, there exists good potential to build upon the base of compliance data collected at WWTPs, rethinking how routine regulatory monitoring may support public health surveillance. Government support, coordination and leadership for WBS are fundamental to exploiting the potential for WBS to contribute to public health protection. In any case, one of the clear benefits from WBS initiatives has been recognition of, and response to, the need for extensive interdisciplinary collaboration among environmental scientists/engineers and public health professionals. The value that WBS has demonstrated toward understanding the progression and dynamics of the COVID-19 pandemic should justify future applications of WBS toward public health surveillance of other pathogens excreted to wastewater.

Finally, addressing ethical considerations will become increasingly important and apparent as WBS is applied to smaller areas and individual buildings or institutions where public health goals and personal privacy issues are more likely to be more obvious and come into conflict.

DATA AVAILABILITY STATEMENT

All relevant data are included in the paper or its Supplementary Information.

ACKNOWLEDGEMENTS

The authors thank the many collaborators of the Canadian COVID-19 Wastewater Coalition established by the Canadian Water Network and the financial support of a Natural Sciences and Engineering Research Council of Canada Discovery Grant to Dr. Hrudey.

REFERENCES

Acosta, N., Bautista, M. A., Hollman, J., McCalder, J., Beaudet, A. B., Man, L., Waddell, B. J., Chen, J., Li, C., Kuzma, D., Bhatnagar, S., Leal, J., Meddings, J., Hu, J., Cabaj, J. L., Ruecker, N. J., Naugler, C., Pillai, D. R., Achari, G., Ryan, M. C., Conly, J. M., Frankowski, K., Hubert, C. R. J. & Parkins, M. D. 2021 A multicenter study investigating SARS-CoV-2 in tertiary-care hospital wastewater. viral burden correlates with increasing hospitalized cases as well as hospital-associated transmissions and outbreaks. *Water Res.* **201**, 117369. doi.org/10.1016/j.watres.2021.117369.

Agrawal, S., Orschler, L. & Lackner, S. 2021 Long-term monitoring of SARS-CoV-2 RNA in wastewater of the Frankfurt metropolitan area in Southern Germany. *Sci. Rep.* **11**, 5372. doi.org/10.1038/s41598-021-84914-2.

Ahmed, W., Angel, N., Edson, J., Bibby, K., Bivins, A., O'Brien, J. W., Choi, P. M., Kitajimae, M., Simpson, S. L., Li, J., Tscharke, B., Verhagen, R., Smith, W. J. M., Zaugg, J., Dierens, L., Hugenholtz, P., Thomas, K. V. & Mueller, J. F. 2020 First confirmed detection of SARS-CoV-2 in untreated wastewater in Australia: a proof of concept for the wastewater surveillance of COVID-19 in the community. *Sci. Total Environ.* **728**, 138764. doi.org/10.1016/j.scitotenv.2020.138764.

Ahmed, F., Islam, M. A., Kumar, M., Hossain, M., Bhattacharya, P., Islam, M. T., Hossen, F., Hossain, M. S., Islam, M. S., Uddin, M. M., Islam, M. N., Bahadur, N. M., Didar-ul-Alam, M., Reza, H. M. & Jakariya, M. 2021a First detection of SARS-CoV-2 genetic material in the vicinity of COVID-19 isolation Centre in Bangladesh: variation along the sewer network. *Sci. Total Environ.* **776**, 145724. doi.org/10.1016/j.scitotenv.2021.145724.

Ahmed, W., Bibby, K., D'Aoust, P. M., Delatolla, R., Gerba, C. P., Haas, C. N., Hamilton, K. A., Hewitt, J., Julian, T. R., Kaya, D., Monis, P., Moulin, L., Naughton, C., Noble, R. T., Shrestha, A., Tiwari, A., Simpson, S. L., Wurtzer, S. & Bivins, A. 2021b Differentiating between the possibility and probability of SARS-CoV-2 transmission associated with wastewater: empirical evidence is needed to substantiate risk. *FEMS Microbes* **2**, xtab007. doi.org/10.1093/femsmc/xtab007.

Ahmed, W., Simpson, S., Bertsch, P. M., Bibby, K., Bivins, A., Blackall, L., Bofill-Mas, S., Bosch, A., Brandao, J., Choi, P., Ciesielski, M., Donner, E., D'Souza, N., Farnleitner, A., Gerrity, D., Gonzalez, R., Griffith, J. F., Gyawali, P., Haas, C. N., Hamilton, K. A., Hapuarachchi, H. C., Harwood, V. J., Haque, R., Jackson, G., Khan, S. J., Khan, W., Kitajima, M., Korajkic, A., La Rosa, G., Layton, B. A., Lipp, E., McLellan, S. L., McMinn, B., Medema, G., Metcalfe, S., Meijer, W. M., Mueller, J. F., Murphy, H., Naughton, C. C., Noble, R. T., Payyappat, S., Petterson, S., Pitkanen, T., Rajal, V. B., Reyneke, B., Roman, F. A., Rose, J. B., Rusinol, M., Sadowsky, M. J., Sala-Comorera, L., Setoh, Y. X., Sherchan, S. P., Sirikanchana, K., Smith, W., Steele, J. A., Sabburg, R., Symonds, E. M., Thai, P., Thomas, K. V., Tynan, J., Toze, S., Thompson, J., Whiteley, A. S., Wong, J. C. C., Sano, D., Wuertz, S., Xagoraraki, I., Zhang, Q., Zimmer-Faust, AG & Shanks, O. C. 2022 Minimizing errors in RT-PCR detection and quantification of SARS-CoV-2 RNA for wastewater surveillance. *Sci. Total Environ.* **805**, 149877. doi.org/10.1016/j.scitotenv.2021.149877.

Albastaki, A., Naji, M., Lootah, R., Almeheiri, R., Almulla, H., Almarri, I., Alreyami, A., Aden, A. & Alghafri, R. 2021 First confirmed detection of SARS-COV-2 in untreated municipal and aircraft wastewater in Dubai, UAE: the use of wastewater based epidemiology as an early warning tool to monitor the prevalence of COVID-19. *Sci. Total Environ.* **760**, 143350. doi.org/10.1016/j.scitotenv.2020.143350.

Alpaslan Kocamemi, B., Kurt, H., Sait, A., Sarac, F., Saatci, A. M. & Pakdemirli, B. 2020 *SARS-CoV-2 Detection in Istanbul Wastewater Treatment Plant Sludges*. medRxiv. doi.org/10.1101/2020.05.12.20099358.

Ampuero, M., Valenzuela, S., Valiente-Echeverria, F., Soto-Rifo, R., Barriga, G. P., Chnaiderman, J., Rojas, C., Guajardo-Leiva, S., Diez, B. & Gaggero, A. 2020 *SARS-CoV-2 Detection in Sewage in Santiago, Chile – Preliminary Results*. medRxiv. doi.org/10.1101/2020.07.02.20145177.

Arora, S., Nag, A., Sethi, J., Rajvanshi, J., Saxena, S., Shrivastava, S. K. & Gupta, A. B. 2020 Sewage surveillance for the presence of SARS-CoV-2 genome as a useful wastewater based epidemiology (WBE) tracking tool in India. *Water Sci. Technol.* **82** (12), 2823–2836. doi.org/10.2166/wst.2020.540.

Balboa, S., Mauricio-Iglesias, M., Rodriguez, S., Martinez-Lamas, L., Vasallo, F. J., Regueiro, B. & Lema, J. M. 2021 The fate of SARS-COV-2 in WWTPS points out the sludge line as a suitable spot for detection of COVID- 19. *Sci. Total Environ.* **772**, 45268. doi.org/10.1016/j.scitotenv.2021.145268.

Bar Or, I., Yaniv, K., Shagan, M., Ozer, E., Erster, O., Mendelson, E., Mannasse, B., Shirazi, R., Kramarsky-Winter, E., Nir, O., Abu-Ali, H., Ronen, Z., Rinott, E., Lewis, Y. E., Friedler, E., Bitkover, E., Paitan, Y., Berchenko, Y. & Kushmaro, A. 2020 *Regressing SARS-CoV-2 Sewage Measurements onto COVID-19 Burden in the Population: A Proof-of-Concept for Quantitative Environmental Surveillance*. medRxiv. doi.10.1101/2020.04.26.20073569.

Bar Or, I., Weil, M., Indenbaum, V., Bucris, E., Bar-Ilan, D., Elul, M., Levi, N., Aguvaev, I., Cohen, Z., Shirazi, R., Erster, O., Sela-Brown, A., Sofer, D., Mor, O., Mendelson, E. & Zuckerman, N. S. 2021 Detection of SARS-CoV-2 variants by genomic analysis of wastewater samples in Israel. *Sci. Total Environ.* **789**, 148002.

Betancourt, W. Q., Schmitz, B. W., Innes, G. K., Prasek, S. M., Brown, K. M. P., Stark, E. R., Foster, A. R., Sprissler, R. S., Harris, D. T., Sherchan, S. P., Gerba, C. P. & Pepper, I. L. 2021 COVID-19 containment on a college campus via wastewater-based epidemiology, targeted clinical testing and an intervention. *Sci. Total Environ.* **779**, 146408. doi.org/10.1016/j.scitotenv.2021.146408.

Bibby, K., Bivins, A., Wu, Z. & North, D. 2021 Making waves: plausible lead time for wastewater based epidemiology as an early warning system for COVID-19. *Water Res.* **202**, 117438.

Bivins, A., North, D., Wu, Z., Shaffer, M., Ahmed, W. & Bibby, K. 2021 *Within-Day Variability of SARS-CoV-2 RNA in Municipal Wastewater Influent During Periods of Varying COVID-19 Prevalence and Positivity*. medRxiv. doi.org/10.1101/2021.03.16.21253652.

Black, J., Aung, P., Nolan, M., Roney, E., Poon, R., Hennessy, D., Crosbie, N. D., Deere, D., Jex, A. R., John, N., Baker, L., Scales, P. J., Usher, S. P., McCarthy, D. T., Schang, C., Schmidt, J., Myers, J. S., Begue, N., Kaucner, C., Thorley, B., Druce, J., Monis, P., Lau, M. & Sarkis, S. 2021 Epidemiological evaluation of sewage surveillance as a tool to detect the presence of COVID-19 cases in a low case load setting. *Sci. Total Environ.* **786**, 147469. doi.org/10.1016/j.scitotenv.2021.147469.

Buonerba, A., Corpuz, M. V. A., Ballesteros, F., Choo, K.-H. C., Hasan, S. W., Korshin, G. V., Belgiorno, V., Barcelo, D. & Naddeo, V. 2021 Coronavirus in water media: analysis, fate, disinfection and epidemiological applications. *J. Hazard. Mater.* **415**, 125580. doi.org/10.1016/j.jhazmat.2021.125580.

Cao, Y. & Francis, R. 2021 On forecasting the community-level COVID-19 cases from the concentration of SARS-CoV-2 in wastewater. *Sci. Total Environ.* **786**, 147451. doi.org/10.1016/j.scitotenv.2021.147451.

Carillo-Reyes, J., Barragan-Trinidad, M. & Buitron, G. 2021 Surveillance of SARS-CoV-2 in sewage and wastewater treatment plants in Mexico. *J. Water Process Eng.* **40**, 101815. doi.org/10.1016/j.jwpe.2020.101815.

Chavarria-Miró, G., Anfruns-Estrada, E., Martinez-Velazquez, A., Vazquez-Portero, A. M., Guix, S., Paraira, M., Galofre, B., Sanchez, G., Pinto, R. M. & Bosch, A. 2021 Time evolution of severe acute respiratory syndrome coronavirus 2 (SARS-CoV-2) in wastewater during the first pandemic wave of COVID-19 in the Metropolitan Area of Barcelona, Spain. *Appl. Environ. Microbiol.* **87** (7), e02750-20. doi.org/10.1128/AEM.02750-20.

Chik, A. H. S., Glier, M. B., Servos, M., Mangat, C. S., Pang, X.-L., D'Aoust, P. M., Burnet, J.-B., Delatolla, R., Dorner, S., Giesy Jr, J. P., McKay, R. M., Mulvey, M. R., Prystajecky, N., Srikanthan, N., Xie, Y.-W., Bernadette Conant, B. & Hrudey, S. E. & Canadian SARS-CoV-2 Inter-Laboratory Consortium 2021 Comparison of approaches to quantify SARS-CoV-2 in wastewater using RT-qPCR: results and implications from a collaborative inter-laboratory study in Canada. *J. Environ. Sci.* **107**, 218–229. doi.org/10.1016/j.jes.2021.01.029.

Curtis, K., Keeling, D., Yetka, K., Larson, A. & Gonzalez, R. 2020 *Wastewater SARS-CoV-2 Concentration and Loading Variability from Grab and 24-hour Composite Samples*. medRxiv. doi.org/10.1101/2020.07.10.20150607.

CWN 2020 *Canadian Coalition on Wastewater-Related COVID-19 Research Ethics Guidance: Ethics and Communications Guidance for Wastewater Surveillance to Inform Public Health Decision-Making about COVID-19*. Canadian Water Network. https://cwn-rce.ca/wp-content/uploads/COVID19-Wastewater-Coalition-Ethics-and-Communications-Guidance-v4-Sept-2020.pdf.

CWN 2021a *Canadian Water Network COVID-19 Wastewater Coalition*. Available from: https://cwn-rce.ca/covid-19-wastewater-coalition/ (accessed 22 November 2021).

CWN 2021b *Canadian Water Network Known COVID-19 Wastewater Sampling Locations*. Available from: https://cwn-rce.ca/covid-19-wastewater-coalition/covid-19-wastewater-coalition-maps/ (accessed 22 November 2021).

Dada, A. C. & Gyawali, P. 2021 Quantitative microbial risk assessment (QMRA) of occupational exposure to SARS-CoV-2 in wastewater treatment plants. *Sci. Total Environ.* **763**, 142989. doi.org/10.1016/j.scitotenv.2020.142989.

D'Aoust, P. M., Graber, T. E., Mercier, E., Montpetit, D., Alexandrov, I., Neault, N., Baig, A. T., Mayne, J., Zhang, X., Alain, T., Servos, M. R., Srikanthan, N., MacKenzie, M., Figeys, D., Manuel, D., Jüni, P., MacKenzie, A. E. & Delatolla, R. 2021a Catching a resurgence: increase in SARS-CoV-2 viral RNA identified in wastewater 48 h before COVID-19 clinical tests and 96 h before hospitalizations. *Sci. Total Environ.* **770**, 145319. doi.org/10.1016/j.scitotenv.2021.145319.

D'Aoust, P. M., Mercier, E., Montpetit, D., Jia, J.-J., Alexandrov, I., Neault, N., Baig, A. T., Mayne, J., Zhang, X., Alain, T., Langlois, M.-A., Servos, M. R., MacKenzie, M., Figeys, D., MacKenzie, A. E. & Delatolla, R. 2021b Quantitative analysis of SARS-CoV-2 RNA from wastewater solids in communities with low COVID-19 incidence and prevalence. *Water Res.* **188**, 116560. doi.org/10.1016/j.watres.2020.116560.

Farkas, K., Hillary, L. S., Malham, S. K., McDonald, J. E. & Jones, D. L. 2020 Wastewater and public health: the potential of wastewater surveillance for monitoring COVID-19. *Current Opinion in Environmental Science & Health* **17**, 14–20. doi.org/10.1016/j.coesh.2020.06.001.

Feng, W., Newbigging, A. M., Le, C., Pang, B., Peng, H., Cao, Y., Wu, J., Abbas, G., Song, J., Wang, D.-B., Cui, M., Tao, J., Lorne Tyrrell, D., Zhang, X.-E., Zhang, H. & Le, X. C. 2020 Molecular diagnosis of COVID-19: challenges and research needs. *Anal. Chem.* **92** (15), 10196–10209. . dx.doi.org/10.1021/acs.analchem.0c02060.

Feng, S., Roguet, A., McClary-Gutierrez, J. S., Newton, R. J., Kloczko, N., Meiman, J. G. & McLellan, S. L. 2021 *Evaluation of Sampling Frequency and Normalization of SARS-CoV-2 Wastewater Concentrations for Capturing COVID-19 Burdens in the Community.* medRxiv. doi.org/10.1101/2021.02.17.21251867.

Fernandez-Cassi, X., Scheidegger, A., Bänziger, C., Cariti, F., Corzon, A. T., Ganesanandamoorthy, P., Lemaitre, J. C., Ort, C., Julian, T. R. & Kohn, T. 2021 Wastewater monitoring outperforms case numbers as a tool to track COVID-19 incidence dynamics when test positivity rates are high. *Water Res.* **200**, 117252. doi.org/10.1016/j.watres.2021.117252.

Fongaro, G., Hermes Stoco, P., Sobral Marques Souza, D., Grisard, E. C., Magri, M. E., Rogovski, P., Schörner, M. A., Barazzetti, F. H., Ana Paula Christoff, A. P., de Oliveira, L. F. V., Bazzo, M. L. & Wagner, G. 2021 The presence of SARS-CoV-2 RNA in human sewage in Santa Catarina, Brazil, November 2019. *Sci. Total Environ.* **778**, 146198. doi.org/10.1016/j.scitotenv.2021.146198.

Fuqua, J. L., Rouchka, E. C., Waigel, S., Sokoloski, K., Chung, D., Zacharias, W., Zhang, M., Chariker, J., Talley, D., Santisteban, I., Varsani, A., Moyer, S., Holm, R. H., Yeager, R. A., Smith, T. & Bhatnagar, A. 2021 *A Rapid Assessment of Wastewater for Genomic Surveillance of SARS-CoV-2 Variants at Sewershed Scale in Louisville, KY.* medRxiv. doi.org/10.1101/2021.03.18.21253604.

Gawlik, B. M., Tavazzi, S., Mariani, G., Skejo, H., Sponar, M., Higgins, T., Medema, G. & Wintgens, T. 2021 *SARS-CoV-2 Surveillance Employing Sewage Towards a Sentinel System – Feasibility Assessment of an EU Approach.* EUR 30684 EN, Publications Office of the European Union, Luxembourg. doi:10.2760/300580, JRC125065.

Gering, E., Colbert, J., Sarah Schmedes, S., Duncan, G., Lopez, J., Motes, J., Weiss, J., Azarian, T., Tekin, O. & Blanton, J. 2021 *ddPCR Reveals SARS-CoV-2 Variants in Florida Wastewater.* medRxiv. doi.org/10.1101/2021.04.08.21255119.

Gerrity, D., Papp, K., Stoker, M., Sims, A. & Frehner, W. 2021 Early-pandemic wastewater surveillance of SARS-CoV-2 in Southern Nevada: Methodology, occurrence, and incidence/prevalence considerations. *Water Res. X* **10**, 100086. doi.org/10.1016/j.wroa.2020.100086.

Giacobbo, A., Rodrigues, M. A. S., Ferreiraa, J. Z., Bernardes, A. M. & de Pinho, M. N. 2021 A critical review on SARS-CoV-2 infectivity in water and wastewater. What do we know? *Sci. Total Environ.* **774**, 145721. doi.org/10.1016/j.scitotenv.2021.145721.

Gibas, C., Lambirth, K., Mittal, N., Juel, M. A. J., Barua, V. B., Brazell, L. R., Hinton, K., Lontai, J., Stark, N., Young, I., Quach, C., Russ, M., Kauer, J., Nicolosi, J. B., Chen, D., Akella, S., Tang, W., Schlueter, J. & Munir, M. 2021 Implementing building-level SARS-CoV-2 wastewater surveillance on a university campus. *Sci. Total Environ.* **782**, 146749. doi.org/10.1016/j.scitotenv.2021.146749.

Goncalves Cabecinhas, A. R., Roloff, T., Stange, M., Bertelli, C., Huber, M., Ramette, A., Chen, C., Nadeau, S., Gerth, Y., Yerly, S., Opota, O., Pillonel, T., Schuster, T., Metzger, C. M. J. A., Sieber, J., Bel, M., Wohlwend, N., Baumann, C., Koch, M. C., Bittel, P., Leuzinger, K., Brunner, M., Suter-Riniker, F., Berlinger, L., Søgaard, K. K., Beckmann, C., Noppen, C., Redondo, M., Steffen, I., Seth-Smith, H. M. B., Mari, A., Lienhard, R., Risch, M., Nolte, O., Eckerle, I., Lucchini, G. M., Hodcroft, E. B., Neher, R. A., Tanja Stadler, T., Hirsch, H. H., Leib, S. L., Risch, L., Kaiser, L., Trkola, A., Greub, G. & Egli, A. 2021 SARS-CoV-2 N501y introductions and transmissions in Switzerland from beginning of October 2020 to February 2021 – implementation of Swiss-wide diagnostic screening and whole genome sequencing. *Microorganisms* **9**, 677. doi.org/10.3390/microorganisms9040677.

Gonzalez, R., Curtis, K., Bivins, A., Bibby, K., Weir, M. H., Yetka, K., Thompson, H., Keeling, D., Mitchell, J. & Gonzalez, D. 2020 COVID-19 surveillance in Southeastern Virginia using wastewater-based epidemiology. *Water Res.* **186**, 116296. doi. org/10.1016/j.watres.2020.116296.

Graham, K. E., Loeb, S. K., Wolfe, M. K., Catoe, D., Sinnott-Armstrong, N., Kim, S., Yamahara, K. M., Sassoubre, L. M., Mendoza Grijalva, L. M., Roldan-Hernandez, L., Langenfeld, K., Wigginton, K. R. & Boehm, A. B. 2021 SARS-CoV-2 RNA in wastewater settled solids is associated with COVID-19 cases in a large urban sewershed. *Environ. Sci. Technol.* **55**, 488–498. doi.org/10.1021/acs.est.0c06191.

Greaves, J., Fischer, R. J., Shaffer, M., Bivins, A., Holbrook, M. G., Munster, V. J. & Bibby, K. 2021 Sodium hypochlorite disinfection of SARS-CoV-2 spiked in water and municipal wastewater. *Sci. Total Environ.* doi.org/10.1016/j.scitotenv.2021.150766.

Green, H., Wilder, M., Collins, M., Fenty, A., Gentile, K., Kmush, B. L., Zeng, T., Middleton, F. A. & Larsen, D. A. 2020 *Quantification of SARS-CoV-2 and Cross-Assembly Phage (crAssphage) from Wastewater to Monitor Coronavirus Transmission Within Communities.* medRxiv. doi.org/10.1101/2020.05.21.20109181.

Guerrero-Latorre, L., Ballesteros, I., Villacres-Granda, I., Granda, M. G., Freire-Paspuel, B. & Rios-Touma, B. 2020 SARS-CoV-2 in river water: implications in low sanitation countries. *Sci. Total Environ.* **743**, 140832. doi.org/10.1016/j.scitotenv.2020.140832.

GWRC 2021 *Global Water Research Coalition.* Available from: http://www.globalwaterresearchcoalition.net/about-us/gwrc-members/ (accessed 22 November 2021).

Hart, O. E. & Halden, R. U. 2020 Computational analysis of SARS-CoV-2/COVID-19 surveillance by wastewater-based epidemiology locally and globally: Feasibility, economy, opportunities and challenges. *Sci. Total Environ.* **730**, 138875. doi.org/10.1016/j.scitotenv.2020.138875.

Hamouda, M., Mustafa, F., Maraqa, M., Rizvi, T. & Hassan, A. A. 2021 Wastewater surveillance for SARS-CoV-2: lessons learnt from recent studies to define future applications. *Sci. Total Environ.* **759**, 143493. doi.org/10.1016/j.scitotenv.2020.143493.

Haramoto, E., Malla, B., Thakali, O. & Kitajima, M. 2020 First environmental surveillance for the presence of SARS-CoV-2 RNA in wastewater and river water in Japan. *Sci. Total Environ.* **737**, 140405. doi.org/10.1016/j.scitotenv.2020.140405.

Harris-Lovett, S., Nelson, K. L., Beamer, P., Bischel, H. N., Bivins, A., Bruder, A., Butler, C., Camenisch, T. D., De Long, S. K., Karthikeyan, S., Larsen, D. A., Meierdiercks, K., Mouser, P. J., Pagsuyoin, S., Prasek, S. M., Radniecki, T. S., Ram, J. L., Roper, D. K., Safford, H., Sherchan, S. P., Shuster, W., Stalder, T., Wheeler, R. T. & Korfmacher, K. T. 2021 Wastewater surveillance for SARS-CoV-2 on college campuses: initial efforts, lessons learned, and research needs. *Int. J. Environ. Res. Public Health.* **18**, 4455. doi.org/10.3390/ijerph18094455.

Hata, A., Honda, R., Hara-Yamamura, H. & Meuchi, Y. 2021 Detection of SARS-CoV-2 in wastewater in Japan by multiple molecular assays – implication for wastewater-based epidemiology (WBE). *Sci. Total Environ.* **758**, 143578. doi.org/10.1016/j.scitotenv.2020.143578.

Hemalatha, M., Kiran, U., Kuncha, S. K., Kopperi, H., Gokulan, C. G., Mohan, S. K. & Mishra, R. K. 2021 Surveillance of SARS-CoV-2 spread using wastewater-based epidemiology: comprehensive study. *Sci Total Environ.* **768**, 144704. doi.org/10.1016/j.scitotenv.2020.144704.

Hillary, L. S., Farkas, K., Maher, K. H., Lucaci, A., Thorpe, J., Distaso, J. M. A., Gaze, W. H., Paterson, S., Burke, T., Connor, T. R., McDonald, J. E., Malham, S. K. & Jones, D. L. 2021 Monitoring SARS-CoV-2 in municipal wastewater to evaluate the success of lockdown measures for controlling COVID-19 in the UK. *Water Res.* **200**, 117214. doi.org/10.1016/j.watres.2021.117214.

Hokajarvi, A.-M., Rytkönena, A., Tiwari, A., Kauppinen, A., Oikarinen, S., Lehto, K.-M., Kankaanpääc, A., Gunnarc, T., Al-Hello, H., Blomqvist, S., Miettinen, I. T., Savolainen-Kopra, C. & Pitkänen, T. 2021 The detection and stability of the SARS-CoV-2 RNA biomarkers in wastewater influent in Helsinki, Finland. *Sci. Total Environ.* **770**, 145274. https://doi.org/10.1016/j.scitotenv.2021.145274.

Hong, P.-Y., Rachmadi, A. T., Mantilla-Calderon, D., Alkahtani, M., Bashawri, Y. M., Al Qarni, H., O'Reilly, K. M. & Zhou, J. 2021 Estimating the minimum number of SARS-CoV-2 infected cases needed to detect viral RNA in wastewater: to what extent of the outbreak can surveillance of wastewater tell us? *Environ. Res.* **195**, 110748. doi.org/10.1016/j.envres.2021.110748.

Hrudey, S. E., Silva, D. S., Shelley, J., Pons, W., Isaac-Renton, J., Chik, A. H.-S. & Conant, B. 2021 Ethics guidance for environmental scientists engaged in surveillance of wastewater for SARS-CoV-2. *Environ. Sci. Technol.* **55**. doi.org/10.1021/acs.est.1c00308.

Huisman, J. S., Scire, J., Caduff, L., Fernandez-Cassi, X., Ganesanandamoorthy, P., Kull, A., Scheidegger, A., Stachler, E., Boehm, A. B., Hughes, B., Knudson, A., Topol, A., Wigginton, K. R., Wolfe, M. K., Kohn, T., Christoph Ort, C., Stadler, T. & Julian, T. R. 2021 *Wastewater-based Estimation of the Effective Reproductive Number of SARS-CoV-2*. medRxiv. doi.org/10.1101/2021.04.29.21255961.

Iglesias, N. G., Gebhard, L. G., Carballeda, J. M., Aiello, I., Recalde, E., Terny, G., Ambrosolio, S., L'Arco, G., Konfino, J. & Brardinelli, J. I. 2021 SARS-CoV-2 surveillance in untreated wastewater: detection of viral RNA in a low-resource community in Buenos Aires, Argentina. *Rev. Panam. Salud Publica* **45**, e137. doi.org/10.26633/RPSP.2021.137.

Jahn, K., Dreifuss, D., Topolsky, I., Kull, A., Ganesanandamoorthy, P., Fernandez-Cassi, X., Banziger, C., Stachler, E., Fuhrmann, L., Jablonski, K. P., Chen, C., Aquino, C., Stadler, T., Ort, C., Kohn, T., Julian, T. R. & Beerenwinkel, N. 2021 *Detection of SARS-CoV-2 Variants in Switzerland by Genomic Analysis of Wastewater Samples*. medRxiv. doi.org/10.1101/2021.01.08.21249379.

Johnson, R., Muller, C. J. F., Ghoor, S., Louw, J., Archer, E., Surujlal-Naicker, S., Berkowitz, N., Volschenk, M., Bröcker, L. H. L., Wolfaardt, G., van der Walt, M., Mutshembele, A. M., Malema, S., Gelderblom, H. C., Mdhuli, M., Gray, G., Mathee, A. & Street, R. A. 2021 Qualitative and quantitative detection of SARS-CoV-2 RNA in untreated wastewater in Western Cape Province, South Africa. *S. Afr. Med. J.* **111** (3), 198–202. doi.org/10.7196/SAMJ.2021.v111i3.15154.

Jones, D. L., Baluja, M. Q., Graham, D. W., Corbishley, A., McDonald, J. E., Malhame, S. K., Hillarya, L. S., Connor, T. R., Gaze, W. H., Mourai, I. B., Wilcox, M. H. & Farkas, K. 2020 Shedding of SARS-CoV-2 in faeces and urine and its potential role in person-to-person transmission and the environment-based spread of COVID-19. *Sci. Total Environ.* **749**, 141364. doi.org/10.1016/j.scitotenv.2020.141364.

Jones, E. R., van Vliet, M. T. H., Qadir, M. & Bierkens, M. F. P. 2021 Country-level and gridded estimates of wastewater production, collection, treatment and reuse. *Earth Syst. Sci. Data* **13**, 237–254. doi.org/10.5194/essd-13-237-2021.

Kang, M. K., Wei, J., Yuan, J., Guo, J., Zhang, Y., Hang, J., Qu, Y., Qian, H., Zhuang, Y., Chen, X., Peng, X., Shi, T., Wang, J., Wu, J., Song, T., He, J., Li, Y. & Zhong, N. 2020 Probable evidence of fecal aerosol transmission of SARS-CoV-2 in a high-rise building. *Ann. Int. Med.* **15**, 2020. doi.org/10.7326/M20-0928.

Kitamura, K., Sadamasu, K., Muramatsu, M. & Yoshida, H. 2021 Efficient detection of SARS-CoV-2 RNA in the solid fraction of wastewater. *Sci. Total Environ.* **763**, 144587. doi.org/10.1016/j.scitotenv.2020.144587.

Kocamemia, B. A., Kurtb, H., Sait, A., Kadid, H., Saracc, F., Aydınd, I., Saatcie, A. M. & Pakdemirlif, B. 2020 *Nationwide SARS-CoV-2 Surveillance Study for Sewage and Sludges of Wastewater Treatment Plants in Turkey*. medRxiv. doi.org/10.1101/2020.11.29.20240549.

Kumar, M., Joshi, M., Patel, A. K. & Joshi, C. G. 2021 Unravelling the early warning capability of wastewater surveillance for COVID-19: a temporal study on SARS-CoV-2 RNA detection and need for the escalation. *Environ. Res.* **196**, 110946. doi.org/10.1016/j.envres.2021.110946.

Kumblathan, T., Liu, Y., Uppal, G. K., Hrudey, S. E. & Li, X.-F. 2021 Wastewater-based epidemiology for community monitoring of SARS-CoV-2: progress and challenges. *ACS Environ. Au* **1**, 18–31. https://doi.org/10.1021/acsenvironau.1c00015.

La Rosa, G., Iaconelli, M., Mancini, P., Bonanno Ferraro, G., Veneri, C., Bonadonna, L., Lucentini, L. & Suffredini, E. 2020 First detection of SARS-CoV-2 in untreated wastewaters in Italy. *Sci. Total Environ.* **736**, 139652. doi.org/10.1016/j.scitotenv.2020.139652.

La Rosa, G., Mancini, P., Bonanno Ferraro, G., Veneri, C., Iaconelli, M., Lucentini, L., Bonadonna, L., Brusaferro, S., Brandtner, D., Fasanella, A., Pace, L., Parisi, A., Galante, D. & Suffredini, E. 2021 Rapid screening for SARS-CoV-2 variants of concern in clinical and environmental samples using nested RT-PCR assays targeting key mutations of the spike protein. *Water Res.* **197**, 117104. doi.org/10.1016/j.watres.2021.117104.

Li, B., Di, D. Y. W., Saingam, P., Jeon, M. K. & Yan, T. 2021 Fine-scale temporal dynamics of SARS-CoV-2 RNA abundance in wastewater during a COVID-19 lockdown. *Water Res.* **197**, 117093. doi.org/10.1016/j.watres.2021.117093.

Liu, P., Ibaraki, M., VanTassell, J., Geith, K., Cavallo, M., Kann, R. & Moe, C. 2020 *A Novel COVID-19 Early Warning Tool: Moore Swab Method for Wastewater Surveillance at an Institutional Level.* medRxiv. doi.org/10.1101/2020.12.01.20238006.

Lo, I. L., Lio, C. F., Cheong, H. H., Lei, C. I., Cheong, T. H., Zhong, X., Tian, Y. & Sin, N. N. 2020 Evaluation of SARS-CoV-2 RNA shedding in clinical specimens and clinical characteristics of 10 patients with COVID-19 in Macau. *Int. J. Biol. Sci.* **16**, 1698–1707. doi.org/10.7150/ijbs.45357.

Lodder, W. & deRoda Huisman, A. M. 2020 SARS-CoV-2 in wastewater: potential health risk, but also data source. *Lancet Gastroenterol. Hepatol.* **5**, 533–534.

Martin, J., Klapsa, D., Wilton, T., Zambon, M., Bentley, E., Bujaki, E., Fritzsche, M., Mate, R. & Majumdar, M. 2021 Tracking SARS-CoV-2 in sewage: evidence of changes in virus variant predominance during COVID-19 pandemic. *Viruses* **12**, 1144. doi:10.3390/v12101144.

Medema, G., Heijnen, L., Elsinga, G. & Italiaander, R. 2020a Presence of SARS-Coronavirus-2 RNA in sewage and correlation with reported COVID-19 prevalence in the early stage of the epidemic in the Netherlands. *Environ. Sci. Technol. Lett.* **7** (7), 511–516. doi.org/10.1021/acs.estlett.0c00357.

Medema, G., Been, F., Heijnen, L. & Petterson, S. 2020b Implementation of environmental surveillance for SARS-CoV-2 virus to support public health decisions: opportunities and challenges. *Curr. Opin. Environ. Sci. Health* **17**, 49–71. doi.org/10.1016/j.coesh.2020.09.006.

Melvin, R. G., Chaudhry, N., Georgewill, O., Freese, R. & Simmons Jr, G. E. 2021 *Predictive Power of SARS-CoV-2 Wastewater Surveillance for Diverse Populations across a Large Geographical Range.* medRxiv. doi.org/10.1101/2021.01.23.21250376.

Miura, F., Kitajima, M. & Omori, R. 2021 Duration of SARS-CoV-2 viral shedding in faeces as a parameter for wastewater-based epidemiology: re-analysis of patient data using a shedding dynamics model. *Sci. Total Environ.* **769**, 144549. doi.org/10.1016/j.scitotenv.2020.144549.

Miyani, B., Fonoll, X., Norton, J., Mehrotra, A. & Xagoraraki, I. 2020 SARS-CoV-2 in detroit wastewater. *J. Environ. Eng.* **146** (11), 06020004. doi.org/10.1061/ (ASCE)EE.1943-7870.0001830.

Mlejnkova, H., Sovova, K., Vasickova, P., Ocenaskova, V., Jasikova, L. & Juranova, E. 2020 Preliminary study of Sars-Cov-2 occurrence in wastewater in the Czech Republic. *Int.J. Environ. Res. Public Health* **17**, 5508. doi.org/10.3390/ijerph17155508.

Napit, R., Manandhar, P., Chaudhary, A., Shrestha, B., Poudel, A., Raut, R., Pradhan, S., Raut, S., Mathema, S., Rajbhandari, R., Dixit, S., Schwind, J. S., Johnson, C. K., Mazet, J. K. & Karmacharya, D. 2021 *Rapid Genomic Surveillance of SARS-CoV-2 in a Dense Urban Community Using Environmental (Sewage) Samples – Genomic Environmental Surveillance of SARS-CoV-2 Using Sewage Samples.* medRxiv. doi.org/10.1101/2021.03.29.21254053.

Nasseri, S., Yavarian, J., Baghani, A. N., Azad, T. M., Nejati, A., Nabizadeh, R., Hadi, M., Jandaghi, N. Z. S., Vakili, B., Vaghefi, S. K. A., Baghban, M., Yousefi, S., Nazmara, S. & Alimohammadi, M. 2021 The presence of SARS-CoV-2 in raw and treated wastewater in 3 cities of Iran: Tehran, Qom and Anzali during coronavirus disease 2019 (COVID-19) outbreak. *J. Environ. Health Sci. Eng.* **19**, 573–584. doi.org/10.1007/s40201-021-00629-6.

Naughton, C. C. 2021 Available from: https://arcg.is/1aummW (accessed 30 November 2021).

Naughton, C. C., Roman Jr., F. A., Alvarado, A. G. F., Tariqi, A. Q., Deeming, M. A., Bibby, K., Bivins, A., Rose, J. B., Medema, G., Ahmed, W., Katsivelis, P., Allan, V., Sinclair, R., Zhang, Y. & Kinyua, M. N. 2021 *Show Us the Data: Global COVID-19 Wastewater Monitoring Efforts, Equity, and Gaps.* medRxiv. doi.org/10.1101/2021.03.14.21253564.

Nemudryi, A., Nemudraia, A., Wiegand, T., Surya, K., Buyukyoruk, M., Cicha, C., Vanderwood, K. K., Wilkinson, R. & Wiedenheft, B. 2020 Temporal detection and phylogenetic assessment of SARS-CoV-2 in municipal wastewater. *Cell Rep. Med.* **1**, 100098. doi.org/10.1016/j.xcrm.2020.100098.

Peccia, J., Zulli, A., Brackney, D. E., Grubaugh, N. D., Kaplan, E. H., Casanovas-Massana, A., Ko, A. I., Malik, A. A., Wang, D., Wang, M., Warren, J. L., Weinberger, D. M., Arnold, W. & Omer, S. B. 2020 Measurement of SARS-CoV-2 RNA in wastewater tracks community infection dynamics. *Nat. Biotechnol.* **38**, 1164–1167. doi.org/10.1038/s41587-020-0684-z.

Pechlivanis, N., Tsagiopoulou, M., Maniou, M. C., Togkousidis, A., Mouchtaropoulou, E., Chassalevris, T., Chaintoutis, S., Dovas, C., Petala, M., Kostoglou, M., Karapantsios, T., Laidou, S., Vlachonikola, E., Chatzidimitriou, A., Papadopoulos, A., Papaioannou, N., Argiriou, A. & Psomopoulos, F. 2021 *Detecting SARS-CoV-2 Lineages and Mutational Load in Municipal Wastewater; a Use-Case in the Metropolitan Area of Thessaloniki, Greece.* medRxiv. doi.org/10.1101/2021.03.17.21252673.

Pecson, B. M., Darby, E., Haas, C. N., Amha, Y., Bartolo, M., Danielson, R., Dearborn, Y., Di Giovanni, G., Ferguson, C., Fevig, S., Gaddis, E., Gray, D., Lukasik, G., Mull, B., Olivas, L., Olivieri, A. & Qu, Y. & SARS-CoV-2 Interlaboratory Consortium 2021 Reproducibility and sensitivity of 36 methods to quantify the SARS-CoV-2 genetic signal in raw wastewater: findings from an interlaboratory methods evaluation in the U.S. *Environ. Sci. Water Res. Technol.* **7**, 504–520. doi.org/10.1039/d0ew00946f.

Philo, S. E., Keim, E. K., Swanstrom, R., Ong, A. Q. W., Burnor, E. A., Kossik, A. L., Harrison, J. C., Demeke, B. A., Zhou, N. A., Beck, N. K., Shirai, J. H. & Meschke, J. S. 2021 A comparison of SARS-CoV-2 wastewater concentration methods for environmental surveillance. *Sci. Total Environ.* **760**, 144215. doi.org/10.1016/j.scitotenv.2020.144215.

Prado, T., Fumian, T. M., Mannarino, C. F., Resende, P. C., Motta, F. C., Eppinghause, A. L. F., Chagas do Vale, V. H., Braz, R. M. S., da Silva Ribeiro de Andrade, J., Maranhão, A. G. & Miagostovich, M. P. 2021 Wastewater-based epidemiology as a useful tool to track SARS-CoV-2 and support public health policies at municipal level in Brazil. *Water Res.* **191**, 116810. doi.org/10.1016/j.watres.2021.116810.

Rafiee, M., Isazadeh, S., Mohseni-Bandpei, A., Mohebbi, S. R., Jahangiri-rad, M., Eslami, A., Dabiri, H., Kasra Roostaei, K., Tanhaei, M. & Amereha, F. 2021 Moore swab performs equal to composite and outperforms grab sampling for SARS-CoV-2 monitoring in wastewater. *Sci. Total Environ.* **790**, 148205. doi.org/10.1016/j.scitotenv.2021.148205.

Randazzo, W., Truchado, P., Cuevas-Ferrando, E., Simon, P., Allende, A. & Sanchez, G. 2020a SARS-CoV-2 RNA in wastewater anticipated COVID-19 occurrence in a low prevalence area. *Water Res.* **181**, 115942. doi.org/10.1016/j.watres.2020.115942.

Randazzo, W., Cuevas-Ferrando, E., Sanjúan, R., Domingo-Calap, P. & Śanchez, G. 2020b Metropolitan wastewater analysis for COVID-19 epidemiological surveillance. *Int. J. Hyg. Environ. Health* **230**, 113621. doi.org/10.1016/j.ijheh.2020.113621.

Reeves, K., Leibig, J., Feula, A., Saldi, T., Lasda, E., Johnson, W., Lilienfeld, J., Maggi, J., Pulley, K., Wilkerson, P. J., Real, B., Zak, G., Davis, J., Fink, M., Gonzalez, P., Hager, C., Ozeroff, C., Tat, K., Alkire, M., Butler, C., Coe, E., Darby, J., Freeman, J., Heuer, H., Jones, J. R., Karr, M., Key, S., Maxwell, K., Nelson, L., Saldana, E., Shea, R., Salveson, L., Tomlinson, K., Vargas-Barriga, J., Vigil, B., Brisson, G., Parker, R., Leinwand, L. A., Bjorkman, K. & Mansfeldt, C. 2021 High-resolution within-sewer SARS-CoV-2 surveillance facilitates informed intervention. *Water Res.* **204**, 117613. doi.org/10.1016/j.watres.2021.117613.

Rimoldi, S. G., Stefani, F., Gigantiello, A., Polesello, S., Comandatore, F., Mileto, D., Maresca, M., Longobardi, C., Mancon, A., Romeri, F., Pagani, C., Cappelli, F., Roscioli, C., Moja, L., Gismondo, M. R. & Salerno, F. 2020 Presence and infectivity of SARS-CoV-2 virus in wastewaters and rivers. *Sci. Total Environ.* **744**, 140911. doi.org/10.1016/j.scitotenv.2020.140911.

Rizak, S. & Hrudey, S. E. 2006 Misinterpretation of drinking water quality monitoring data with implications for risk management. *Environ. Sci. Technol.* **40** (17), 5244–5250. doi.org/10.1021/es0520417.

Roka, E., Khayer, B., Kis, Z., Kovacs, L. B., Schuler, E., Magyar, N., Malnasi, T., Oravecz, O., Palyi, B., Pandics, T. & Vargha, M. 2021 Ahead of the second wave: early warning for COVID-19 by wastewater surveillance in Hungary. *Sci. Total Environ.* **786**, 147398. doi.org/10.1016/j.scitotenv.2021.147398.

Rusinol, M., Zammit, I., Itarte, M., Fores, E., Martinez-Puchol, S., Girones, R., Borrego, C., Corominas, L. & Bofill-Mas, S. 2021 Monitoring waves of the COVID-19 pandemic: inferences from WWTPs of different sizes. *Sci. Total Environ.* **787**, 147463. doi.org/10.1016/j.scitotenv.2021.147463.

Saguti, F., Magnil, E., Enache, L., Churqui, M. P., Johansson, A., Lumely, D., Davidson, F., Dotevall, L., Mattsson, A., Trybala, E., Lagging, M., Lindh, M., Gisslen, M., Brezicka, T., Nystrom, K. & Norder, H. 2021 Surveillance of wastewater revealed peaks of SARS-CoV-2 preceding those of hospitalized patients with COVID-19. *Water Res.* **189**, 116620. doi.org/10.1016/j.watres.2020.116620.

Saththasivam, J., El-Malaha, S. S., Gomeza, T. A., Jabbar, K. A., Remanana, R., Krishnankutty, A. K., Ogunbiyi, O., Rasool, K., Ashhab, S., Rashkeev, S., Bensaad, M., Ahmed, A. A., Mohamoud, Y. A., Malek, J. A., Raddad, L. J. A., Jeremijenko, A., Halaweh, H. A. A., Lawler, J. & Mahmoud, K. A. 2021 COVID-19 (SARS-CoV-2) outbreak monitoring using wastewater-based epidemiology in Qatar. *Sci. Total Environ.* **774**, 145608. doi.org/10.1016/j.scitotenv.2021.145608.

Schang, C., Crosbie, N. D., Nolan, M., Poon, R., Wang, M., Jex, A., John, N., Baker, L., Scales, P., Schmidt, J., Thorley, B. R., Hill, K., Zamyadi, A., Tseng, C.-W., Henry, R., Kolotelo, P., Langveld, J., Schilperoot, R., Shi, B., Einsiedel, S., Thomas, M., Black, J., Wilson, S. & McCarthy, D. T. 2021 Passive sampling of SARS-CoV-2 for wastewater surveillance. *Environ. Sci. Technol.* **55** (15), 10432–10441. doi.org/10.1021/acs.est.1c01530.

Sharif, S., Ikram, A., Khurshid, A., Salman, M., Mehmood, N., Arshad, Y., Ahmed, J., Safdar, R. M., Rehman, L., Mujtaba, G., Hussain, J., Ali, J., Angez, M., Alam, M. M., Akthar, R., Malik, M. W., Baig, M. Z. I., Rana, M. S., Usman, M., Ali, M. Q., Ahad, A., Badar, N., Umair, M., Tamim, S., Ashraf, A., Tahir, F. & Ali, N. 2021 Detection of SARs-CoV-2 in wastewater using the existing environmental surveillance network: a potential supplementary system for monitoring COVID-19 transmission. *PLoS ONE* **16** (6), e0249568. doi.org/10.1371/journal.pone.0249568.

Sherchan, S. P., Shahin, S., Ward, L. M., Tandukar, S., Aw, T. G., Schmitz, B., Ahmed, W. & Kitajima, M. 2020 First detection of SARS-CoV-2 RNA in wastewater in North America: a study in Louisiana, USA. *Sci. Total Environ.* **743**, 140621. doi.org/ 10.1016/j.scitotenv.2020.140621.

Stadler, L. B., Ensor, K. B., Clark, J. R., Kalvapalle, P., LaTurner, Z. W., Mojica, L., Terwilliger, A., Zhuo, Y., Ali, P., Avadhanula, V., Bertolusso, R., Crosby, T., Hernandez, H., Hollstein, M., Weesner, K., Zong, D. M., Persse, D., Piedra, P. A., Maresso, A. W. & Hopkins, L. 2020 *Wastewater Analysis of SARS-CoV-2 as a Predictive Metric of Positivity Rate for a Major Metropolis*. medRxiv. doi.org/10.1101/2020.11.04.20226191.

Street, R., Malemaa, S., Mahlangeni, N. & Mathee, A. 2020 Wastewater surveillance for Covid-19: an African perspective. *Sci. Total Environ.* **743**, 140719. doi.org/10.1016/j.scitotenv.2020.140719.

Tomasino, M. P., Semedo, M., Vieira, P., Ferraz, E., Rocha, A., Carvalho, M. F., Magalhaes, C. & Mucha, A. P. 2021 SARS-CoV-2 RNA detected in urban wastewater from Porto, Portugal: method optimization and continuous 25-week monitoring. *Sci. Total Environ.* **792**, 148467. doi.org/10.1016/j.scitotenv.2021.148467.

Trottier, J., Darques, R., Ait Mouheb, N., Partiot, E., Bakhache, W., Deffieu, M. S. & Gaudin, R. 2020 Post-lockdown detection of SARS-CoV-2 RNA in the wastewater of Montpellier, France. *One Health* **10**, 100157. doi.org/10.1016/j.onehlt.2020.100157.

Wannigama, D. L., Amarasiri, M., Hurst, C., Phattharapornjaroen, P., Abe, S., Hongsing, P., Rad, S. M. A. H., Pearson, L., Saethang, T., Luk-in, S., Kueakulpattana, N., Storer, R. J., Ounjai, P., Jacquet, A., Leelahavanichkul, A. & Chatsuwan, T. 2021 Tracking COVID-19 with wastewater to understand asymptomatic transmission. *Int. J. Infect. Dis.* **108**, 296–299. doi.org/10.1016/j.ijid.2021.05.005.

Weidhaas, J., Aanderud, J. T., Keith Roper, D., VanDerslice, J., Erica Brown Gaddis, E. B., Ostermiller, J., Hoffman, K., Jamal, R., Heck, P., Zhang, Y., Torgersen, K., Vander Laan, J. & LaCross, N. 2021 Correlation of SARS-CoV-2 RNA in wastewater with COVID-19 disease burden in sewersheds. *Sci. Total Environ.* **775**, 145790. doi.org/10.1016/j.scitotenv.2021.145790.

Westhaus, S., Weber, F.-A., Schiwy, S., Linnemann, V., Brinkmann, M., Widera, M., Greve, C., Janke, A., Hollert, H., Wintgens, T. & Ciesek, S. 2021 Detection of SARS-CoV-2 in raw and treated wastewater in Germany – suitability for COVID-19 surveillance and potential transmission risks. *Sci. Total Environ.* **751**, 141750. doi.org/10.1016/j.scitotenv.2020.141750.

WHO 2017 *WHO Guidelines on Ethical Issues in Public Health Surveillance*. World Health Organization. Available from: https://www.who.int/publications/i/item/who-guidelines-on-ethical-issues-in-public-health-surveillance (accessed 24 November 2021).

WHO 2021a *A Timeline of WHO's Response to COVID-19 in the WHO European Region: A Living Document*. Version 2.0 from 31 December 2019 to 31 December 2020. WHO Regional Office for Europe, Copenhagen. Available from: https://apps.who.int/iris/bitstream/handle/10665/339983/WHO-EURO-2021-1772-41523-56652-eng.pdf?sequence=1&isAllowed=y (accessed 24 November 2021).

WHO 2021b *Tracking SARS-CoV-2 Variants*. Available from: https://www.who.int/en/activities/tracking-SARS-CoV-2-variants/ (accessed 26 November 2021).

Wilder, M. L., Middleton, F., Larsen, D. A., Du, Q., Fenty, A., Zeng, T., Insaf, T., Kilaru, P., Collins, M., Kmush, B. & Green, H. C. 2021 Co-quantification of crAssphage increases confidence in wastewater-based epidemiology for SARS-CoV-2 in low prevalence areas. *Water Res. X* **11**, 100100. doi.org/10.1016/j.wroa.2021.100100.

Wong, J. C. C., Tan, J., Lim, Y. X., Arivalan, S., Hapuarachchi, H. C., Mailepessov, D., Griffiths, J., Jayarajah, P., Seto, Y. X., Tien, W. P., Lowa, S. L., Koo, C., Yenamandra, S. P., Kong, M., Lee, V. J. M. & Ng, L. C. 2021 Non-intrusive wastewater surveillance for monitoring of a residential building for COVID-19 cases. *Sci. Total Environ.* **786**, 147419. doi.org/10.1016/j.scitotenv.2021.147419.

Wu, F., Zhang, J., Xiao, A., Gu, X., Lee, W. L., Armas, F., Kauffman, K., Hanage, W., Matus, M., Ghaeli, N., Endo, N., Duvallet, C., Poyet, M., Moniz, K., Washburne, A. D., Erickson, T. B., Chai, P. R., Thompson, J. & Alm, E. J. 2020 SARS-CoV-2 titers in wastewater are higher than expected from clinically confirmed cases. *mSystems* **5**, e00614-20. doi.org/10.1128/mSystems.00614-20.

Wurtzer, S., Marechal, V., Mouchel, J. M., Maday, Y., Teyssou, R., Richard, E., Almayrac, J. L. & Moulin, L. 2020 Evaluation of lockdown effect on SARS-CoV-2 dynamics through viral genome quantification in waste water, Greater Paris, France, 5 March to 23 April 2020. *Eurosurveillance* **25** (50), 2000776. doi.org/10.2807/1560-7917.ES.2020.25.50.2000776.

Xu, X., Zheng, X., Li, S., Lama, N. S., Wang, Y., Chu, D. K. W., Poon, L. L. M., Tun, H. M., Peiris, M., Deng, Y., Leung, G. M. & Zhang, T. 2021 The first case study of wastewater-based epidemiology of COVID-19 in Hong Kong. *Sci. Total Environ.* **790**, 148000. doi.org/10.1016/j.scitotenv.2021.148000.

Yaniv, K., Ozer, E., Shagan, M., Lakkakula, S., Plotkin, N., Bhandarkar, N. S. & Kushmaro, A. 2021 Direct RT-qPCR assay for SARS-CoV-2 variants of concern (alpha, B.1.1.7 and beta, B.1.351) detection and quantification in wastewater. *Environ. Res.* **201**, 111653. doi.org/10.1016/j.envres.2021.111653.

Yaqub, T., Nawaz, M., Shabbir, M. Z., Ali, M. A., Altaf, I., Raza, S., Shabbir, M. A. B., Ashraf, M. A., Aziz, S. Z., Fazal, S., Hussain, N., Khan, M. T., Atique, M. M., Asif, A., Anwar, M., Awan, N. A., Younis, M. U., Bhattee, M. A., Tahir, Z., Mukhtar, N., Sarwar, H., Rana, M. S. & Farooq, O. 2021 A longitudinal survey for genome-based identification of SARS-CoV-2 in sewage water in selected lockdown areas of Lahore City, Pakistan: a potential approach for future smart lockdown strategy. *Biomed. Environ. Sci.* **34** (9), 729–733. doi.org/10.3967/bes2021.101.

Zaneti, R. N., Girardi, V., Spilki, F. R., Mena, K., Westphalen, A. P. C., Colares, E. R. C., Pozzebon, A. G. & Etchepare, R. G. 2021 Quantitative microbial risk assessment of SARS-CoV-2 for workers in wastewater treatment plants. *Sci. Total Environ.* **754**, 142163. doi.org/10.1016/j.scitotenv.2020.142163.

Zhang, D., Ling, H., Huang, X., Li, J., Li, W., Yi, C., Zhang, T., Jiang, Y., He, Y., Deng, S., Zhang, X., Wang, X., Liu, Y., Li, G. & Qu, J. 2020a Potential spreading risks and disinfection challenges of medical wastewater by the presence of severe acute respiratory syndrome coronavirus 2 (SARS-CoV-2) viral RNA in septic tanks of Fangcang hospital. *Sci. Total Environ.* **741**. doi.org/10.1016/j.scitotenv.2020.140445.

Zhang, S., Diaob, M. Y., Yuc, W., Peic, L., Lind, Z. & Chen, D. 2020b Estimation of the reproductive number of novel coronavirus (COVID-19) and the probable outbreak size on the Diamond Princess cruise ship: a data-driven analysis. *Int. J. Infect. Dis.* **93**, 201–204. doi.org/10.1016/j.ijid.2020.02.033.

Zhao, L., Atoni, E., Du, Y., Zhang, H., Donde, O., Huang, D., Xiao, S., Ren, N., Ma, T., Shu, Z., Yuan, Z., Tong, L. & Xia, H. 2021 Environmental surveillance of SARS-CoV-2 RNA in wastewater systems and related environments in Wuhan: April to May of 2020. *J. Environ. Sci.* **112**, 115–120. doi.org/10.1016/j.jes.2021.05.005.

Zhou, N. A., Tharpe, C., Meschke, J. S. & Ferguson, C. M. 2021 Survey of rapid development of environmental surveillance methods for SARS-CoV-2 detection in wastewater. *Sci. Total Environ.* **769**, 144852.

Zhu, Y., Oishi, W., Maruo, C., Saito, M., Chend, R., Kitajima, M. & Sano, D. 2021 Early warning of COVID-19 via wastewater-based epidemiology: potential and bottlenecks. *Sci. Total Environ.* **767**, 145124. doi.org/10.1016/j.scitotenv.2021.145124.

First received 31 July 2021; accepted in revised form 6 December 2021. Available online 28 December 2021

SARS-CoV-2 in Water doi: 10.2166/wh.2022.210

One-year surveillance of SARS-CoV-2 in wastewater from vulnerable urban communities in metropolitan São Paulo, Brazil

Mikaela Renata Funada Barbosa [a,*], Suzi Cristina Garcia[a], Antonio de Castro Bruni[a], Flávio Silva Machado[b], Roberto Xavier de Oliveira[a], Milena Dropa[c], Antonio Charlys da Costa[d], Elcio Leal[e], Carlos Jesus Brandão[a], Renan Lourenço Oliveira da Silva [a], Beatriz Yukie Iko [a], Victor Kenji Matsuoka Kondo[a], Ronalda Silva de Araújo[a], Vanessa Barbosa da Silveira [f], Tatiane Montes de Andrade [f], Danilo Rosa Nunes [f], Luiz Mário Ramos Janini[f], Carla Torres Braconi [f], Juliana Terzi Maricato[f] and Maria Inês Zanoli Sato [a]

[a] Department of Environmental Analysis, Division of Microbiology and Parasitology, Environmental Company of the São Paulo State (CETESB), Av. Prof. Frederico Hermann Jr., 345, São Paulo, CEP 05459-900, Brazil
[b] Basic Sanitation Company of the State of São Paulo (SABESP), São Paulo, Brazil
[c] School of Public Health, Universidade de São Paulo (USP), São Paulo, Brazil
[d] Instituto de Medicina Tropical, Universidade de São Paulo (IMT/USP), São Paulo, Brazil
[e] Institute of Biological Sciences, Universidade Federal do Pará (UFPA), Belém, Brazil
[f] Department of Microbiology, Immunology and Parasitology, Escola Paulista de Medicina, (UNIFESP), São Paulo, Brazil
*Corresponding author. E-mail: mrbarbosa@sp.gov.br

MRFB, 0000-0002-7730-9247; RLOd, 0000-0003-1985-6549; BYI, 0000-0003-4273-4021; VBd, 0000-0002-9699-0875; TMd, 0000-0002-3600-261X; DRN, 0000-0003-4079-2914; CTB, 0000-0002-5370-8980

ABSTRACT

The current COVID-19 pandemic has emphasized the vulnerability of communities living in the urban outskirts and informal settlements. The lack of reliable COVID-19 case data highlights the importance and application of wastewater-based epidemiology. This study aimed to monitor the COVID-19 trends in four vulnerable urban communities (slums and low-income neighborhoods) in metropolitan São Paulo by assessing the SARS-CoV-2 RNA viral load in wastewater. We analyzed 160 samples from May 2020 to June 2021 with weekly or fortnightly samplings. The samples were ultracentrifuged with glycine elution and quantified by N1/N2 SARS-CoV-2 RT-qPCR. The results of positivity were 100% (Paraisópolis, Heliópolis and Cidade Tiradentes) and 76.9% (Vila Brasilândia). The new case numbers of COVID-19, counted from the onset of symptoms, positively correlated with SARS-CoV-2 N1 viral loads from the two largest communities ($p<0.001$). SARS-CoV-2 infectivity was tested in Vero E6 cells after concentration with the two techniques, ultrafiltration (Centricon® Plus-70 10 kDa) and sucrose cushion ultracentrifugation, but none of the evaluated samples presented positive results. Next-generation sequencing (NGS) analysis from samples collected in March and August 2021 revealed the presence of the clade 20 J (lineage P.1) belonging to the most prevalent circulating variant in the country. Our results showed that wastewater surveillance data can be used as complementary indicators to monitor the dynamics and temporal trends of COVID-19. The infectivity test results strengthened the evidence of low risk of infection associated with SARS-CoV-2 in wastewater.

Key words: environmental surveillance, infectivity, SARS-CoV-2, sequencing, vulnerable communities, wastewater

HIGHLIGHTS

- SARS-CoV-2 WBE can be a valuable tool to follow trends of COVID-19 cases in vulnerable communities (slums and low-income neighborhoods).
- Positive samples for SARS-CoV-2 RNA presented negative results in the infectivity assays performed with Vero E6 culture cell.
- Preliminary results of NGS sequencing analysis showed the circulation of lineage P.1, the predominant SARS-CoV-2 variant in clinical genomic surveillance

GRAPHICAL ABSTRACT

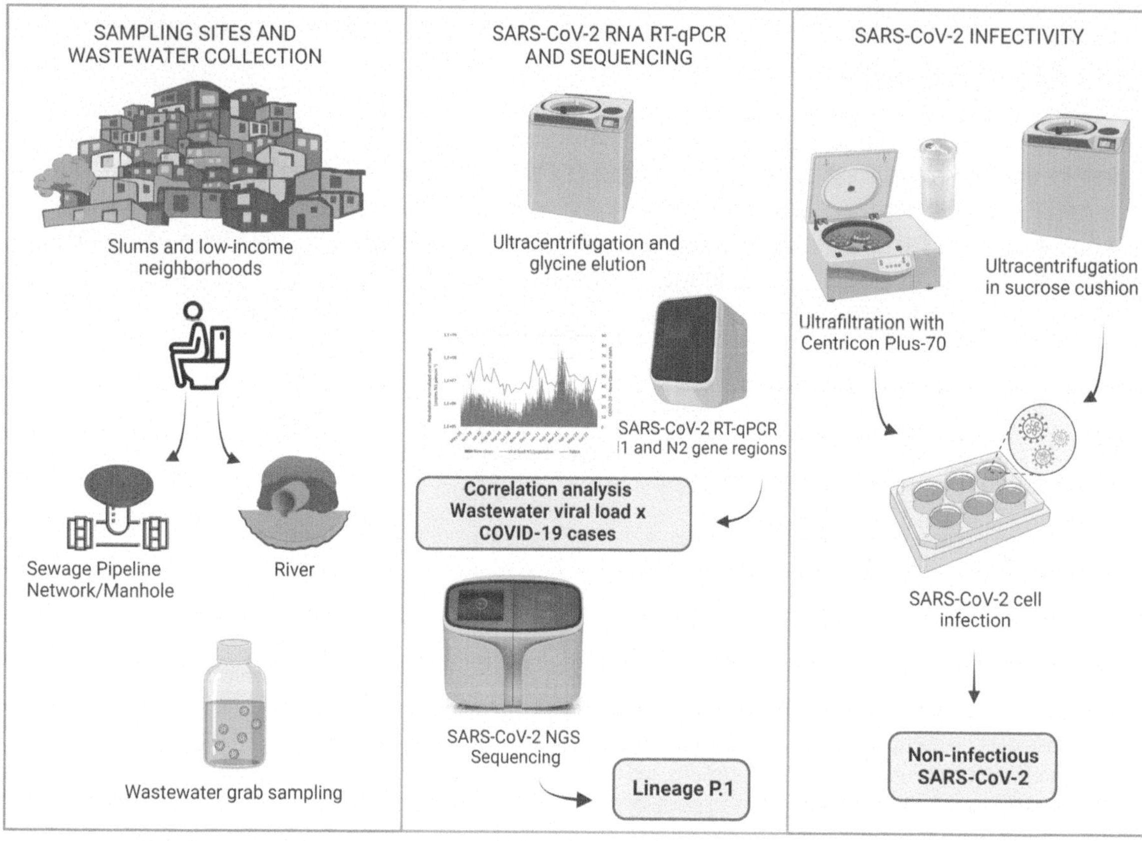

1. INTRODUCTION

The first occurrence of COVID-19 in Latin America was confirmed in São Paulo, Brazil, on February 26, 2020, two weeks before the World Health Organization declared COVID-19 a pandemic (Cucinotta & Vanelli 2020). Since then, the world has been under a global crisis with a severe impact on the public health system and economy (CRS 2021). The challenges the world has faced with the pandemic triggered an accelerated scientific and technological development in the areas of diagnosis, vaccine, logistics, modeling and epidemiological surveillance.

Environmental surveillance has been used for decades to evidence the circulation of waterborne pathogens in the population, particularly poliovirus, and more recently antimicrobial-resistant microorganisms, as a complementary approach to infectious diseases surveillance (Asghar *et al.* 2014; GPEI 2016; Sims & Kasprzyk-Hordern 2020; WHO 2020; Xagoraraki & O'Brien 2020). The first report of SARS-CoV-2 RNA detection in sewage in March 2020, in the Netherlands (Medema *et al.* 2020b), indicated that wastewater surveillance was also a sensitive tool to assess the circulation of SARS-Cov-2 in the community and the COVID infection trends. Since then, several research groups worldwide have been using this approach as an epidemiological device and as an early warning system (Haramoto *et al.* 2020; La Rosa *et al.* 2020; Wu *et al.* 2020; Ahmed *et al.* 2021; https://www.covid19wbec.org/collaborators). It has also been used as a valuable instrument to provide information about the emergence and abundance of SARS-CoV-2 strains (Nemudryi *et al.* 2020; Rimoldi *et al.* 2020; Crits-Christoph *et al.* 2021; Jahn *et al.* 2021). Many countries have already implemented a national wastewater surveillance system or program to support the government in formulating epidemic prevention policies for COVID-19, including the Netherlands (Rijksoverheid 2021), the USA (CDC 2021; Kirby *et al.* 2021), the UK (Wade *et al.* 2022), South Africa (NICD 2021), Japan (Takeda *et al.* 2021) and others.

Brazil has been severely affected by the COVID pandemic. By July 1, 2021, it was the second country in a total number of deaths (520,000), only ranking behind the USA (605,000). It was also ranked ninth in the number of deaths per million inhabitants (Ritchie *et al.* 2020), with daily death rates ranging from 3,000 to 4,000 between March and April 2021. Given the testing limitations and the challenges faced in obtaining accurate clinical and epidemiological national data, several wastewater-based epidemiology (WBE) programs have been implemented by local governments and/or research projects mostly driven to assess spatial and temporal SARS-CoV-2 load fluctuations in the sewer system and wastewater treatment plants (WWTPs) of large metropolitan regions, supporting decision-makers on COVID-19 infection trends at the population level (ANA 2021; Claro *et al.* 2021; Fongaro *et al.* 2021; Mota *el al.* 2021; Prado *et al.* 2021).

Assessing the circulation of SARS-CoV-2 in the wastewater of a huge city as São Paulo has been a major challenge. With around 13 million people, the city is served by four WWTPs, responsible for the treatment of around 75% of the sewage generated. The remaining 25% are ultimately released into surface waters.

The city of São Paulo is responsible for the highest number of cases and deaths by COVID-19 among Brazilian cities, since the beginning of the pandemic. The city has the largest number of slums (favelas), with more than 1,700 reported by the Municipal Housing Secretariat (SEHAB, http://www.habitasampa.inf.br/habitacao/). It is estimated that there are more than 390,000 homes with more than two million inhabitants living in slums in São Paulo, about 11% of the city's population. The mortality rate of the new coronavirus in São Paulo can be up to 10 times higher in neighborhoods with the worst social conditions that coincide with the areas with the highest slum concentrations (PMSP 2020; Figueiredo 2021). However, cases confirmed by laboratory examination were not predominant, which may show greater difficulty in obtaining confirmation of suspected cases in these areas (PMSP 2020).

Poor sanitation conditions, water scarcity, hyper-dense dwellings, crowded households and lack of health care access, among other factors, make urban slums and other vulnerable communities major COVID-19 hotspots and relevant areas for the application of WBE programs; however, few studies have been reported in these communities. Mota *et al.* (2021) and Prado *et al.* (2021) in decentralized monitoring of SARS-CoV-2 RNA in sewages from Belo Horizonte and Rio de Janeiro, two other large Brazilian cities, showed that wastewater monitoring data is more sensitive for identifying hotspots in vulnerable areas than clinical/epidemiological data, allowing early intervention actions by public health authorities. Razzolini *et al.* (2021) followed the evolution of SARS-CoV-2 RNA concentration for 7 months in a contaminated stream that receives raw sewage from an urban underprivileged settlement in the city of São Paulo and observed a statistically significant correlation between SARS-CoV-2 concentration in water and COVID-19 cases in the community. The authors concluded that virus concentration in the environment reflects the epidemiological status of the community. Iglesias *et al.* (2021) in a similar study conducted in a low-resource community in Buenos Aires noted that SARS-CoV-2 measurements in the lagoon that receives sewage from the community could be applied to estimate the changes in the COVID-19 prevalence. Another frequent concern in these highly vulnerable regions is the possibility of fecal–oral and fecal–respiratory transmissions, although the viability of SARS-CoV-2 in sewage samples has not been demonstrated to date (Rimoldi *et al.* 2020; Westhaus *et al.* 2021).

The present study is part of the São Paulo State wastewater surveillance program for COVID-19 initiated in April 2020 and focused on assessing trends in SARS-CoV-2 RNA concentrations in wastewater from the two main favelas of the city of São Paulo (Paraisópolis and Heliópolis) and two low-income neighborhoods (Vila Brasilândia and Cidade Tiradentes), areas of high vulnerability considered as a priority for the fight against COVID-19 by the city Health Service System. SARS-CoV-2 concentration methods, genetic diversity and viability of SARS-CoV-2 were also evaluated.

2. MATERIAL AND METHODS

2.1. Sampling

Four communities with different population sizes were selected for SARS-CoV-2 monitoring, including Paraisópolis and Heliópolis, the two largest favelas in the City of São Paulo and two communities located in low-income neighborhoods (Cidade Tiradentes and Vila Brasilândia), the most affected by COVID-19 at the start of the study (Figure 1). The selection of the sampling sites had the support of the São Paulo State Sanitation Company (SABESP), which provided information about the sub-sewershed localization, flow rate and the respective population that contributes to the sewage discharge at the collection point, enabling the definition of sampling sites of each community. Paraisópolis does not have a sewage network infrastructure and monitoring was carried out by sampling surface water at a point in the stream that receives sewage discharge from the community. The other communities were monitored at manhole points of the collection network in sewer systems. Except for Cidade Tiradentes, where the collection point represents the entire sanitary sewage basin of the neighborhood, in the other locations, due to the complexity of the sanitary sewage system, points of easier and safer access,

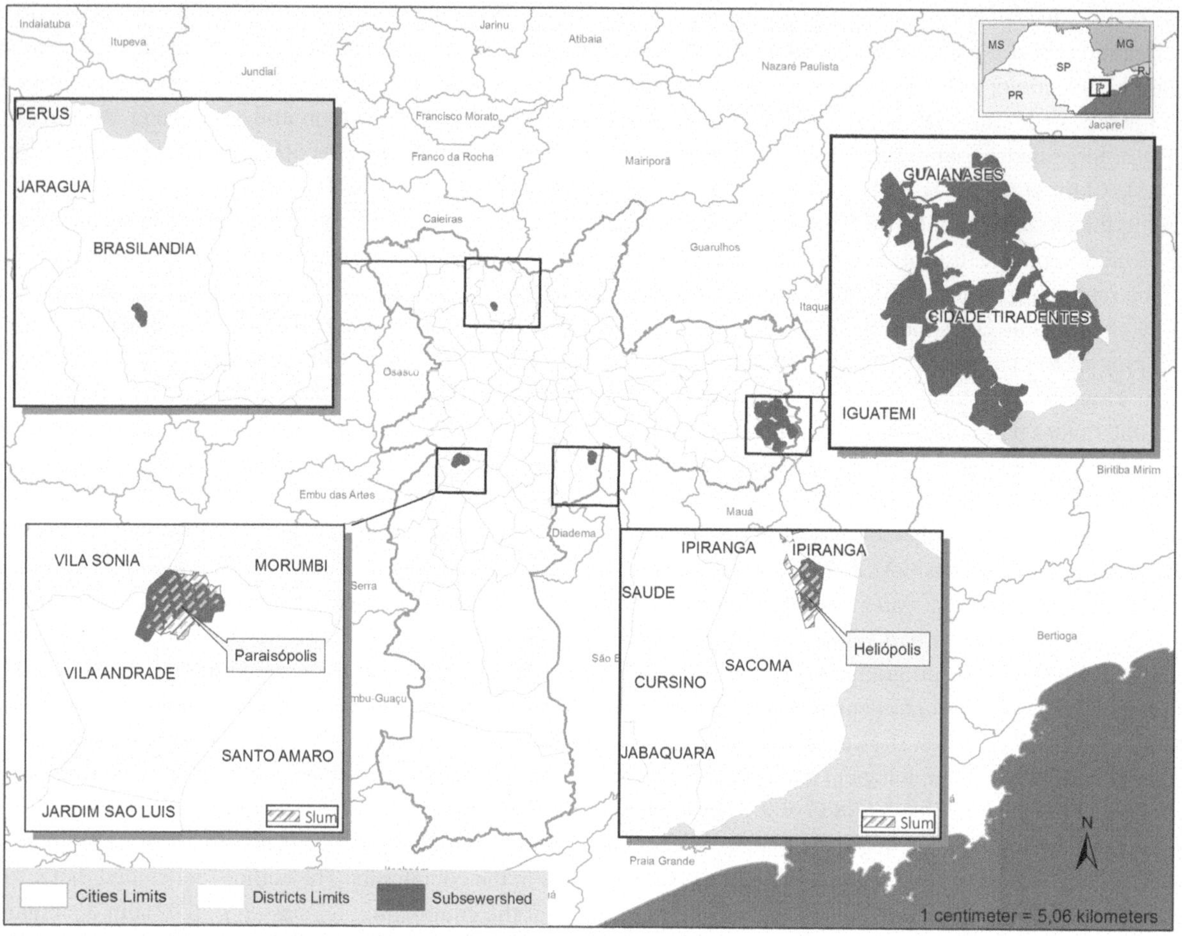

Figure 1 | Sampling locations and associated sub-sewersheds for the four selected vulnerable urban communities in São Paulo city.

representative of the community, were established. The geographic and population characteristics of each area are shown in Table 1. The average flow rates were calculated from the micromeasured flow data of treated water.

Grab samples were collected once a week, during the period from May to September 2020. From October 2020 onwards, the frequency of sampling was reduced to fortnightly.

For viruses' analyses, wastewater samples were manually collected in stainless steel buckets and transferred to sterile 1 L polypropylene bottles (APHA; AWWA; WEF 2017a). Temperature and pH were measured during sampling. Samples were transported to the laboratory on ice and kept refrigerated (4 °C) before analyses for a maximum of 24 h. Additionally, volumes of 500 mL were collected for chemical assay of total suspended solids (TSS) (APHA; AWWA; WEF 2017).

Table 1 | Wastewater sampling sites in low-income communities and estimated population residing in the area corresponding to each sub-sewershed

Sampling site	Collection point	Geographic coordinate	Average flow rate (L/day)	Estimated population
Paraisópolis (slum)	Stream	23° 36′28″ S 46° 43′11″ W	9,070,000	51,000
Heliópolis (slum)	Sewer pipeline (manhole)	23° 36′38″ S 46° 35′11″ W	2,060,000	9,800
Cidade Tiradentes	Sewer pipeline (manhole)	23° 33′36″ S 46° 25′06″ W	23,100,000	130,000
Vila Brasilândia	Sewer pipeline (manhole)	23° 27′54″ S 46° 41′30″ W	170,000	1,700

2.2. COVID-19 and SARI data

Correlation analyses were performed between confirmed cases of COVID-19 and SARI (Severe Acute Respiratory Infection) in the resident population of the area corresponding to the sub-sewersheds and SARS-CoV-2 RNA viral loads in wastewater samples. Confirmed cases of COVID-19 were defined according to the following criteria: laboratory tests, clinical (flu-like syndrome cases associated with anosmia or acute ageusia), clinical–epidemiological (flu-like syndrome cases and SARI that had close contact with a confirmed COVID-19 patient during the 14 days before the appearance of the symptoms) and clinical imaging (flu-like syndrome cases and SARI with tomography diagnose) results. A unified database with geocoded information of the residence addresses of all cases was provided by the Municipal Health Department of São Paulo, which gathered data from the SIVEP-Gripe and e-SUS Notifica databases (https://www.prefeitura.sp.gov.br/cidade/secretarias/saude/vigilancia_em_saude/doencas_e_agravos/coronavirus/index.php?p=313773). Confirmed cases of COVID-19 and SARI were counted considering the symptom onset date.

2.3. Temporal evaluation of SARS-CoV-2 RNA viral load

2.3.1. Virus concentration by ultracentrifugation and glycine elution

The ultracentrifugation and glycine elution method was used to concentrate the wastewater samples (Pina *et al.* 1998). Volumes of 40 mL of each concentrated sample were aliquoted and contaminated with 10^6 copies of bovine coronavirus (BCoV). After incubation of 1 h at 4 °C, samples were ultracentrifuged at 110,000 g for 1 h at 4 °C using a Sorvall RC 90 ultracentrifuge.

The supernatant was discarded, and the pellet was eluted with 4 mL of 0.25 N glycine buffer (pH 9.5) in an ice bath for 30 min, mixing by vortex each 5 min. After adding 4 mL of 2×phosphate-buffered saline (PBS, pH 7.2), the samples were centrifuged at 3,000 g for 20 min at 4 °C. Then, the supernatant was transferred to a new tube, previously weighed. After ultracentrifugation at 110,000 g for 1 h at 4 °C, the supernatant was carefully discarded, and the tube weight was re-recorded.

The pellet eluted in the remaining supernatant (approximately 300 μL) was transferred to 1.5 mL microtubes. The samples were kept under refrigeration (2–8 °C) until RNA extraction.

2.3.2. RNA extraction and RT-qPCR for SARS-CoV-2 and BCoV

The concentrated sample (140 μL) was extracted with the QIAamp® Viral RNA Mini kit (Qiagen, Germany) according to the manufacturer's instructions. Elution was performed in two steps with 40 μL of elution buffer (AVE), adding up to nearly 80 μL. Nucleic acid extracts were stored at −70 °C until viral quantification by RT-qPCR.

The RT-qPCR for SARS-CoV-2 was performed by quantifying the N1 and N2 gene regions according to primers, probes and reaction parameters described by the CDC protocol (CDC 2020). In addition, BCoV RNA concentrations were determined using primers and the probe previously described by Decaro *et al.* (2008). The BCoV probe was labeled with HEX at the 5′ and Black Hole Quencher 1 (BHQ1) at the 3′ end.

RT-qPCR assays for N1 and N2 were performed separately in 20 μL reactions containing 5 μL TaqPath™ 1-Step RT-qPCR Master Mix, CG (ThermoFisher, MA, USA), 1.5 μL of N1 or N2 primer and probe (2019-nCoV RUO kit, IDT, IA, USA) and 5 μL of extracted RNA. The thermal cycling conditions of one-step RT-qPCR were: 25 °C for 2 min, reverse transcription at 50 °C for 15 min, preheating at 95 °C for 2 min, 45 cycles of amplification at 95 °C for 3 s and 55 °C for 30 s. BCoV RNA was quantified using 900 nM of each primer and 200 nM of probe at the same conditions of N1 and N2.

The N1 assay was performed in a multiplex reaction with an RNA internal positive control (IPC), with the addition of 0.8 μL of VetMAX™ Xeno™ Internal Positive Control VIC Assay (Applied Biosystems, ThermoFisher) and 0.1 μL of Xeno RNA Control (Applied Biosystems, Thermo Fisher Scientific, MA, USA). If the *Cq* value of a wastewater sample increases more than $1-Cq$ compared with the reference *Cq* value, the sample is considered to have PCR inhibitors.

Standard quantification curves for N1 and N2 SARS-CoV-2 regions were constructed using serial dilutions of the 2019-nCoV_N_Positive Control kit (IDT, IA, USA), consisting of 2×10^4 to 2 copies/μL of a plasmid cloned with the N1 and N2 sequences of SARS-CoV-2. BCoV standard curves were constructed with serial dilutions of synthetic fragments (GeneArt, Thermo Fisher Scientific, MA, USA) containing BCoV sequences ranging from 2.5×10^5 to 2.5 copies/μL. Negative template control was included in all qPCR assays. All reactions were performed in duplicate in a StepOne Plus instrument (ABI, CA, USA), and data were extracted from ABI StepOne Software version 2.3.

The assay limits of detection (LODs) were defined as the minimum copy number/reaction with a 95% probability of detection. Samples that had at least one RT-qPCR replicate amplified with a *Cq* value of >40 were considered detected but not quantifiable.

The concentrations of N1 and N2 of SARS-CoV-2 and BCoV were obtained by the following equation:

$$\text{virus genome copies per L} = \text{genome copy number} * (\text{RNA}^{\text{total}}/\text{RNA}^{\text{PCR}}) * (\text{concentrate}^{\text{total}}/\text{concentrate}^{\text{extracted}}) * (1,000\ \text{mL}/\text{sample})$$

where $\text{RNA}^{\text{total}}$ is the total volume of RNA eluted (0.08 mL), RNA^{PCR} is the volume of purified RNA tested in RT-qPCR (0.005 mL), $\text{concentrate}^{\text{total}}$ is the total volume of water concentrate: Estimated volume of final concentrate=weight of tube with sample (g)−weight of tube (g), $\text{concentrate}^{\text{extracted}}$ is the volume of wastewater concentrate from which RNA was extracted (0.140 mL) and sample is the volume of original wastewater sample processed with ultracentrifugation procedure (40 mL).

To calculate the recovery efficiency of the total method, BCoV gene copies detected in the initial volume of 40 mL were divided by the BCoV gene copies spiked, and the result was expressed as a percentage.

Table 2 summarizes the RT-qPCR parameters obtained in the N1 and N2 SARS-CoV-2 and BCoV assays, for temporal evaluation of SARS-CoV-2 RNA concentrations in wastewater samples.

2.4. Infectivity assay of SARS-CoV-2

Wastewater samples with N1 and N2 concentrations greater than 10^4 GC/L in the RT-qPCR assays were submitted to the SARS-CoV-2 viability assay in cell culture.

The samples were processed using two different methods: ultrafiltration with Centricon® Plus-70 and ultracentrifugation in sucrose cushion (Summer & Smith 1987).

2.4.1. Ultrafiltration – Centricon® Plus-70

Approximately 200 mL of water or wastewater sample were filtered through a 0.22 µm PES filter (Stericup, Merck Millipore, MA, USA). Then, 60 mL of the filtered sample were centrifuged at 3,000 g for 20 min in a Centricon® Plus-70 filter (10 kDa) (Merck Millipore, MA, USA). The procedure was repeated until the volume of concentrate was reduced to approximately 1 mL. The filter was inverted and the concentrate was recovered after centrifugation (1,000 g for 2 min) and treated with 10 µl of 10 mg/mL gentamicin (Gibco, China).

2.4.2. Ultracentrifugation in sucrose cushion

The volume of 60 mL of 0.22 µm filtered sample was carefully added to the ultracentrifuge tube containing 6 mL of sucrose solution (25% w/w sucrose, 5 mM NaCl, 10 mM EDTA). After ultracentrifugation at 55,000 g for 75 min at 4 °C, the supernatant was discarded, and the pellet was resuspended in 1 mL of PBS (pH 7.2). The sample was treated with 10 µl of 10 mg/mL gentamicin (Gibco, China).

An aliquot of 140 µL of the concentrated samples was subjected to nucleic acid extraction and N1/N2 RT-qPCR, and the remaining samples were stored at −80 °C and posteriorly sent to the Laboratory of the Federal University of São Paulo.

Table 2 | Performance of N1 and N2 SARS-CoV-2 and M gene BCoV RT-qPCR assays from a composite of seven standard curves

	Range				Genome copies/reaction
Assay target gene	Efficiency (*E*) (%)	Linearity (R^2)	Slope	*Y*-intercept	LOD
SARS-CoV-2 N1	91.3–101.8	0.994–0.998	−3.290 to −3.572	36.213–38.089	10
SARS-CoV-2 N2	91.2–101.1	0.996–1.000	−3.296 to −3.552	37.608–39.304	10
BcoV M gene	91.6–102.5	0.991–0.998	−3.262 to −3.511	45.593–48.782	12

2.4.3. Cell culture for viral isolation

All experiments were conducted using two biological replicates and two technical duplicates. The experiments for viral isolation and initial passages were performed in a biosafety level 3 laboratory (BLS3), in accordance with WHO recommendations and under the laboratory biosafety guidance required for the SARS-CoV-2 at the BLS3 facilities at the Federal University of São Paulo.

Concentrated samples which previously tested positive for SARS-CoV-2 RNA were managed for viral isolation and cytopathic effect (CPE) observation in cell culture (Araujo *et al.* 2020). We used for the experiments the Vero E6 cell line (ATCC® CRL-1586™) maintained in Minimum Essential Medium (MEM; Gibco), supplemented with 10% fetal bovine serum (FBS) (Gibco) and 1% penicillin/streptomycin (Gibco). The Vero E6 cells were kept in a humidified 37 °C incubator in an atmosphere of 5% CO_2. After reaching 80% confluent monolayer, we seeded cells in 24-well plates within a concentration of 5×10^5 cells for each well and in 6-well plates within a concentration of 7.5×10^5 cells for each well. After 24 h, for the 24-well plates, we removed the culture medium, washed the wells three times with PBS 1× and inoculated them with aliquots of 150 μL of MEM medium supplemented with 150 μL of concentrated wastewater sample. For the 6-well plates, 200 μL of the concentrated wastewater sample were added to 800 μL of supplemented MEM medium. For both experiments, after 3 h of incubation, the supernatants were removed to perform the N RT-qPCR assay (Corman *et al.* 2020) to evaluate eventually remaining virus in the supernatants after incubation. Then, the medium volumes in each well in 24 or 6-well plates were completed to 500 μL and 3 mL of MEM supplemented with 2.5% FBS and 1% penicillin–streptomycin, respectively. The Vero E6 cells were incubated under humidified 37 °C with a 5% CO_2 atmosphere and were observed daily for CPE up to 6 days. The supernatant was collected, and virus replication was confirmed through CPE with an optical microscope and RT-qPCR for SARS-CoV-2. We used a positive control of SARS-CoV-2 for all experiments, kindly given by Prof. José Luiz Proença-Módena (University of Campinas – UNICAMP, SP, Brazil).

2.5. Statistical analysis

For statistical analysis, concentrations below the LODs were assigned a value equal to the LOD divided by the square root of 2 (Croghan & Egeghy 2003). The population-normalized viral loads (gene copies/capita/day) were estimate by multiplying (gene copies/L) × (L wastewater/day) × (1/sub-sewershed population).

The values of N1 and N2 RNA concentrations were transformed using the log10 function. The Wilcoxon test for paired data (Bauer 1972) was used to compare mean N1 concentrations for the two methods of concentration of SARS-CoV-2. The significance level adopted for the statistical tests was 5%.

Scatterplot diagrams were created to assess correlations between quantitative variables (recovery percentage and TSS). Linear models were fitted to assess the significance of correlations and trends. Confidence intervals of 95 and 99% were fitted for linear models. Multiple comparisons were tested using the statistics proposed by Dunn (1961).

Spearman correlations of population-normalized viral loads with new cases of COVID-19 (7-day cumulative) were calculated using IBM® SPSS® Statistics version 27.0, with a *p*-value statistically significant of <0.05. A cross-correlation analysis (Haugh 1976; Polanco-Martinez *et al.* 2019) was used to assess whether a time displacement of the COVID-19/SARI case dataset (7-day cumulative) impacts the strength of the correlation between N1 SARS-CoV-2 viral loads dataset quantified in wastewater samples. For viral loads, the frequency was fortnightly and when more than one data was available in the fortnight, the mean values were used. Analyses were performed using the R language with RStudio.

2.6. SARS-CoV-2 sequencing

Whole viral genome sequencing was carried out by a Brazilian reference laboratory located at the Instituto Adolfo Lutz (IAL, Center for Interdisciplinary Procedures, Strategic Laboratory). The partnership for sample sequencing was only made possible from March 2021, and therefore, only samples with concentrations above 10^5 GC N1 L^{-1}, collected from that date onwards, were sequenced. An aliquot of extracted RNA underwent a retrotranscription (RT) step where cDNA was synthesized using the SuperScript IV VILO Master Mix according to the manufacturer's instructions (Thermo Fisher Scientific, MA, USA). Fifteen microliters of cDNA were used to amplify SARS-CoV-2 full-length genome using an AmpliSeq SARS-CoV-2 panel (Thermo Fisher Scientific, MA, USA) (Alessandrini *et al.* 2020). The library was adjusted to 30 pM and then loaded onto an Ion Chef instrument (Thermo Fisher Scientific, MA, USA) for emulsion PCR, enrichment and loading onto an Ion 530 chip on the Ion GeneStudio S5 Prime Series system (Thermo Fisher Scientific, MA, USA). Reads from the library were aligned with the Wuhan-Hu-1 NCBI Reference Genome (Accession No. MN908947.3) in Torrent Suite v. 5.10.1.

For mutation calling, the following plugins were used: Coverage Analysis (v. 5.10.0.3), Variant Caller (v.5.10.1.19) and COVID19AnnotateSnpEff (v.1.0.0), a plugin specifically developed for SARS-CoV-2 that can predict the effect of base substitution (Alessandrini *et al.* 2020). The software Geneious R9 was used to visualize each sample's torrent variant caller (TVC) bam files to check the consistency of nucleotide calls and analyze sequencing data. Raw sequence reads were aligned to the complete genome of the SARS-CoV-2 Wuhan-Hu-1 isolate (Genbank Accession No. NC_045512.2).

The resulting contigs were subjected to a blast search using Ugene software (Okonechnikov *et al.* 2012) to identify members of the betacoronavirus genera. To classify and determine the mutational pattern of our sequences, NextClade v1.5.0 clade assigner (https://clades.nextstrain.org/) was used.

3. RESULTS

3.1. SARS-CoV-2 RNA quantification

We detected N1 or N2 of SARS-CoV-2 RNA in 153 (95.6%) of the 160 wastewater samples analyzed over the study period. Of these 153 samples, 150 (93.8%) had quantifiable concentrations of SARS-CoV-2 RNA. The frequency of positive samples in each community and the SARS-CoV-2 RNA concentration ranges are summarized in Table 3.

The N2 RT-qPCR detected SARS-CoV-2 RNA in 152 (95%) samples, while for the N1 RT-qPCR assay, 149 (93%) samples were positive. No significant differences were observed between the concentrations (genome copies/L) of N1 and N2 in wastewater samples, although N1 values were systematically higher than N2. The linear model (GAMLSS) relating N1 to N2 concentrations presented an R^2 value of 99.85%. Therefore, only results related to N1 data were presented for statistical analyses.

3.1.1. Concentration methods

Preliminary tests were performed to select the most effective concentration method for recovering SARS-CoV-2 RNA from wastewater samples (n=13) naturally contaminated. The ultracentrifugation and glycine elution method was compared with the ultrafiltration method using Centricon® Plus-70 (10 kDa). Statistical analysis (Wilcoxon test) showed that the ultracentrifugation results were statistically superior to ultrafiltration (W=85.0; p=0.003418; Figure 2).

3.1.2. Recovery efficiency

The recovery efficiency using BCoV control was evaluated only in samples ultracentrifuged with glycine elution. Considering all the results obtained in the laboratory including other sampling locations, with the analysis of wastewater samples (n=267), the average recovery percentage was 8.3% (s.d. 11.7%). The characteristics of the different sample matrices significantly influenced the results. Statistical analysis showed that these differences in recovery efficiencies might be associated with variations in TSS. Samples with higher amounts of TSS showed lower recovery rates of BCoV (t=−4,933 with g.l.=335; p<0.001; 95% confidence interval (−0.3570 to −0.1580); Figure 3).

Table 3 | Frequency, average and range concentrations of SARS-CoV-2 RNA (N1 and N2) in wastewater samples at four vulnerable communities of São Paulo city, Brazil

Sampling site	Number of analyzed samples	Frequency of positive samples (%)		SARS-CoV-2 RNA Average (range) genome copies/L	
		N1	N2	N1	N2
Paraisópolis	40	97.5	100	9.7×10^4 (<LOD – 7.2×10^5)	5.5×10^4 (DNQ – 5.8×10^5)
Heliópolis	41	97.6	100	1.5×10^5 (<LOD – 1.7×10^6)	6.3×10^4 (2.1×10^3–8.4×10^5)
Cidade Tiradentes	40	100	100	1.2×10^5 (1.3×10^4–5.8×10^5)	6.0×10^4 (4.9×10^3–5.8×10^5)
Brasilândia	39	76.9	76.9	4.1×10^4 (<LOD – 2.7×10^5)	2.6×10^4 (<LOD – 2.0×10^5)

DNQ, detected but not quantifiable; LOD, limit of detection.

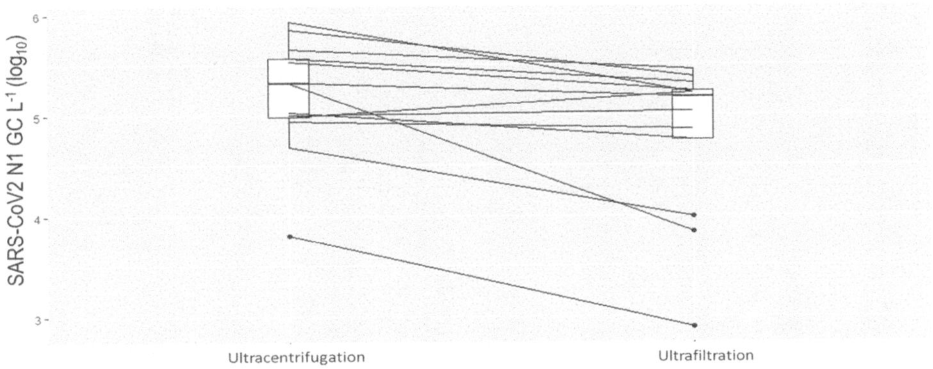

Figure 2 | Comparison between log-transformed N1 concentration (copies/L) results in wastewater samples concentrated by ultracentrifugation (with glycine elution) and ultrafiltration with Centricon® Plus-70 methods.

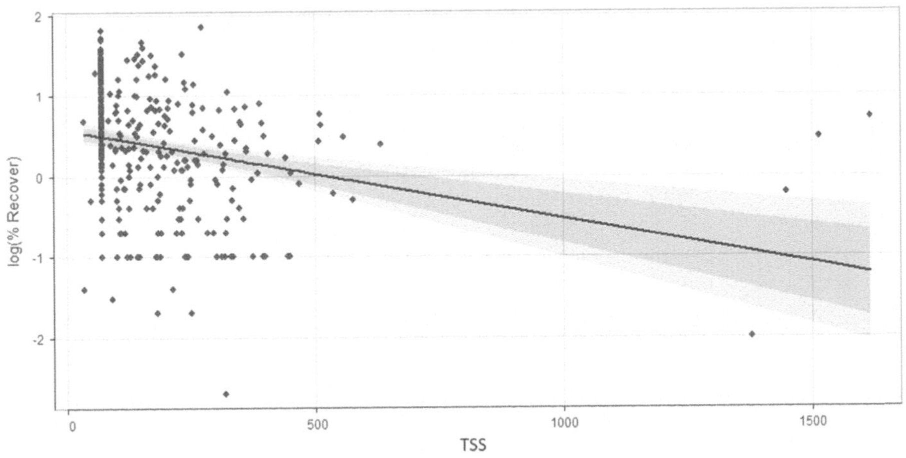

Figure 3 | Influence of the variable total suspended solids amount (TSS) on the recovery efficiencies of SARS-CoV-2 RNA by ultracentrifugation and the glycine elution method.

3.1.3. RT-qPCR inhibition

All the N1 SARS-CoV-2 assays were performed in multiplex RT-qPCR with an internal RNA positive control to ascertain inhibitors' interference present in the RNA samples. The presence of inhibitors in the samples was not observed since the IPC amplification results did not show differences greater than 1 *Cq* compared to the reference control IPC. RNA extracts (n=9) were also analyzed in different volumes (5, 1 and 0.5 μL) by N1 and N2 RT-qPCR, and no significant differences were observed between the values of copies/L concentrations.

3.1.4. SARS-CoV-2 RNA in wastewater samples and COVID-19/SARI cases

The evolution of population-normalized SARS-CoV-2 loads observed in wastewater from May 2020 to June 2021 in the four communities is shown in Figure 4(a). The periods of highest concentrations (May to June 2020, December 2020 to January 2021 and March 2021) coincided with records of an increase in new cases of COVID-19 in the city of São Paulo (Figure 4(b)). The alert levels of the São Paulo Coronavirus Control Plan that were in force during the study period are also plotted in Figure 4(b) (see Supplementary material for classifying criteria).

Considering the population-normalized viral load data in the wastewater of each community (Figure 5), we observed a greater similarity to the new cases of temporal variations only in the Cidade Tiradentes and Paraisópolis communities, which correspond to the sub-sewersheds of the largest contributing population. This observation was confirmed by the

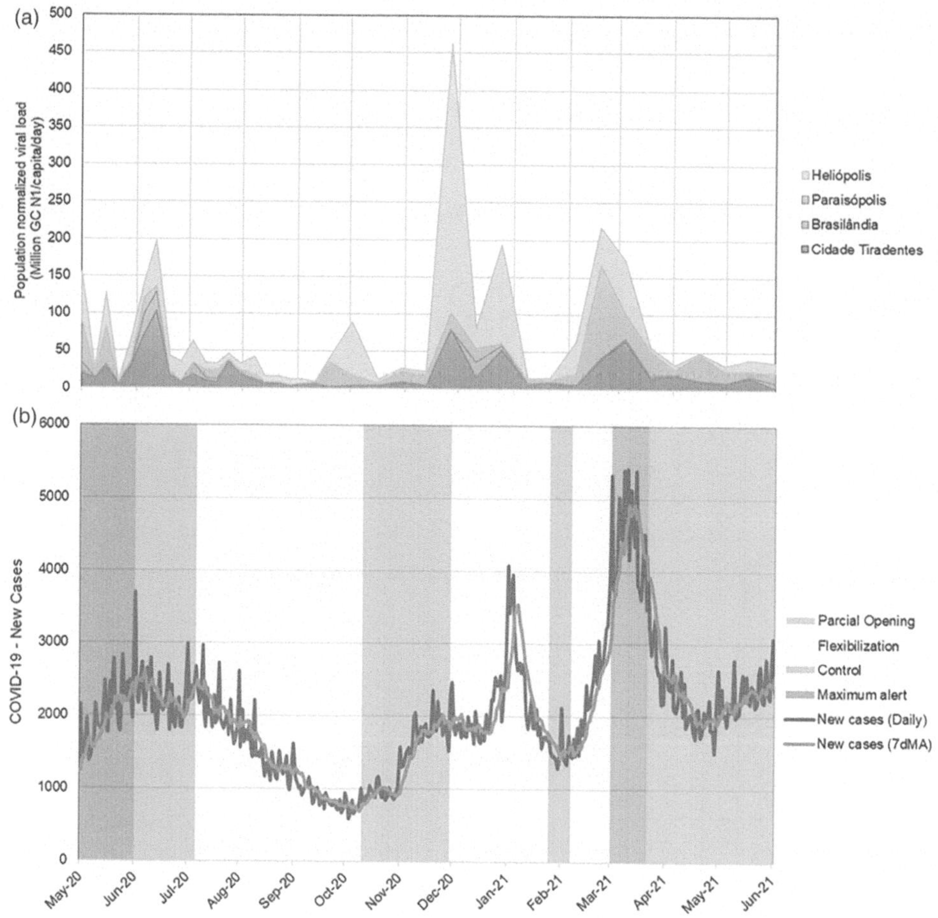

Figure 4 | Population-normalized SARS-CoV-2 loads for the four communities (a) and new cases and 7-day moving average new cases (7dMA) of COVID-19 (b).

statistical analysis of the Spearman correlation, in which significant positive correlations were observed in both communities ($p<0.001$) (Figure 6).

To assess the differences in the behavior of the temporal curves, the viral loads of SARS-CoV-2 RNA were correlated to the curves of the number of 7-day cumulative cases for each community using a cross-correlation analysis. Although the viral load curves in wastewater and reported cases of COVID-19/SARI presented similar trends in different periods, no significant correlation was observed considering the entire study period (Figure 7). Only time series of Cidade Tiradentes, for lag 0, the correlation coefficient was close to the limit of 95%; however, the representativeness was not guaranteed, since it did not reach robust CI.

3.1.5. Virus isolation

The positive concentrated samples for N1 and N2 SAR-CoV-2 RT-qPCR were inoculated in the previously prepared plates of Vero E6 cells. The samples were frozen and then used for virus isolation with daily observation for CPE by optical microscopy. Twenty-four hours post-infection (24 h.p.i.), all samples were compared with the positive control, and no morphological changes were observed. Three days post-infection (3 d.p.i), the cultures continued without any sign of CPE. After 6 d.p.i, no virus propagation was observed by CPE in the cultures, but all supernatants were collected. Samples of wastewater tested in the Vero E6 cell line had no sign of virus infectivity (Figure 8(a) and 8(d)). For all independent experiments performed in cell culture, CPE was observed in positive controls for the SARS-CoV-2, and negative controls had no signs of morphological changes (Figure 8(e) and 8(f), respectively). RT-qPCR further confirmed these results, corroborating the lack of

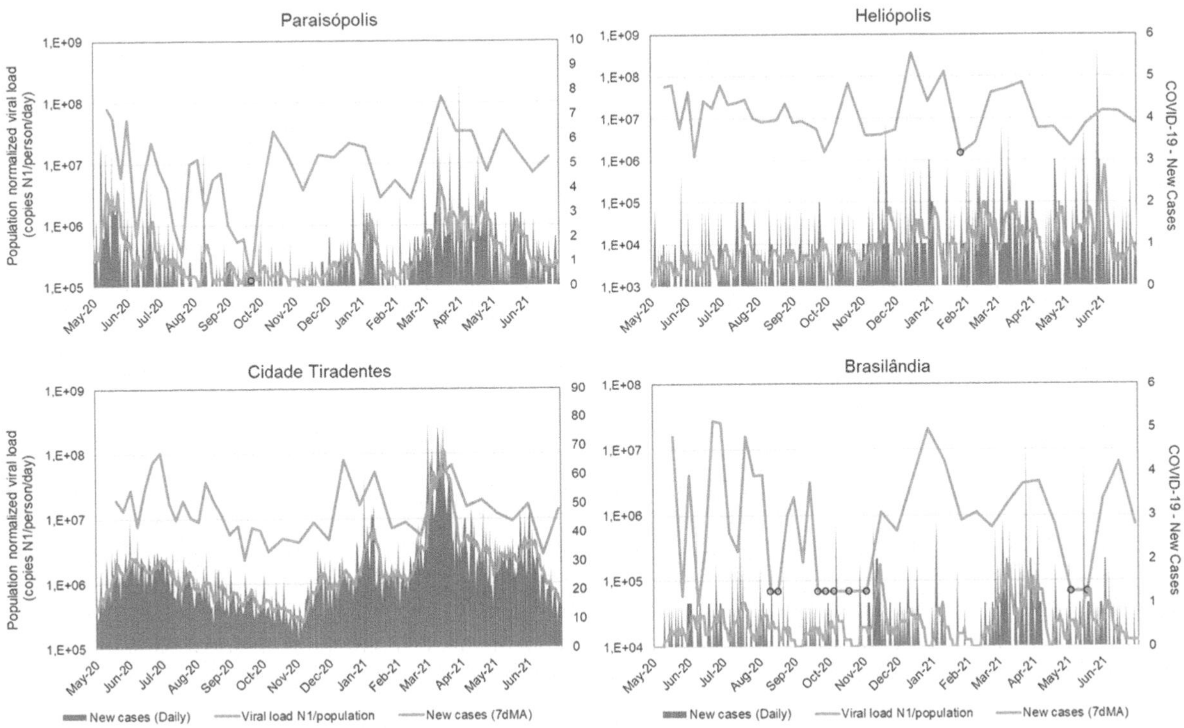

Figure 5 | Population-normalized SARS-CoV-2 RNA viral loads in wastewater and number of COVID-19/SARI cases in vulnerable communities from São Paulo city, sampled between May 2020 and June 2021. Hollow dots indicate samples below the limit of detection.

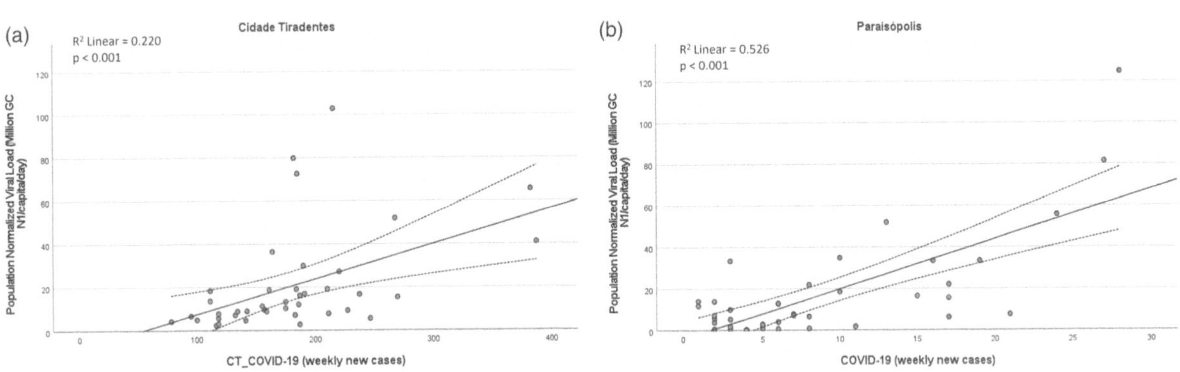

Figure 6 | Spearman's correlation between population-normalized SARS-CoV-2 viral load and weekly new cases of COVID-19 from (a) Cidade Tiradentes and (b) Paraisópolis data.

SARS-CoV-2 detection in the 6 d.p.i culture supernatants (Table 4). Similar results were observed for negative control samples. Positive controls used in the Vero E6 cells presented clear CPE and SARS-CoV-2 RNA detection in the supernatant by RT-qPCR using gene N as target.

3.1.6. SARS-CoV-2 genomes in wastewater samples

We performed NGS on three ultracentrifuged (with glycine elution) wastewater samples, and we were able to recover only one nearly complete genome of SARS-CoV-2. The results of the BlastN search using these genomes indicated that our sequences presented similarities ranging from 96.0 to 99.9% with SARS-CoV-2 reference sequences. Based on the SARS-CoV-2 Spike region (3,822 bp), we performed genotyping of the sequences identified in this study (Table 5). According to

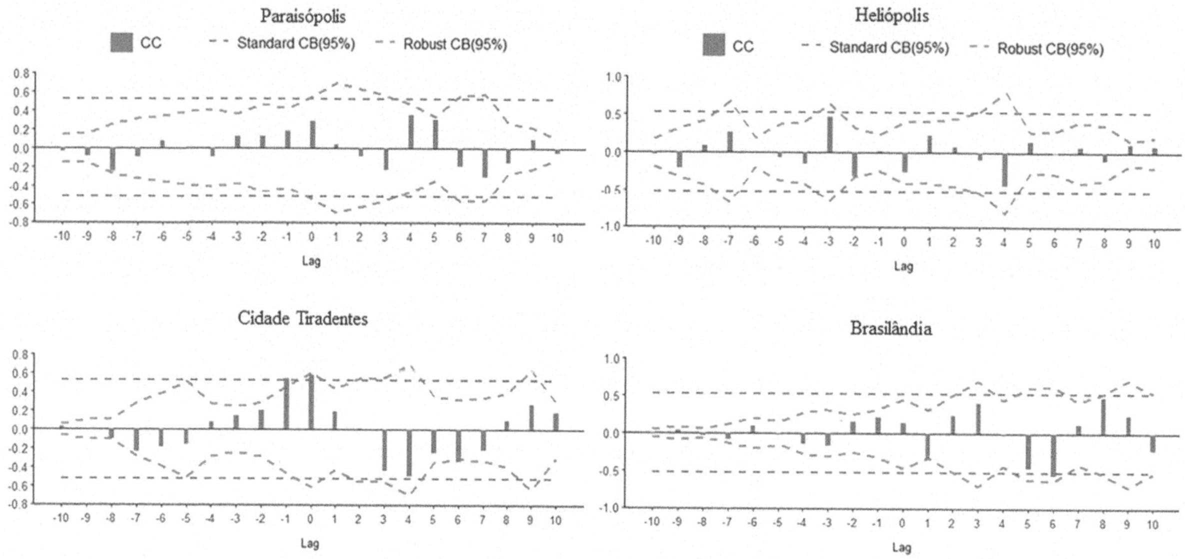

Figure 7 | Cross-correlation analysis between SARS-CoV-2 RNA viral loads and 7-day cumulative cases of COVID-19/SARI time series, in four vulnerable communities of São Paulo city, from May 2020 to June 2021.

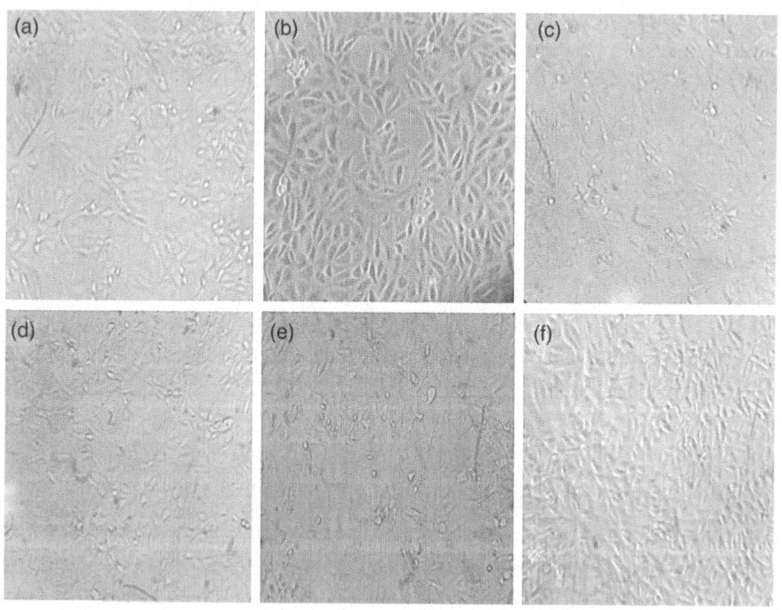

Figure 8 | Wastewater samples tested in Vero E6 cells, observed under optical microscopy. (a,b) Cells infected with the wastewater samples concentrated by ultrafiltration with Centricon® Plus-70 and (c,d) ultracentrifugation in sucrose cushion. The samples of wastewater tested in the cell culture had no sign of virus infectivity. (e,f) Cells infected with the SARS-CoV-2 virus and uninfected cells, respectively.

the clade assignment of the NextClade application, sequences 915 and 920 belong to the clade 20 J (also known as lineage P.1 or variant gamma).

4. DISCUSSION

This study presents data from SARS-CoV-2 monitoring in wastewater samples from vulnerable communities (slums and low-income neighborhoods) located in São Paulo city. During the 1-year study from May 2020 to June 2021, SARS-CoV-2 RNA (N1 or N2) was detectable in all samples from Cidade Tiradentes (n=40, ~130,000 individuals), Paraisópolis (n=40, ~51,000

Table 4 | SARS-CoV-2 RNA concentrations in wastewater samples submitted to infectivity assay in Vero E6 cells, and the corresponding results of evaluating the presence of CPE and the detection of SARS-CoV-2 replication by RT-qPCR

Sample ID	Sampling site	Sampling date	Concentration method	SARS-CoV-2, N1 copies/L (*Cq*)[a]	Cell culture (CPE-SARS-CoV-2)	SARS-CoV-2 N gene detection post-infection
2006032	Paraisópolis	26/05/2020	Ultrafiltration	2.1×10^4 (35.4)	Negative	Negative
2006073	Cidade Tiradentes	02/06/2020	Ultrafiltration	4.6×10^4 (32.9)	Negative	Negative
2006075	Heliópolis	02/06/2020	Ultrafiltration	8.9×10^4 (34.4)	Negative	Negative
2006199	Heliópolis	16/06/2020	Ultrafiltration	1.1×10^6 (30.4)	Negative	Negative
2006201	Cidade Tiradentes	15/06/2020	Ultrafiltration	1.6×10^6 (30.0)	Negative	Negative
2006250	Cidade Tiradentes	22/06/2020	Ultrafiltration	1.5×10^5 (33.9)	Negative	Negative
2006248	Heliópolis	23/06/2020	Ultrafiltration	7.1×10^4 (35.8)	Negative	Negative
2006251	Paraisópolis	23/06/2020	Ultrafiltration	1.1×10^5 (33.2)	Negative	Negative
2011128	Cidade Tiradentes	14/12/2020	Ultrafiltration	2.8×10^5 (31.7)	Negative	Negative
			Sucrose cushion ultracentrifugation	6.2×10^4 (34.1)	Negative	Negative
2011132	Heliópolis	15/12/2020	Ultrafiltration	8.8×10^5 (30.1)	Negative	Negative
			Sucrose cushion ultracentrifugation	1.1×10^5 (33.2)	Negative	Negative
2103497	Paraisópolis	15/03/2021	Ultrafiltration	2.4×10^5 (32.0)	Negative	Negative
			Sucrose cushion ultracentrifugation	2.6×10^5 (31.8)	Negative	Negative

[a]Quantification prior to inoculation in cell culture.

Table 5 | Genotyping of SARS-CoV-2 detected in wastewater samples by illumina sequencing

Sample ID	Sampling site	Sampling date	N1 (copies/L) *Cq*	Genome coverage (%)	NextClade[a]		
					Complete genome	Orf1ab	Spike
915	Paraisópolis	10/05/2021	2.0×10^5 32.4	99.6	20 J (Gamma, V3)	20B	20 J (Gamma, V3)
918	Heliópolis	09/03/2021	2.4×10^5 32.1	97.1	NA	NA	NA
920	Paraisópolis	29/03/2021	1.9×10^5 32.4	80	NA	NA	20 J (Gamma, V3)

NA, genotype not assigned due to poor quality of the sequence.
[a]NextClade v1.5.0 clade assigner (https://clades.nextstrain.org/).

individuals) and Heliópolis (n=41, ~10,000 individuals), and in 30 of a total of 39 from Vila Brasilândia, with an estimated population of 1,700 individuals contributing to sub-sewershed. The temporal variations of SARS-CoV-2 RNA were evaluated in ultracentrifuged wastewater samples, and the N1 and N2 concentrations ranged from 1×10^3 to 1×10^6 gene copies/L (Table 3). These values are within the range of SARS-CoV-2 RNA concentrations reported in wastewater around the world (https://www.covid19wbec.org/covidpoops19; https://sphere.waterpathogens.org/map).

Ultracentrifugation and glycine elution was the method of choice to determine the time series of SARS-CoV-2 RNA concentrations as it resulted in significantly higher SARS-CoV-2 N1 concentrations than the Centricon® Plus-70 ultrafiltration method ($p<0.001$) (Figure 2). Some authors, when comparing the recovery results of SARS-CoV-2 or enveloped viruses (MHV, BCoV, PMMoV) obtained by different concentration methods, observed that ultrafiltration using Centricon® Plus-70 had better performance than other methods such as CP-Select™ ultrafiltration (Forés *et al.* 2021), PEG 8000 precipitation (Gerrity *et al.* 2021), Amicon, Macrosep, Vivaspin ultrafiltration, PEG 8000 precipitation (Boogaerts *et al.* 2021), PEG 6000 precipitation (Bertrand *et al.* 2021) and the adsorption–elution method using electronegative membranes (Sherchan *et al.*

2020). Our comparative study was based on a few samples (n=13) previously filtered on 0.22 μm membranes (not centrifuged) before the Centricon concentration step, a difference that may have reduced the effectiveness of the method but was necessary for the viability test in cell culture.

Few studies have comparatively evaluated the ultracentrifugation method. Ahmed *et al.* (2020) compared seven different concentration methods and had better recovery efficiency of MHV in ultracentrifugation (33.5 ± 12.1%) than Centricon® Plus-70 ultrafiltration (28.0 ± 9.10%), although the most efficient was the adsorption–extraction method with $MgCl_2$ pretreatment (65.7 ± 23.8%). Prado *et al.* (2021) reported mean recoveries of 27.4 ± 8.64% of BRSV, using the same ultracentrifugation method in the wastewater sample concentration. The viral recoveries reported within this study (8.3 ± 11.7%) are in line with those reported by other comparative studies with recoveries of BCoV of no greater than 20% for the majority of samples (Jafferali *et al.* 2021; Philo *et al.* 2021; Wilder *et al.* 2021).

The ultracentrifugation technique employed in this work was originally developed to concentrate non-enveloped enteric viruses that exhibit lower partitioning to solids present in wastewater compared to enveloped viruses (Ye *et al.* 2016). Due to the strong interaction of enteric viruses with solid particles, the considerable viral loss probably occurs during the glycine elution step followed by centrifugation. This could explain the lower recovery rates in samples with higher amounts of TSS.

SARS-CoV-2 (N1 and N2 gene regions) was detected consistently throughout the monitoring period, except in Vila Brasilândia, where the virus was not found in 23% of the samples, probably as a consequence of false-negative results due to the low flow rate (Weidhaas *et al.* 2021).

The SARS-CoV-2 densities were greater than 4 $\log_{10}$ GC/L for about 80% of the samples, with 36% of these showing values greater than 5 log GC/L. Heliópolis was the place that showed the highest concentrations of viral RNA, 39% of the values were above 5 log GC/L, with a peak in December (1.7×10^6 GC/L), followed by Cidade Tiradentes (32.5%), Paraisópolis (30%) and Vila Brasilândia (12.8%) (Supplementary Fig. S1). These data are in line with those reported in other Brazilian cities (Claro *et al.* 2021; Mota *et al.* 2021; Prado *et al.* 2021).

Comparing the population-normalized viral load in the four monitored sites, both slums, Heliópolis and Paraisópolis, presented higher values of N1/capita/day over the course of the 1-year study (Figure 4(a)). Trends in viral load showed an increase in three different periods, at the beginning of the monitoring (May/June 2020), at Christmas time (December 2020/January 201) and after the carnival holiday (March 2021), which coincide with the peaks of the curve of COVID-19 new cases (daily and 7-day moving average (dMA)) reported by the São Paulo city (Figure 4(b)). After July, new COVID-19 cases and deaths decreased, but the facilitation of preventive measures in mid-October 2020 (partial opening) increased the number of daily deaths cases, exceeding 1,000 per day in mid-December (Ministério da Saúde 2020). The emergency of the new SARS-CoV-2 variant lineage P.1 in January 2021 in the city of Manaus, with more significant transmission potential, added to a greater people circulation in Carnival (mid-February) led to a new peak of cases and deaths in March.

However, when we look at the sub-sewershed level, trends in SARS-CoV-2 wastewater loads versus COVID-19 cases are distinct (Figure 5). At Cidade Tiradentes and Paraisópolis, the wastewater viral loads followed a trend similar to new cases of COVID-19 (daily and 7 dMA) reported for the population of these watersheds over the monitoring time period ($p<0.001$) (Figure 6); the same was not observed for the other two communities. Also, the cross-correlation analysis, considering the viral loads of SARS-CoV-2 RNA and the 7-day cumulative COVID-19 cases for each community, showed no significant temporal dependence among the COVID-19 cases and the respective virus load for the four communities. Although it was not possible to establish a model for predicting cases of COVID-19 in these communities based on the SARS-CoV-2 data in the respective sewersheds, following the trends of SARS-CoV-2 wastewater loads in the four vulnerable communities allows the Public Health Service to have a more refined picture of infected individuals circulating within these sewersheds.

It is important to mention some limitations in our study. The monitoring was conducted with weekly or fortnightly sampling with weekly or biweekly frequency, which may have reduced the representativeness of SARS-CoV-2 RNA concentrations, especially in the Brasilândia and Heliópolis communities. The sampling frequency is one important aspect to identifying the COVID-19 incidence dynamics. Some studies indicated that the least resource-intensive sampling scheme that maintained a high degree of confidence in capturing case trends was two nonconsecutive days per week (Feng *et al.* 2021; Graham *et al.* 2021). Another important aspect is that the lack of reliable data on COVID-19 cases does not allow a more accurate comparison with SARS-CoV-2 wastewater surveillance results. This issue is mainly due to the high rate of underreported cases, resulting from the limitation in the number of tests for COVID-19, with rates much lower than in developed countries (https://www.worldometers.info/coronavirus/). Veiga Silva *et al.* (2020) estimated an underreporting rate of deaths of 35.28% in the city of São Paulo, considering data up to May 2020. Kupek (2021) estimated an underreporting rate of deaths by COVID-19 of 22.62% in 2020 in Brazil. Furthermore, the underreporting rate is more accentuated in low-income communities due to the limited availability of tests in the public health system (de Souza *et al.* 2020). Another limiting factor for comparing data is the considerable percentage of missing data (date of initial symptoms, ZIP code), especially in slum residents.

A few studies have evaluated the infectivity in cell cultures of SARS-CoV-2 isolated from water and sewage samples (Rimoldi *et al.* 2020; Razzolini *et al.* 2021; Westhaus *et al.* 2021), and as reported in the present study, the presence of infectious SARS-CoV-2 was also not evidenced. Factors like the propagation process, susceptibility of cells and incubation time may affect the sensitivity of viral culture. Furthermore, pretreatment processes by filtration and concentration of samples (Giacobbo *et al.* 2021), as well as storage and freezing–thawing procedures, can also contribute to the inactivation of SARS-CoV-2 particles. Therefore, regarding viral culture, there is a lack of standardized methods for the recovery of viable SARS-CoV-2.

Although studies have shown that viral particles of SARS-CoV-2 remain viable at 20–24 °C in river water samples (1.9–2.3 days, T90) and sewage (1.2–1.6 days, T90) experimentally contaminated (Bivins *et al.* 2020; de Oliveira *et al.* 2021; Sala-Comorera *et al.* 2021), many factors contributed to the inactivation of SARS-CoV-2 in environmental waters such as temperature, pH, exposure to UV, antagonistic microorganisms and presence of disinfectants (La Rosa *et al.* 2020; Giacobbo *et al.* 2021; Paul *et al.* 2021; Tran *et al.* 2021). However, due to the high viral load released into the environment, especially in communities without access to a sewage collection network, the low rate of sewage treatment in Brazil (SNIS 2020) and the low rate of virus removal in conventional sewage treatments from activated sludge processes, there is a clear need for further investigation to clarify the real microbiological risk associated with waterborne transmission of COVID-19 (Kumar *et al.* 2021).

The preliminary sequencing analyses detected the circulating variants, identified as the 20 J (P.1) variant, in two of the three samples evaluated. However, the complete virus genome was achieved in only one sample. Additional tests are needed to assess the sequencing approach's feasibility, including the RNA concentration and RNA purification steps. The results agreed with the SARS-CoV-2 genomic surveillance data in clinical samples conducted during the same period of collection of wastewater samples. In São Paulo city, the P.1 variant was identified in 64.4% of the 73 samples analyzed in the first week of March 2021 (http://www.prefeitura.sp.gov.br/cidade/secretarias/upload/saude/situacao_-covid19_03_26_03_2021.pdf). Considering the variants circulating in the metropolitan São Paulo in 2021, P.1 was predominant in March (84.6%), April (94.7%) and May (96.3%) (https://butantan.gov.br/assets/files/Covid/Boletim_epidemiologico/relatorio.pdf).

SARS-CoV-2 wastewater surveillance proved to be an efficient tool to detect and monitor trends and the prevalence of COVID-19 cases in communities, as demonstrated in several studies worldwide carried out from wastewater analysis (WWTP, sub-sewershed and facilities) and surface water (Medema *et al.* 2020a). In low-income communities with high rates of transmission and underreporting cases, SARS-CoV-2 WBE has an even more critical role because it can provide information on infection trends without being influenced by the availability and access to clinical testing resources or data on healthcare-seeking behavior.

5. CONCLUSIONS

This study demonstrated that in vulnerable communities such as low-income neighborhoods and slums, where underreporting rates are pronounced, wastewater surveillance could be an essential tool to provide reliable indicators of COVID-19 prevalence, despite the difficulty of validating the strategy and demonstrating its value to clinical surveillance.

The ultracentrifugation with the glycine method showed greater sensitivity in quantifying SARS-CoV-2 compared to the Centricon® Plus-70 ultrafiltration method. We observed that the different characteristics of wastewater matrices influenced the concentration method recovery. The total solids amount negatively influenced the recovery efficiency.

Wastewater-based genomic epidemiology can provide complementary and more comprehensive data on the circulation of variants in a community. Preliminary results of NGS sequencing evidenced the circulation of the P.1 lineage, the predominant variant in the area and periods analyzed. SARS-CoV-2 RNA positive sewage samples showed no replication of

SARS-CoV-2 in Vero E6 cell culture, which reinforces the evidence suggesting a low risk of transmission of SARS-CoV-2 by wastewater contamination.

ACKNOWLEDGEMENTS

The authors thank the Fundação de Amparo à Pesquisa do Estado de São Paulo – FAPESP: Brazil (2020/08943-5) (C.T.B, J.T.M. and L.M.R.J) and to the Conselho Nacional de Desenvolvimento Científico e Tecnológico – CNPq: 405691/2018-1 (C.T.B); we thank the CETESB team from Sampling Division and Inorganic Chemistry Laboratory; José Augusto de Areas, Ivo Freitas de Oliveira and Ana Maria Brokelmann (CETESB) for assistance in the wastewater samples processing; Paulo Brandão (Faculty of Veterinary Medicine and Animal Science of University of São Paulo) for kindly providing the BCoV control; Cláudio Tavares Sacchi (IAL – Center for Interdisciplinary Procedures, Strategic Laboratory) for sequencing and Ana Carolina Aguiar de Carvalho (City Hall of São Paulo, Acute Communicable Diseases Nucleus, Health Surveillance Coordination) for sharing the SIVEP/e-SUS database.

CONFLICT OF INTERESTS

The authors declare that there are no competing interests in this article.

DATA AVAILABILITY STATEMENT

All relevant data are included in the paper or its Supplementary Information.

REFERENCES

Ahmed, W., Bertsch, P. M., Bivins, A., Bibby, K., Farkas, K., Gathercole, A., Haramoto, E., Gyawali, P., Korajkic, A., McMinn, B. R., Mueller, J. F., Simpson, S. L., Smith, W. J. M., Symonds, E. M., Thomas, K. V., Verhagen, R. & Kitajima, M. 2020. Comparison of virus concentration methods for the RT-qPCR-based recovery of murine hepatitis virus, a surrogate for SARS-CoV-2 from untreated wastewater. *Science of Total Environment* **739**, 139960. https://doi.org/10.1016/j.scitotenv.2020.139960

Ahmed, W., Tscharke, B., Bertsch, P. M., Bibby, K., Bivins, A., Choi, P., Clarke, L., Dwyer, J., Edson, J., Nguyen, T. M. H., O'Brien, J. W., Simpson, S. L., Sherman, P., Thomas, K. V., Verhagen, R., Zaugg, J. & Mueller, J. F. 2021 SARS-CoV-2 RNA monitoring in wastewater as a potential early warning system for COVID-19 transmission in the community: a temporal case study. *Science of the Total Environment* **761**, 144216. https://doi.org/10.1016/j.scitotenv.2020.144216.

Alessandrini, F., Caucci, S., Onofri, V., Melchionda, F., Tagliabracci, A., Bagnarelli, P., Di, S. L., Turchi, C. & Menzo, S. 2020 Evaluation of the Ion AmpliSeq SARS-CoV-2 Research Panel by massive parallel sequencing. *Genes* **11** (8), 1–14. https://doi.org/10.3390/genes11080929.

ANA Agência Nacional de Águas 2021 *Monitoramento COVID Esgotos – Painel Monitoramento COVID em Esgotos (COVID Sewage Monitoring – COVID Sewage Monitoring Dashboard).* Report for Agency of Water and Sanitation (ANA), Brasília, Brazil. Available from: https://www.gov.br/ana/pt-br/assuntos/acontece-na-ana/monitoramento-covid-esgotos (accessed 28 December 2021).

APHA; AWWA; WEF 2017a *Standard Methods for the Examination of Water and Wastewater: 1060 Collection and Preservation of Samples*, 23rd edn. Washington, DC.

APHA; AWWA; WEF 2017b *Standard Methods for the Examination of Water and Wastewater: 2540 D Total Suspended Solids Dried from 103 to 105 °C*, 23rd edn. Washington, DC.

Araujo, D. B., Machado, R. R. G., Amgarten, D. E., Malta, F. d. M., de Araujo, G. G., Monteiro, C. O., Candido, E. D., Soares, C. P., de Menezes, F. G., Pires, A. C. C., Santana, R. A. F., Viana, A. d. O., Dorlass, E., Thomazelli, L., Ferreira, L. C. d. S., Botosso, V. F., Carvalho, C. R. G., Oliveira, D. B. L., Pinho, J. R. R. & Durigon, E. L. 2020 SARS-CoV-2 isolation from the first reported patients in Brazil and establishment of a coordinated task network. *Memorias Do Instituto Oswaldo Cruz* **115** (12), 1–8. https://doi.org/10.1590/0074-02760200342.

Asghar, H., Diop, O. M., Weldegebriel, G., Malik, F., Shetty, S., El Bassioni, L., Akande, A. O., Al Maamoun, E., Zaidi, S., Adeniji, A. J., Burns, C. C., Deshpande, J., Oberste, M. S. & Lowther, S. A. 2014 Environmental surveillance for polioviruses in the global polio eradication initiative. *Journal of Infectious Diseases* **210** (suppl 1), S294–S303. https://doi.org/10.1093/infdis/jiu384.

Bauer, D. F. 1972 Constructing confidence sets using rank statistics. *Journal of the American Statistical Association* **67** (339), 687–690. https://doi.org/10.2307/2284469.

Bertrand, I., Challant, J., Jeulin, H., Hartard, C., Mathieu, L., Lopez, S., Schvoerer, E., Courtois, S. & Gantzer, C. 2021 Epidemiological surveillance of SARS-CoV-2 by genome quantification in wastewater applied to a city in the northeast of France: comparison of ultrafiltration- and protein precipitation-based methods. *International Journal of Hygiene and Environmental Health* **233**. https://doi.org/10.1016/j.ijheh.2021.113692

Bivins, A., Greaves, J., Fischer, R., Yinda, K. C., Ahmed, W., Kitajima, M., Munster, V. J. & Bibby, K. 2020 Persistence of SARS-CoV-2 in water and wastewater. *Environmental Science and Technology Letters* **7** (12), 937–942. https://doi.org/10.1021/acs.estlett.0c00730.

Boogaerts, T., Jacobs, L., De Roeck, N., Van den Bogaert, S., Aertgeerts, B., Lahousse, L., van Nuijs, A. L. N. & Delputte, P. 2021 An alternative approach for bioanalytical assay optimization for wastewater-based epidemiology of SARS-CoV-2. *Science of the Total Environment* **789**, 148043. https://doi.org/10.1016/J.SCITOTENV.2021.148043.

CDC 2020 *US CDC RT-PCR-Panel-for-Detection-Instructions V20.01.27.* pp. 1–12.

CDC 2021 *National Wastewater Surveillance System (NWSS).* Waterborne Disease & Outbreak Surveillance. Available from: https://www.cdc.gov/healthywater/surveillance/wastewater-surveillance/wastewater-surveillance.html.

Claro, I. C. M., Cabral, A. D., Augusto, M. R., Duran, A. F. A., Graciosa, M. C. P. G., Fonseca, F. L. A., Speranca, M. A. & Bueno, R. F. 2021 Long-term monitoring of SARS-COV-2 RNA in wastewater in Brazil: a more responsive and economical approach. *Water Research* **203**, 117534. https://doi.org/10.1016/j.watres.2021.117534.

Corman, V. M., Landt, O., Kaiser, M., Molenkamp, R., Meijer, A., Chu, D. K. W., Bleicker, T., Schneider, J., Schmidt, M. L., Mulders, D. G. J. C., Haagmans, B. L., Veer, B. V. D., Den, S. V., Wijsman, L., Goderski, G., Ellis, J., Zambon, M., Peiris, M., Goossens, H., Reusken, C., Koopmans, M. P. G. & Drosten, C. 2020 Detection of 2019 novel coronavirus (2019-nCoV) by real-time RT-PCR. *Eurosurveillance* **25** (3). https://doi.org/10.2807/1560-7917.ES.2020.25.3.2000045

Crits-Christoph, A., Kantor, R. S., Olm, M. R., Whitney, O. N., Al-Shayeb, B., Lou, Y. C., Flamholz, A., Kennedy, L. C., Greenwald, H., Hinkle, A., Hetzel, J., Spitzer, S., Koble, J., Tan, A., Hyde, F., Schroth, G., Kuersten, S., Banfield, J. F. & Nelson, K. L. 2021 Genome sequencing of sewage detects regionally prevalent SARS-CoV-2 variants. *mBio* **12** (1), 1–9. https://doi.org/10.1128/mBio.02703-20.

Croghan, C. W. & Egeghy, P. P. 2003 *Methods of Dealing with Values below the Limit of Detection Using SAS.* US-EPA, Research Triangle Park, NC.

CRS Report 2021 *Global Economic Effects of COVID-19.* November 10, 2021. R46270. Available from: https://crsreports.congress.gov.

Cucinotta, D. & Vanelli, M. 2020 WHO declares COVID-19 a pandemic. *Acta Biomedica* **91** (1), 157–160. doi:10.23750/abm.v91i1.9397.

Decaro, N., Elia, G., Campolo, M., Desario, C., Mari, V., Radogna, A., Colaianni, M. L., Cirone, F., Tempesta, M. & Buonavoglia, C. 2008 Detection of bovine coronavirus using a TaqMan-based real-time RT-PCR assay. *Journal of Virological Methods* **151** (2), 167–171. https://doi.org/10.1016/J.JVIROMET.2008.05.016.

de Oliveira, L. C., Torres-Franco, A. F., Lopes, B. C., Santos, B. S. Á. d. S., Costa, E. A., Costa, M. S., Reis, M. T. P., Melo, M. C., Polizzi, R. B., Teixeira, M. M. & Mota, C. R. 2021 Viability of SARS-CoV-2 in river water and wastewater at different temperatures and solids content. *Water Research* **195**, 117002. https://doi.org/10.1016/J.WATRES.2021.117002.

de Souza, W. M., Buss, L. F., Candido, D. d. S., Carrera, J. P., Li, S., Zarebski, A. E., Pereira, R. H. M., Prete, C. A., de Souza-Santos, A. A., Parag, K. V., Belotti, M. C. T. D., Vincenti-Gonzalez, M. F., Messina, J., da Silva Sales, F. C., Andrade, P. d. S., Nascimento, V. H., Ghilardi, F., Abade, L., Gutierrez, B., Kraemer, M. U. G., Braga, C. K. V., Aguiar, R. S., Alexander, N., Mayaud, P., Brady, O. J., Marcilio, I., Gouveia, N., Li, G., Tami, A., de Oliveira, S. B., Porto, V. B. G., Ganem, F., de Almeida, W. A. F., Fantinato, F. F. S. T., Macário, E. M., de Oliveira, W. K., Nogueira, M. L., Pybus, O. G., Wu, C. H., Croda, J., Sabino, E. C. & Faria, N. R. 2020 Epidemiological and clinical characteristics of the COVID-19 epidemic in Brazil. *Nature Human Behaviour* **4** (8), 856–865. https://doi.org/10.1038/s41562-020-0928-4.

Dunn, O. J. 1961 Multiple comparisons among means. *Journal of the American Statistical Association* **56** (293), 52–64. https://doi.org/10.1080/01621459.1961.10482090.

Feng, S., Roguet, A., McClary-Gutierrez, J. S., Newton, R. J., Kloczko, N., Meiman, J. G. & McLellan, S. L. 2021 Evaluation of Sampling, Analysis, and Normalization Methods for SARS-CoV-2 Concentrations in Wastewater to Assess COVID-19 Burdens in Wisconsin Communities. *ACS ES&T Water* **1**, 1955–1965. https://doi.org/10.1021/acsestwater.1c00160

Figueiredo, A. C. C. 2021 Social housing and the Favela Urbanization Program: the Paraisópolis complex, in São Paulo. *Risco – Journal of Architeture and Urbanisms (Online)* **19**, 1–15. https://doi.org/10.11606/1984-4506.risco.2021.159431.

Fongaro, G., Rogovski, P., Pereira Savi, B., Dorighello Cadamuro, R., Virgínia Faria Pereira, J., Hashimoto Sant Anna, I., Henrique Rodrigues, I., Sobral Marques Souza, D., Gregory Torres Saravia, E., Rodríguez-Lázaro, D. & Célia da Silva Lanna, M. 2021 SARS-CoV-2 in human sewage and river water from a remote and vulnerable area as a surveillance tool in Brazil. *Food and Environmental Virology* **1**, 3. https://doi.org/10.1007/s12560-021-09487-9.

Forés, E., Bofill-Mas, S., Itarte, M., Martínez-Puchol, S., Hundesa, A., Calvo, M., Borrego, C. M., Corominas, L. L., Girones, R. & Rusiñol, M. 2021 Evaluation of two rapid ultrafiltration-based methods for SARS-CoV-2 concentration from wastewater. *Science of the Total Environment* **768**, 144786. doi:10.1016/j.scitotenv.2020.144786.

Gerrity, D., Papp, K., Stoker, M., Sims, A. & Frehner, W. 2021 Early-pandemic wastewater surveillance of SARS-CoV-2 in Southern Nevada: Methodology, occurrence, and incidence/prevalence considerations. *Water Research* **10**, 100086. https://doi.org/10.1016/J.WROA.2020.100086.

Giacobbo, A., Rodrigues, M. A. S., Zoppas Ferreira, J., Bernardes, A. M. & de Pinho, M. N. 2021 A critical review on SARS-CoV-2 infectivity in water and wastewater. What do we know? *Science of the Total Environment* **774**, 145721. https://doi.org/10.1016/j.scitotenv.2021.145721.

GPEI 2016 *Explaining Environmental Surveillance.* Available from: https://polioeradication.org/news-post/explaining-environmental-surveillance/ (accessed 28 August 2021).

Graham, K. E., Loeb, S. K., Wolfe, M. K., Catoe, D., Sinnott-Armstrong, N., Kim, S., Yamahara, K. M., Sassoubre, L. M., Mendoza Grijalva, L. M., Roldan-Hernandez, L., Langenfeld, K., Wigginton, K. R. & Boehm, A. B. 2021 SARS-CoV-2 RNA in Wastewater Settled Solids Is Associated with COVID-19 Cases in a Large Urban Sewershed. *Environmental Science and Technology* **55**, 488–498. https://doi.org/10.1021/acs.est.0c06191.

Haramoto, E., Malla, B., Thakali, O. & Kitajima, M. 2020 First environmental surveillance for the presence of SARS-CoV-2 RNA in wastewater and river water in Japan. *Science of the Total Environment* **737**, 140405. https://doi.org/10.1016/j.scitotenv.2020.140405.

Haugh, L. D. 1976 Checking the independence of two covariance-stationary time series: a univariate residual cross-correlation approach. *Journal of the American Statistical Association* **71** (354), 378–385. https://doi.org/10.2307/2285318.

Hewett, P. & Ganser, G. H. 2007 A comparison of several methods for analyzing censored data. *Annals of Occupational Hygiene* **51**, 611–632. https://doi.org/10.1093/annhyg/mem045.

Iglesias, N. G., Gebhard, L. G., Carballeda, J. M., Aiello, I., Recalde, E., Terny, G., Ambrosolio, S., L'Arco, G., Konfino, J. & Brardinell, J. I. 2021 SARS-CoV-2 surveillance in untreated wastewater: detection of viral RNA in a low-resource community in Buenos Aires, Argentina. *Rev Panam Salud Publica* **45**, e137. https://doi.org/10.26633/RPSP.2021.137.

Jafferali, M. H., Khatami, K., Atasoy, M., Birgersson, M., Williams, C. & Cetecioglu, Z. 2021 Benchmarking virus concentration methods for quantification of SARS-CoV-2 in raw wastewater. *Science of Total Environment* **755**, 142939. https://doi.org/10.1016/J.SCITOTENV.2020.142939.

Jahn, K., Dreifuss, D., Topolsky, I., Kull, A., Ganesanandamoorthy, P., Fernandez-Cassi, X., Bänziger, C., Devaux, A. J., Stachler, E., Caduff, L., Cariti, F., Corzón, A. T., Fuhrmann, L., Chen, C., Jablonski, K. P., Nadeau, S., Feldkamp, M., Beisel, C., Aquino, C., Stadler, T., Ort, C., Kohn, T., Julian, T. R. & Beerenwinkel, N. 2021 Detection and surveillance of SARS-CoV-2 genomic variants in wastewater. MedRxiv. https://doi.org/10.1101/2021.01.08.21249379.

Jensen, S., Nguyen, C. & Jewett, J. 2016 A gradient-free method for the purification of infective dengue virus for protein-level investigations. *Journal of Virological Methods* **235**, 125–130. https://doi.org/10.1016/J.JVIROMET.2016.05.017.

Kirby, A. E., Walters, M. S., Jennings, W. C., Fugitt, R., LaCross, N., Mattioli, M., Marsh, Z. A., Roberts, V. A., Mercante, J. W., Yoder, J. & Hill, V. R. 2021 Using wastewater surveillance data to support the COVID-19 response – United States, 2020–2021. *MMWR – Morbidity and Mortality Weekly Report* **70**, 1242–1244. http://dx.doi.org/10.15585/mmwr.mm7036a2.

Kumar, M., Joshi, M., Patel, A. K. & Joshi, C. G. 2021 Unravelling the early warning capability of wastewater surveillance for COVID-19: a temporal study on SARS-CoV-2 RNA detection and need for the escalation. *Environmental Research* **196**, 110946. https://doi.org/10.1016/j.envres.2021.110946.

Kupek, E. 2021 How many more? Under-reporting of the COVID-19 deaths in Brazil in 2020. *Tropical Medicine and International Health*. https://doi.org/10.1111/tmi.13628

La Rosa, G., Bonadonna, L., Lucentini, L., Kenmoe, S. & Suffredini, E. 2020 Coronavirus in water environments: occurrence, persistence and concentration methods – a scoping review. *Water Research* **179**, 115899. https://doi.org/10.1016/j.watres.2020.115899.

Medema, G., Been, F., Heijnen, L. & Petterson, S. 2020a Implementation of environmental surveillance for SARS-CoV-2 virus to support public health decisions: opportunities and challenges. *Current Opinion in Environmental Science & Health* **17**, 49–71. https://doi.org/10.1016/J.COESH.2020.09.006.

Medema, G., Heijnen, L., Elsinga, G., Italiaander, R. & Brouwer, A. 2020b Presence of SARS-Coronavirus-2 RNA in sewage and correlation with reported COVID-19 prevalence in the early stage of the epidemic in the Netherlands. *Environmental Science and Technology Letters* **7** (7), 511–516. https://doi.org/10.1021/ACS.ESTLETT.0C00357.

Ministério da Saúde 2020 Brasil confirma primeiro caso da doença (Brazil confirms first case of the disease) – Português (Brasil). Available from: https://www.gov.br/saude/pt-br/assuntos/noticias/brasil-confirma-primeiro-caso-de-novo-coronavirus (accessed 28 August 2021).

Mota, C. R., Bressani-Ribeiroa, T., Araújo, J. C., Leal, C. D., Leroy-Freitas, D., Machado, E. C., Espinosa, M. F., Fernandes, L., Leão, T. L., Chamhum-Silva, L., Azevedo, L., Morandia, T., Freitas, G. T. O., Costa, M. S., Carvalho, B. O., Reis, M. T. P., Melo, M. C., Ayrimoraes, S. R. & Chernicharo, C. A. L. 2021 Assessing spatial distribution of COVID-19 prevalence in Brazil using decentralised sewage monitoring. *Water Research* **202**, 117388. https://doi.org/10.1016/j.watres.2021.117388.

Nemudryi, A., Nemudraia, A., Wiegand, T., Surya, K., Buyukyoruk, M., Cicha, C., Vanderwood, K. K., Wilkinson, R. & Wiedenheft, B. 2020 Temporal detection and phylogenetic assessment of SARS-CoV-2 in municipal wastewater. *Cell Reports Medicine* **1** (6), 100098. https://doi.org/10.1016/J.XCRM.2020.100098.

NICD (National Institute for Communicable Diseases) 2021 *COVID 19: Wastewater-Based Epidemiology for SARS-CoV-2 in South Africa*. Available from: https://www.nicd.ac.za/diseases-a-z-index/disease-index-covid-19/surveillance-reports/weekly-reports/wastewater-based-epidemiology-for-sars-cov-2-in-south-africa/ (accessed 30 december2021).

Okonechnikov, K., Golosova, O. & Fursov, M. 2012 Genome analysis Unipro UGENE: a unified bioinformatics toolkit. *Bioinformatics Applications Note* **28** (8), 1166–1167. https://doi.org/10.1093/bioinformatics/bts091.

Paul, D., Kolar, P. & Hall, S. G. 2021 A review of the impact of environmental factors on the fate and transport of coronaviruses in aqueous environments. *npj Clean Water* **4** (1), 1–13. https://doi.org/10.1038/s41545-020-00096-w.

Philo, S. E., Keim, E. K., Swanstrom, R., Ong, A. Q. W., Burnor, E. A., Kossik, A. L., Harrison, J. C., Demeke, B. A., Zhou, N. A., Beck, N. K., Shirai, J. H. & Meschke, J. S. 2021 A comparison of SARS-CoV-2 wastewater concentration methods for environmental surveillance. *Science of the Total Environment* **760**, 144215. https://doi.org/10.1016/J.SCITOTENV.2020.144215.

Pina, S., Jofre, J., Emerson, S. U., Purcell, R. H. & Girones, R. 1998 Characterization of a strain of infectious hepatitis E virus isolated from sewage in an area where hepatitis E is not endemic. *Applied And Environmental Microbiology* **64**, 4485–4488.

PMSP 2020 *COVID-19 Relatório Situacional (COVID-19 Situation Report)*. São Paulo. Available from: https://www.prefeitura.sp.gov.br/cidade/secretarias/upload/saude/COVID19_Relatorio_Situacional_SMS_20201203.pdf.

Polanco-Martinez, J. M., Medina-Elizalde, M. A., Sanchez Goni, M. F. & Mudelsee, M. 2019 BINCOR: an r package for estimating the correlation between two unevenly spaced time series. *R Journal* **11** (1), 1–14. https://doi.org/10.32614/rj-2019-035.

Prado, T., Fumian, T. M., Mannarino, C. F., Resende, P. C., Motta, F. C., Eppinghaus, A. L. F., Chagas do Vale, V. H., Braz, R. M. S., de Andrade, J. d. S. R., Maranhão, A. G. & Miagostovich, M. P. 2021 Wastewater-based epidemiology as a useful tool to track SARS-CoV-2 and support public health policies at municipal level in Brazil. *Water Research* **191**. https://doi.org/10.1016/j.watres.2021.116810

Razzolini, M. P., Funada Barbosa, M., Silva de Araújo, R., Freitas de Oliveira, I., Mendes-Correa, M., Sabino, E., Garcia, S., de Paula, A., Villas-Boas, L., Costa, S., Dropa, M., Brandão de Assis, D., Levin, B., Pedroso de Lima, A. & Levin, A. 2021 SARS-CoV-2 in a stream running through an underprivileged, underserved, urban settlement in São Paulo, Brazil: a 7-month follow-up. *Environmental Pollution (Barking, Essex: 1987)* **290**, 118003. https://doi.org/10.1016/J.ENVPOL.2021.118003.

Rijksoverheid (Government of the Netherlands) 2021 *Coronavirus Dashboard. Early Indicators: Virus Particle in Wastewater*. Available from: https://coronadashboard.government.nl/landelijk/rioolwater (accessed 30 December 2021).

Rimoldi, S. G., Stefani, F., Gigantiello, A., Polesello, S., Comandatore, F., Mileto, D., Maresca, M., Longobardi, C., Mancon, A., Romeri, F., Pagani, C., Cappelli, F., Roscioli, C., Moja, L., Gismondo, M. R. & Salerno, F. 2020 Presence and infectivity of SARS-CoV-2 virus in wastewaters and rivers. *Science of the Total Environment* **744**, 140911. https://doi.org/10.1016/j.scitotenv.2020.140911.

Ritchie, H., Mathieu, E., Rodés-Guirao, L., Appel, C., Giattino, C., Ortiz-Ospina, E., Hasell, J., Macdonald, B., Beltekian, D. & Roser, M. 2020 Coronavirus pandemic (COVID-19). Our World in Data. Available from: https://ourworldindata.org/coronavirus

Sala-Comorera, L., Reynolds, L. J., Martin, N. A., O'Sullivan, J. J., Meijer, W. G. & Fletcher, N. F. 2021 Decay of infectious SARS-CoV-2 and surrogates in aquatic environments. *Water Research* **201**. https://doi.org/10.1016/j.watres.2021.117090.

Sherchan, S. P., Shahin, S., Ward, L. M., Tandukar, S., Aw, T. G., Schmitz, B., Ahmed, W. & Kitajima, M. 2020 First detection of SARS-CoV-2 RNA in wastewater in North America: a study in Louisiana, USA. *Science of the Total Environment* **743**. https://doi.org/10.1016/j.scitotenv.2020.140621.

Sims, N. & Kasprzyk-Hordern, B. 2020 Future perspectives of wastewater-based epidemiology: monitoring infectious disease spread and resistance to the community level. *Environmental International* **139**. https://doi.org/10.1016/j.envint.2020.105689.

SNIS–Sistema Nacional de Informações em Saneamento 2020 Série Histórica (Historical Series). http://app4.mdr.gov.br/serieHistorica/ . (accessed 10 May 2020).

Summer, M. D. & Smith, D. E. A. 1987 *Manual of Methods of Baculovirus Vectors and Insect Cell Culture Procedures*. Texas A&M University, College Station, TX.

Takeda, T., Kitajuma, M., Abeynayaka, A., Huong, N. T. T., Dinh, N. Q., Sirikanchana, K., Navia, M., Sam, A. A., Tsudaka, M., Setiadi, T., Hung, D. T. & Haramoto, E. 2021 Governance of wastewater surveillance systems to minimize the impact of COVID-19 and future epidemics: cases across Asia-Pacific. In: *Environmental Resilience and Transformation in Times of COVID-19* (Ramanathan, A. L., Shidambaram, S., Jonathan, M. P., Prasanna, M. V., Kumar, P. & Arriola, F. M., eds). Elsevier Radarweg, Amsterdam, Netherlands, Chapter 11, pp. 115–126. https://doi.org/10.1016/B978-0-323-85512-9.00010-3.

Tran, H. N., Le, G. T., Nguyen, D. T., Juang, R. S., Rinklebe, J., Bhatnagar, A., Lima, E. C., Iqbal, H. M. N., Sarmah, A. K. & Chao, H. P. 2021 SARS-CoV-2 coronavirus in water and wastewater: a critical review about presence and concern. *Environmental Research* **193**, 110265. https://doi.org/10.1016/j.envres.2020.110265.

Veiga Silva, L., Da Penha de Andrade Abi Harb, M., Milene Teixeira Barbosa dos Santos, A., André de Mattos Teixeira, C., Hugo Macedo Gomes, V., Helena Silva Cardoso, E., da Silva, M. S., Vijaykumar, N. L., Venâncio Carvalho, S., Ponce de Leon Ferreira de Carvalho, A. & Renato Lisboa Frances, C. 2020 COVID-19 mortality underreporting in Brazil: analysis of data from government internet portals. *Journal of Medical Internet Research* **22** (8), e21413. https://doi.org/10.2196/21413.

Wade, M. J., Jacomo, A. L., Armenise, L., Brown, M. R., Bunce, J. T., Cameron, G. J., Zhou Fang, Z., Farkas, K., Gilpin, D. F., Graham, D. W., Grimsley, J. M. S., Alwyn Hart, A., Hoffmann, T., Katherine, J., Jackson, K. J., Jones, D. L., Lilley, C. L., McGrath, J. W., McKinley, J. M., McSparron, C., Behnam, F., Nejad, B. F., Mario Morvan, M., Quintela-Baluja, M., Roberts, A. M. I., Singer, A. C., Souque, C., Speigh, V. L., Sweetapple, C., Walker, D., Watts, G., Weightman, A. & Kasprzyk-Hordern, B. 2022 Understanding and managing uncertainty and variability for wastewater monitoring beyond the pandemic: lessons learned from the United Kingdom national COVID-19 surveillance programmes. *Journal of Hazardous materials* **424** (Pt B), 127456. https://doi.org/10.1016/B978-0-323-85512-9.00010-3.

Weidhaas, J., Aanderud, Z. T., Roper, D. K., VanDerslice, J., Gaddis, E. B., Ostermiller, J., Hoffman, K., Jamal, R., Heck, P., Zhang, Y., Torgersen, K., Laan, J. V. & LaCross, N. 2021 Correlation of SARS-CoV-2 RNA in wastewater with COVID-19 disease burden in sewersheds. *Science of the Total Environment* **775**, 145790. doi:10.1016/j.scitotenv.2021.145790.

Westhaus, S., Weber, F. A., Schiwy, S., Linnemann, V., Brinkmann, M., Widera, M., Greve, C., Janke, A., Hollert, H., Wintgens, T. & Ciesek, S. 2021 Detection of SARS-CoV-2 in raw and treated wastewater in Germany – suitability for COVID-19 surveillance and potential transmission risks. *Science of the Total Environment* **751**, 141750. https://doi.org/10.1016/j.scitotenv.2020.141750.

WHO 2020 *Status of Environmental Surveillance for SARS-CoV-2 Virus: Scientific Brief*. WHO/2019-nCoV/Sci_Brief/EnvironmentalSampling/2020.1 (5 August), pp. 1–4.

Wilder, M. L., Middleton, F., Larsen, D. A., Du, Q., Fenty, A., Zeng, T., Insaf, T., Kilaru, P., Collins, M., Kmush, B. & Green, H. C. 2021 Co-quantification of crAssphage increases confidence in wastewater-based epidemiology for SARS-CoV-2 in low prevalence areas. *Water Research X* **11**, 100100. https://doi.org/10.1016/J.WROA.2021.100100.

Wu, F., Zhang, J., Xiao, A., Gu, X., Lin Lee, W., Armas, F., Kauffman, K., Hanage, W., Matus, M., Ghaeli, N., Endo, N., Duvallet, C., Poyet, M., Moniz, K., Washburne, A. D., Erickson, T. B., Chai, P. R., Thompson, J. & Alm, E. J. 2020 SARS-CoV-2 titers in wastewater are higher than expected from clinically confirmed cases. *Applied and Environmental Science* **5** (4), e00614–20. https://doi.org/10.1128/mSystems.00614-20.

Xagoraraki, I. & O'Brien, E. 2020 Wastewater-based epidemiology for early detection of viral outbreaks. In: *Women in Water Quality* (O'Bannon, D. J., ed.). Springer Nature, Switzerland, pp. 75–97. Available from: https://link.springer.com/book/10.1007/978-3-030-17819-2.

Ye, Y., Ell enberg, R. M., Graham, K. E. & Wigginton, K. R. 2016 Survivability, partitioning, and recovery of enveloped viruses in untreated municipal wastewater. *Environmental Science and Technology* **50** (10), 5077–5085. https://doi.org/10.1021/ACS.EST.6B00876.

First received 31 August 2021; accepted in revised form 4 February 2022. Available online 11 February 2022

SARS-CoV-2 in Water doi: 10.2166/wh.2022.179

SARS-CoV-2 variant detection from wastewater: rapid spread of B.1.1.7 lineage in Hungary

Eszter Róka[a], Dániel Déri[b], Bernadett Khayer[a], Zoltán Kis[b], Eszter Schuler[a], Nóra Magyar[b], Bernadett Pályi[b], Tamás Pándics[a] and Márta Vargha[a,*]

[a] Department of Public Health Laboratory, National Public Health Center, Albert Flórián út 2-6., H-1097 Budapest, Hungary
[b] National Biosafety Laboratory, Division of Microbiological Reference Laboratories, National Public Health Center, Albert Flórián út 2-6., H-1097 Budapest, Hungary
*Corresponding author. E-mail: vargha.marta@nnk.gov.hu

DD, 0000-0002-9078-4923; BK, 0000-0003-3484-9232; MV, 0000-0002-6426-2056

ABSTRACT

Wastewater-based epidemiology (WBE) is a recognised tool for tracking community transmission of COVID-19. From the second half of 2020, the emergence of new, highly infective, more pathogenic or vaccine-escape SARS-CoV-2 variants is the major public health concern. Variant analysis in sewage might assist the early detection of new mutations. Weekly raw sewage samples from 22 wastewater treatment plants (WWTPs) in Hungary (representing 40% of the population) were analysed between December 2020 and March 2021 for signature mutations N501Y and del H69/V70 of B.1.1.7 lineage by melting point genotyping and RT-digital droplet PCR (RT-ddPCR). The latter method proved to be more efficient in parallel detection of different variants and also provides quantitative information. Wastewater surveillance indicated that the B.1.1.7 variant first emerged in Budapest in early January 2021 and rapidly became dominant in the entire country. Results are in close agreement with the available clinical data (Pearson's correlation coefficient, $R = 0.9153$). RT-ddPCR was confirmed to be a reliable tool for tracking emerging variant ratios in wastewaters. It is a rapid and cost-effective method compared to whole-genome sequencing, but only applicable for the detection of known mutations. Efficient variant surveillance might require the combination of multiple methods.

Key words: digital droplet PCR, SARS-CoV-2, variants of concern, wastewater-based epidemiology

HIGHLIGHTS

- RT-digital droplet PCR was used to detect B.1.1.7 SARS-CoV-2 variant in sewage.
- Signature mutations N501Y and del H69/V70 were detected in equal concentrations.
- Signature mutations of B.1.1.7 became dominant within 6 weeks in Hungarian wastewater treatment plants.
- Sewage data closely correlated with clinical emergence of B.1.1.7 cases.
- Wastewater-based epidemiology is an efficient tool in tracing emerging SARS-CoV-2 variants of concern.

INTRODUCTION

The COVID-19 pandemic is a severe public health threat and its management is a serious challenge to governments. The need for rapid outbreak detection accelerated the introduction of new methods. Wastewater-based epidemiology (WBE) has been used for decades to monitor the health of the population (Sims & Kasprzyk-Hordern 2020), but it gained global attention in the current COVID-19 pandemic as an important complementary method to clinical surveillance.

Although it is a respiratory virus, SARS-CoV-2 is shed in faeces and (rarely) in urine by both symptomatic and asymptomatic individuals. Shedding may start before clinical symptoms and can extend up to several weeks, especially in faeces (Walsh *et al.* 2020). The presence of SARS-CoV-2 in the sewage and the correlation of its concentration to COVID-19 incidence rates were investigated in many countries by several research groups (e.g. Medema *et al.* 2020; Weidhaas *et al.* 2021; Wu *et al.* 2020). Most studies find a correlation between the SARS-CoV-2 concentration in wastewater and the number of clinically confirmed new COVID-19 cases with an offset of 4–10 days after wastewater sampling. The role of WBE in the COVID-19 pandemic and in possible subsequent epidemics is acknowledged by the WHO (2020). The European Commission issued a recommendation to its Member States to introduce wastewater surveillance for SARS-CoV-2, covering at least the agglomerations over 150,000 inhabitants (European Commission 2021). The recommendation also calls for the determination of virus variants from sewage.

The emergence of new SARS-CoV-2 variants is the main driver of consecutive new waves of the pandemic. The variants that pose an increased risk to global public health are termed variants of interest (VOIs) and variants of concern (VOCs) by the WHO (2021). The WHO in February 2021 defined VOIs as genetic changes that are likely to affect virus characteristics, such as transmissibility, disease severity, immune or treatment escape, and its epidemiological impacts suggest an emerging risk to global public health. VOCs are VOIs that have been demonstrated to be associated with significantly increased transmissibility, virulence or decreased effectiveness of diagnostics, vaccines or therapeutics. According to the WHO, the threshold for the determination of VOCs is high in order to focus attention and resources on the variants with the highest public health implications.

B.1.1.7 lineage or Alpha variant was the first VOC emerging in Europe from December 2020 (Davies *et al.* 2021a). The first Hungarian case was confirmed on 13 January. The lineage was defined with 17 non-synonymous mutations. The most important ones include N501Y (related to increased ACE2 receptor binding affinity) and deletion 69H/70 V in the S gene (Rambaut *et al.* 2020) encoding the surface spike protein of the virus. Increased transmissibility and increased disease severity were both demonstrated for this lineage (Davies *et al.* 2021b).

While lineage B.1.1.7 became dominant by the beginning of February 2021 in most European countries, B.1.351 and P.1 variants also appeared in the continent, but did not reach high prevalence in the first half of 2021 (ECDC 2021). These lineages also contain the N501Y amino acid change, but no 69H/70 V deletion. Both were declared VOC as they were considered vaccine-escape variants since they harbour the E484 K mutation (Wang *et al.* 2021a, 2021b).

Estimating the proportion of variants among the COVID-19 cases requires continuous and representative clinical investigation, which can be very resource-intensive. WBE was suggested as a cost-effective complementary method for the surveillance of virus variants. Whole-genome sequencing (WGS) is the generally accepted method of variant identification. Jahn *et al.* (2021) could detect 10 out of the 17 B.1.1.7 mutations with this method in a ski resort in Switzerland, 1 day before the first reported human case. Other methods are also proposed for variant detection from wastewaters. Allele-specific PCR methods screen for one or few signature mutations. Heijnen *et al.* (2021) used digital droplet PCR for selective quantification for N501Y, a key mutation in B.1.1.7, B.1.351 and P.1. Allele-specific RT-qPCR (Lee *et al.* 2021) and nested RT-PCR (La Rosa *et al.* 2021) were also suggested to be applicable for this purpose.

The aim of the present study is to identify a suitable method to assess the prevalence and proportion of SARS-CoV-2 VOC lineages in Hungary. Hungarian wastewater samples from December 2020 to mid-March 2021 – the period when the lineage B.1.1.7 variant became dominant in Hungary – were screened by multiple methods. A method applicable for the rapid detection of variants from sewage will be an important asset in the preparation for further pandemic waves.

MATERIALS AND METHODS

Sample collection

Raw sewage samples were collected in the wastewater treatment plants (WWTPs) of Budapest ($n = 3$), of all Hungarian county seats ($n = 18$), and a composite sample from five cities in agglomeration of Budapest (Tököl, Biatorbágy, Szigetszentmiklós, Budakeszi and Százhalombatta). Six sampled WWTPs (Budapest Central, Kaposvár, Miskolc, Nyíregyháza, Székesfehérvár and Szombathely) were equipped with built-in composite auto-sampler; from the other sampling points, grab samples were taken at the peak load of the WWTP in the morning hours. Samples were collected by accredited personnel of the WWTPs and transported within 24 h at 4 °C to the laboratory where it was processed within 6 h. Weekly samples from 1 December 2020 (Budapest) or from 15 January 2021 (other sampling points) to 15 March 2021 were investigated in this study (51 composite and 158 grab samples in total). The population served by each WWTP is included in Supplementary Material, Table S1.

Sample preparation and concentration

Sewage samples were concentrated by an in-house ultrafiltration method as described previously (Róka *et al.* 2021). Briefly, samples were sedimented by centrifugation (4,500 g, 30 min, 4 °C), followed by membrane ultrafiltration (Suez Water Technologies & Solutions, Membrane Research Center, Tatabánya, Hungary). In this process, 50 ml sewage is concentrated to 1 ml. External process control (wastewater sample spiked with inactivated SARS-CoV-2) was used to ensure consistency (Róka *et al.* 2021).

RNA extraction

For melting curve genotyping and RT-digital droplet PCR (RT-ddPCR), RNA was isolated using a commercial kit (QIAamp Viral RNA Mini Kit, Qiagen, Germany) in accordance with the manufacturer's instructions. The method yields 30 μl RNA from 140 μl concentrate.

Detection of SARS-CoV-2

For the detection of SARS-CoV-2 RNA, quantitative RT-PCR was conducted on the LightCycler 480 Instrument II (Roche, Switzerland) platform using the LightCycler Multiplex RNA Virus Master kit (Roche, Switzerland). Primers and probes used were specific for the nucleocapsid 1 (N1) gene of the virus (2019-nCoV-N1-F: GAC CCC AAA ATC AGC GAA AT; 2019-nCoV-N1-R: TCT GGT TAC TGC CAG TTG AAT CTG; 2019-nCoV-N1-P: FAM-ACC CCG CAT TAC GTT TGG TGG ACC-BHQ1) (CDC, Atlanta, https://www.cdc.gov/coronavirus/2019-ncov/lab/rt-pcr-panel-primer-probes.html). Virus copy number was calculated based on the standard curve of a commercially available copy control (EURM-019) obtained from the European Commission Joint Research Center, Geel, measured with an automatic second derivative maximum method threshold. The limit of detection (LOD) of the RT-PCR was measured using log10 serial dilution of the EURM-19 copy control.

Melting curve genotyping

For qualitative detection of spike mutations N501Y and del H69/V70, the VirSNiP SARS-CoV-2 Spike N501Y and Spike del H69/V70 assays were used (TIB MOLBIOL, Berlin, Germany). The genotyping RT-PCR and melting analysis were conducted on the LightCycler 480 Instrument II platform using LightCycler Multiplex RNA Virus Master kit in 20 μL final reaction volume and using simple probe detection format. For the melting curve analysis, Tm calling was performed using the Standard ModularDX Program for Roche LC480 Instruments. The Tm temperature of the samples was compared to the N501/501Y and del H69/V70 positive and negative controls and to the RT-PCR negative controls.

RT-ddPCR for mutation detection

For simultaneous quantitative detection of the signature mutations of the B.1.1.7 variant and wild-type of the same positions, ddRT-PCR mutation detection assays were used (Bio-Rad, CA, USA). For 501Y and the wild-type variant (N501) (assay ID: dMDS731762551), for the del69H/70 V and the wild-type variant (no deletion) (assay ID: dMDS284738817), primers-probes were used. The 22 μl reaction volume from One-Step RT-ddPCR Advanced Kit for Probes contained 5.5 μl one-step RT-ddPCR Supermix, 2.2 μl reverse transcriptase, 1.1 μl 300 mM DTT, 1 μl mutation detection assay, 7.2 μl water and 5.0 μl purified RNA. Droplets were generated using the Automated Droplet Generator (Bio-Rad, CA, USA), and RT-ddPCR was carried out on the Bio-Rad T100 Thermal Cycler platform (25 °C 30 min pre-incubation, 50 °C 60 min reverse transcription, 95 °C 10 min enzyme activation, 40 cycles of 95 °C 30 s denaturation and 55 °C 60 s annealing/extension; 98 °C 10 min enzyme deactivation and overnight droplet stabilisation at 4 °C; ramp rate was decreased to 2.0 °C/s in all steps). The RT-ddPCR product read-out was performed using the QX200 system (Bio-Rad) and analysed using the QuantaSoft™ Analysis Pro 1.0.596 software (Bio-Rad). Wild-type, B.1.1.7 lineage and non-template control were used in every reaction.

The LOD and the limit of quantification (LOQ) were determined using replicates of 10-fold dilutions of purified SARS-CoV-2 RNA (same applied as positive amplification control in RT-qPCR). The LOD was defined as the last serial dilution detected in 95% of replicates and LOQ as the lowest dilution showing a coefficient of variation percentage below the threshold (CV% = 25) (Rocchigiani *et al.* 2020). The calculated value both for the LOD and the LOQ was 0.703 copies/μl (Supplementary Material, Table S3).

Sewage data

The *Enterococcus* count of the sewage samples was determined according to ISO 7899-1 standard methods. MUD/SF plates were obtained from Bio-Rad (CA, USA). Plates were incubated at 44 °C for 48 h and read under UV light.

Sewage flow data (daily volume) were provided by the WWTP operators (data shown in Supplementary Material, Table S2).

Epidemiological data

The total number of new COVID-19 cases was obtained from the National Surveillance Database of National Public Health Centre, and the number of VOC-infected cases (B.1.1.7, B.1.351 and P.1) was obtained from the database of the COVID Laboratory Working Group of the National Public Health Centre.

Statistical analysis

Microsoft Excel 2010 (version: 14.0.7232.5000) was used for the statistical analysis of data. Pearson's correlation coefficient was calculated for comparing the concentration of the two investigated mutations and investigating the association of clinical B.1.1.7 cases and the variant's concentration in sewage. Linear regression was used to confirm the association between the SARS-CoV-2 gene copy numbers in sewage and the weekly number of new cases.

RESULTS

SARS-CoV-2 concentration

During the monitored period, the SARS-CoV-2 concentration in the wastewater samples decreased slowly until the third week of 2021, parallel to the decline of the second wave. From week 4 in 2021, a rapid increase was observed (Figure 1; Supplementary Material, Table S2). The comparison of the SARS-CoV-2 genome copy numbers and the weekly average of new cases by linear regression confirmed that trends sewage data precede the changes in the case number by approximately 1 week ($R = 0.843$, $p = 0.0002$).

Melting curve genotyping

Melting curve genotyping is only suitable for the qualitative detection of mutations. However, if different variants are present, multiple melting temperatures are also detectable. Wastewater samples spiked with heat-inactivated SARS-CoV-2 (using confirmed B.1.1.7 lineage and wild-type viruses in a final concentration of approximately 10^6 GC/l) were used to confirm the suitability of the method for wastewaters. During melting curve analysis, multiple peaks of melting temperatures were observed in sewage samples spiked with both isolates (B.1.1.7 and wild-type in 1:1 proportion).

Twenty-one raw sewage samples (6 composite and 15 grab samples) from the 6–7th week of 2021 were analysed with this method. Wild-type N501 was detected in two samples, N501Y genotype in nine samples and both variants in two samples. In eight cases, N501Y detection PCR was unsuccessful. Screening for del H69/V70 was only successful in five cases, of which two samples

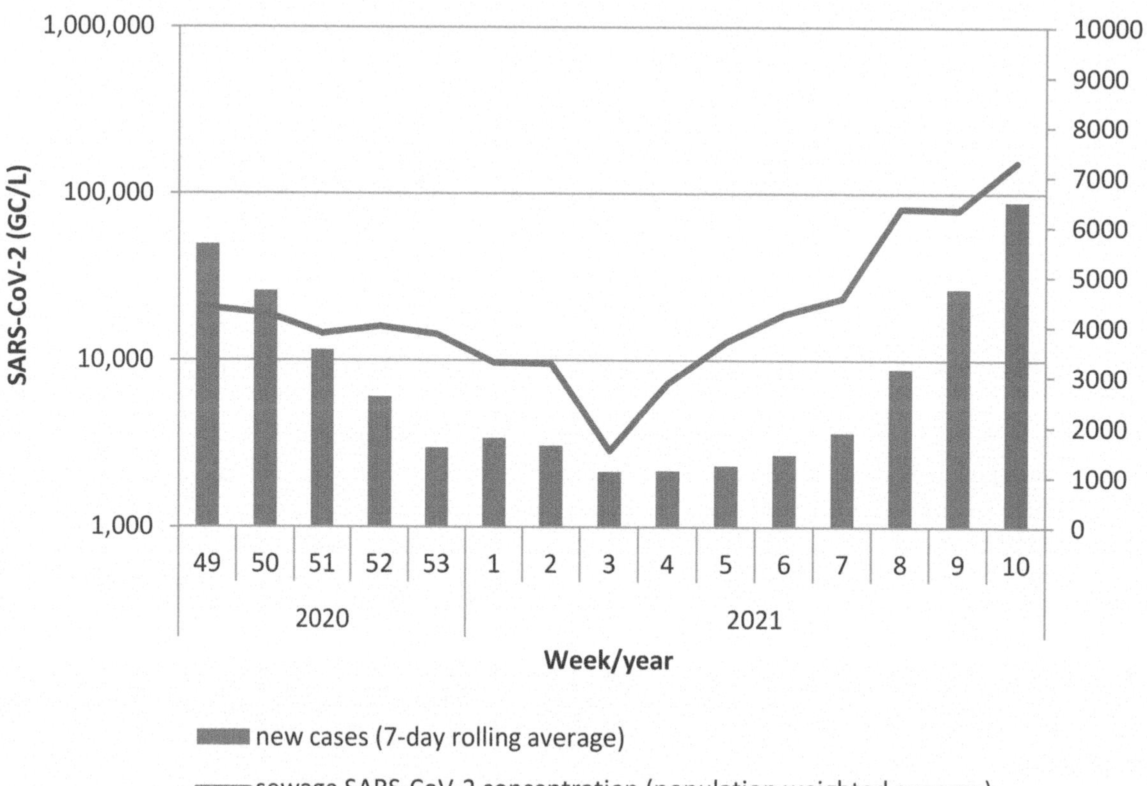

Figure 1 | Population weighted average of SARS-CoV-2 concentrations and 7-day rolling average (from Thursday to Wednesday) of new COVID-19 cases.

showed wild-type genotype, two samples contained the deletion and one sample contained both variants (Supplementary Material, Table S2).

RT-digital droplet PCR

The method allows parallel quantification of the mutant and wild-type of a genomic locus in the S gene (N501Y or delH69/V70). 212 samples (47 composite and 165 grab samples) obtained between weeks 49/2020 and 10/2021 (in the case of Budapest) and between weeks 3/2021 and 10/2021 (at other sampling sites) were analysed by RT-ddPCR.

The LOD for different variants – as a percentage of the total concentration – was estimated based on the process control samples. To determine the virus concentration of the process control in the wastewater, a natural wastewater sample spiked with heat-inactivated, wild-type SARS-CoV-2 was used. By using this method, we were able to efficiently recover other lineages present naturally in low concentrations (0.5–5%) in the spiked sample.

Wastewater samples from Budapest were investigated for a longer period, as earlier emergence of new types was expected in the capital based on clinical data. The results show high levels of the wild-type virus at the beginning of December. The concentration of the wild-type strain started to decrease from mid-December and continued to decrease over the study period. The first part (until week 4 of 2021) of the observed trend is associated with the decline of the second wave of the outbreak but after that, the cumulated SARS-CoV-2 concentration started to increase (see Figure 1). Hence, the further decrease of the wild-type virus is due to the emergence of other variants. Both of the monitored spike mutations occurred at low (but not zero) concentrations at the beginning of December and increased rapidly from the beginning of February (Figure 2(a)). Comparing the proportion of the mutant and wild-type variants, the first emergence of N501Y and del H69/V70 can be observed from the third week of 2021 (Figure 2(b)), and by the end of February, it became the variant almost exclusively present in wastewaters (Figure 2(b)).

Similar trends were observed in every WWTP sampled in the national surveillance (Table 1). The average proportion of variants N501Y and del H69/V70 was nearly zero in mid-January and become almost exclusively present by mid-March. However, several weeks of discrepancy are observed in the spread of the new variant across the country.

B.1.1.7 cases and signature mutations in the wastewater

The ratio of N501Y+del69H/V70 mutations in the clinical samples analysed by the National Public Health Center COVID Laboratory Working Group was similar to that observed in wastewater in the investigated period (Figure 3). Clinical samples positive for N501Y+del69H/V70 mutations by the VirSNiP Mutation Assays were also confirmed partially by WGS, indicating the B.1.1.7 variant present in 100% of the tested samples. Presumptive lineage B.1.1.7 was accountable for approximately 10% of all cases in mid-January 2021 and reached 90% by mid-March.

DISCUSSION

In the present study, two methods were tested for the detection of SARS-CoV-2 VOCs in wastewater, melting curve genotyping and RT-ddPCR. The detection of N501Y or wild-type N501 was successful with both methods from 13 samples. The two samples showing wild-type N501 according to melting curve genotyping contained 6–13% N501Y according to RT-ddPCR, while N501Y positive samples contained 56–93% N501Y. The two samples with multiple maximums of melting temperature showed 64 and 94% N501Y according to RT-ddPCR. Similar results can be observed by the H69/V70 deletion screening methods (Supplementary Material, Table S2).

The advantages of the melting curve genotyping method are the low cost, rapid identification and easy access to the materials. For these reasons, investigations were carried out nearly in real-time in February, from fresh samples. However, low sensitivity and lack of quantitative data are strong limitations of melting curve genotyping.

The results of RT-ddPCR indicate a rapid dispersion of the two signature mutations of the B.1.1.7 variant in the country. The variants first emerged in Budapest and its agglomeration but appeared in all WWTPs within 4 weeks. The onset of the emergence of the variants did not show a clear correlation either with the size of the population serviced by the WWTP or the geographical location. The increasing trend was consistent in most WWTPs, though there were occasional outliers. The reason behind unexpectedly high (e.g. Szeged, 3rd week or Győr, 4th week) or unexpectedly low (e.g. Debrecen, 9th week, Veszprém, 8th week or Zalaegerszeg, 10th week) VOC ratios is not clear. One potential explanation is the inherent heterogeneity of the wastewater which may lead to a bias, especially in the case of the grab sampling. Outlying results were more frequent in those WWTPs where an automated sampler was not available.

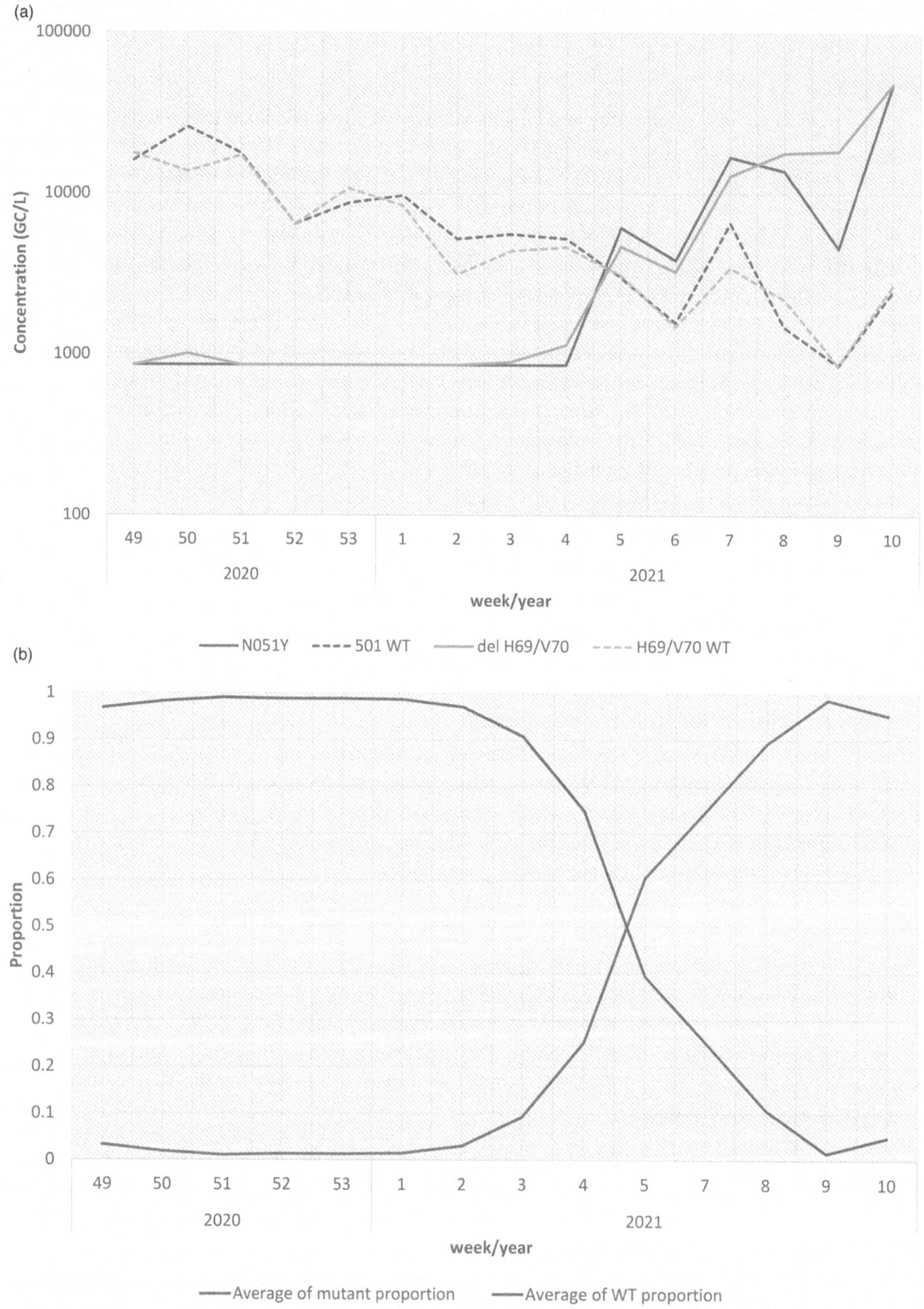

Figure 2 | (a) Weekly changes of the emergence of signature mutations of the S gene specific for the B.1.1.7 variant in Budapest (average if three WWTPs). (b) Changes in the ratio of the wild-type and presumptive B.1.1.7 SARS-CoV-2 variants in Budapest wastewater (average of three WWTPs) between December 2020 and March 2021. Presumptive B.1.1.7 genome copy numbers were calculated as the average of N501Y and del H69/V70 copy numbers, while wild-type was expressed as the average of the non-mutated copy numbers of the two sites.

Table 1 | Average of N501Y and del H69/V70 proportion in the investigated WWTPs

week	Békéscsaba	Budapest	Debrecen	Eger	Győr	Kaposvár	Kecskemét	Miskolc	Nyíregyháza	Pécs	Salgótarján	Szeged	Székesfehérvár	Szekszárd	Szolnok	Szombathely	Tatabánya	Budapest agglomeration*	Veszprém	Zalaegerszeg
3	0.00	0.09	0.32		0.00	0.16	0.00	0.00	0.06	0.00	0.08	0.09	0.00	0.10	0.00	0.00	0.23	0.00	0.00	0.09
4	0.00	0.25	0.03	0.00	0.53	0.23	0.00	0.05	0.34	0.00	0.03	0.39	0.03	0.00	0.00	0.00	0.29	0.58	0.00	0.00
5	0.00	0.61	0.31	0.00	0.18	0.50	0.00	0.03		0.18	0.48	0.14	0.00	0.27	0.08	0.00	0.50	0.71	0.00	0.68
6			0.20	0.00				0.00	0.75			0.67	0.11	0.06	0.20	0.44	0.83	0.92	0.00	0.45
7	0.75	0.75	0.83	0.83	0.90	0.66	0.00	0.44	0.95	0.91	0.40	0.77	0.56	0.20	0.03	0.71	0.76	0.84	0.75	
8		0.89	0.94	0.59	0.82	0.88	0.85	0.66	0.71	0.85	0.89	0.89	0.64	0.81	0.43	0.82	0.55	0.99	0.43	0.91
9	0.78	0.99	0.63	1.00	0.94	0.91		0.66	0.99	0.90	0.91	0.96	0.95	0.92	0.66	0.91	0.98	1.00	0.91	
10	0.75	0.95	0.93	1.00	0.92	1.00	0.93	0.90	0.95	0.95	0.91	0.96	0.97	0.98	0.82	0.97	0.94	0.97	0.84	0.41

□ no data. The proportion of presumptive B.1.1.7 lineage in sewage sample (average ratio of N501 and delH69/V70).

*The mixed sample from five cities (Tököl, Biatorbágy, Szigetszentmiklós, Budakeszi and Százhalombatta).

The cells are coloured on a colour scale indicating the presence of different variants (green, only wild type is present; from light green to orange, both types are present, with increasing ratio of mutants; red, only mutants are present).

Please refer to the online version of this paper to see this table in colour: http://dx.doi.10.2166/wh.2022.179.

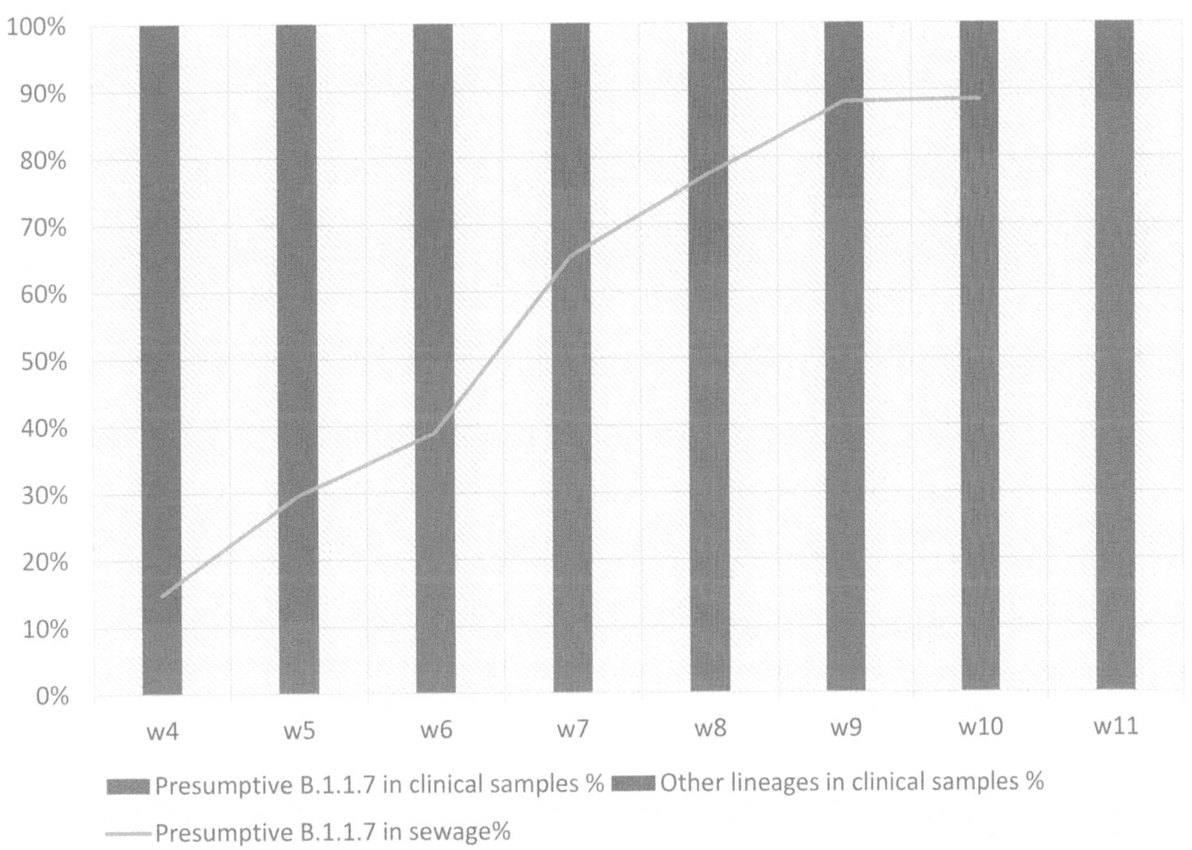

Figure 3 | Percentage of presumptive lineage B.1.1.7 (based on N501Y and del H69/V70 mutations screening) in clinical samples and the same mutations measured in wastewater (national average).

The frequency of B.1.1.7 variant cases and the ratio of its signature mutations in sewage increased rapidly between mid-January and mid-March 2021 (see Figure 3). Though clinical results are not representative for the entire country (figures are based on clinical samples tested in NPHC, which represents approximately 3–15% of the new daily cases in the country), the correlation is statistically significant (Pearson's correlation coefficient, $R = 0.921$, $p < 0.05$). Other studies using RT-ddPCR (Heijnen *et al.* 2021) or sequencing (Jahn *et al.* 2021) for variant analysis in sewage have also indicated that the ratio of different SARS-CoV-2 variants represents well the distribution of different lineages in clinical cases.

The two investigated mutations are key changes in the B.1.1.7 lineage but it is not defined exclusively by them: N501Y can also be found in B.1.351 and P1 lineages, while B.1.1.7 also harbours several other non-synonymous mutations (Rambaut *et al.* 2020). The RT-ddPCR method detects the two mutation sites in two separate reactions; therefore, it is not automatically evident that they are in present the same viral genomes. However, the prevalence of N501Y and del 69H/70 V was closely linked in the investigated samples (Pearson's correlation, $R = 0.9153$, $p < 0.05$; Figure 4).

A less clear correlation was observed between the quantitative RT-qPCR (CDC N1 primer-probe set) and RT-ddPCR (S gene) results (Pearson's correlation, $R = 0.7173$, $p < 0.05$; Figure 5). Heijnen *et al.* (2021) found a very similar correlation between the two datasets (RT-qPCR with N2 primer set compared to RT-ddPCR with N501Y/N501 WT primer set). The total concentration of different variants measured by RT-ddPCR was lower than the qPCR results by an average of 62% (average of N501Y+N501 WT and del H69/V70+wild-type without deletion). The difference is probably due to the different quantification methods and the different target genes of the two types of PCR. A similar difference was seen in RT-ddPCR and RT-qPCR comparison with the same primer set (N1, data not shown).

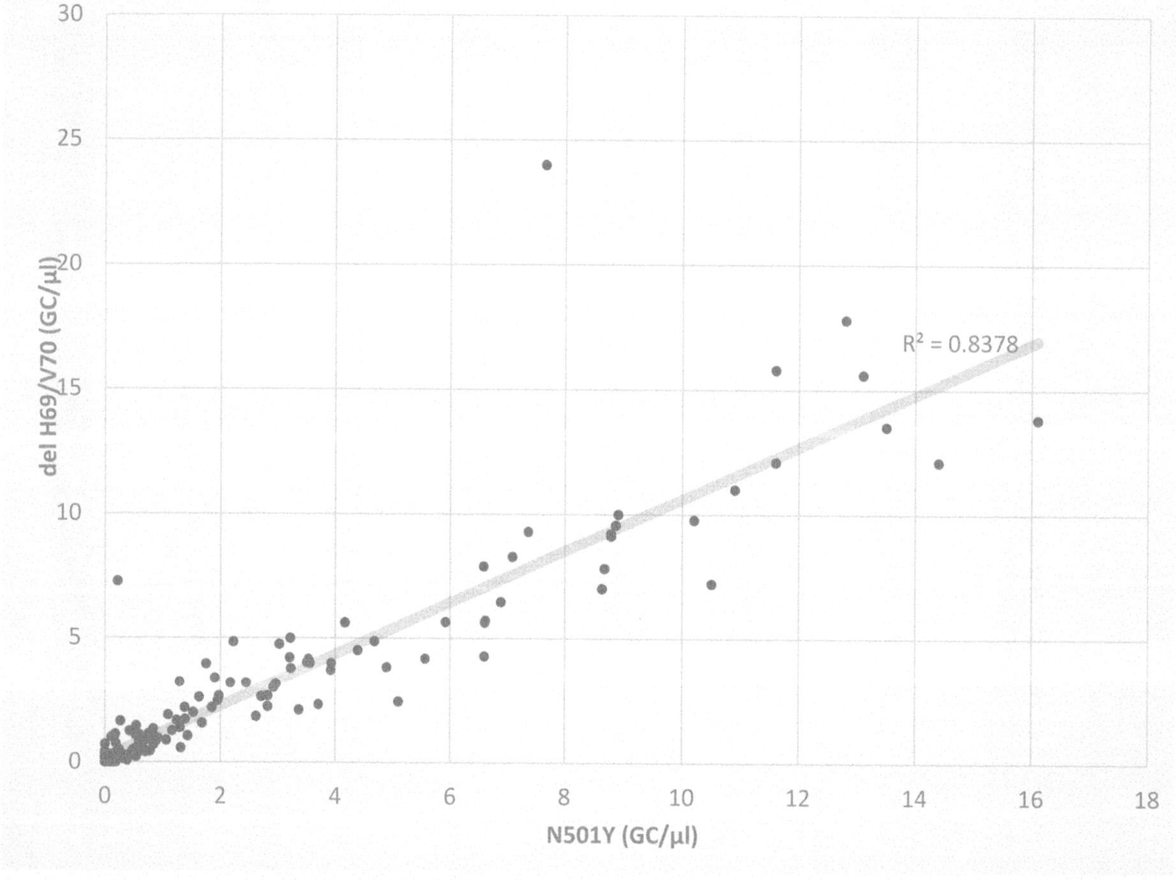

Figure 4 | Correlation between N501Y and del H69/V70 concentrations (GC/µl, in RT-ddPCR reaction).

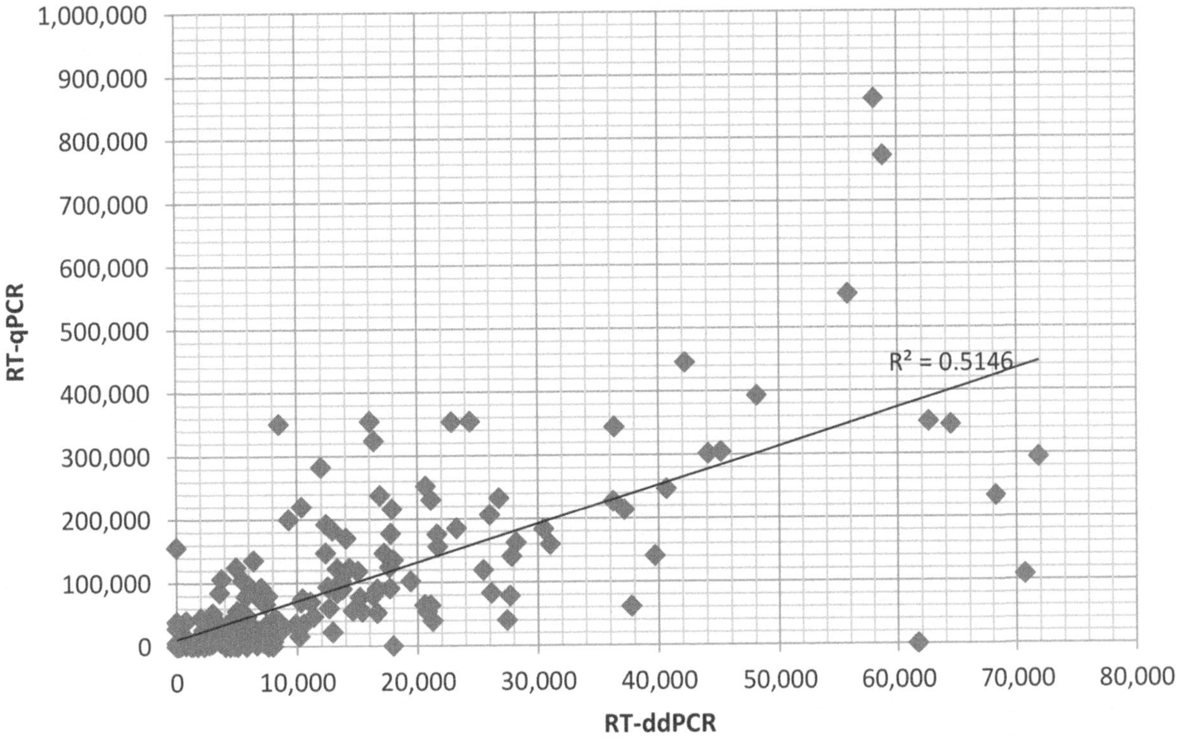

Figure 5 | Correlation between RT-qPCR results (N1 primer set) and RT-ddPCR (average of N501Y+N501WT and del H69/V70+WT without deletion).

CONCLUSION

Results of the present study confirmed that RT-ddPCR-based methods are rapid and reliable tools for tracking VOCs in sewage, and the results are also indicative for community circulation of different variants. It is more convenient for surveillance than sequencing due to its lower time demand (6 h vs. several days) and cost-efficiency, especially in resource-limited settings.

Though RT-ddPCR uses signature mutations for variant detection, and not the full mutation profile, the close correlation of the two mutations' concentration and the association with the clinical data strongly suggest that their presence was due to the emergence of the B.1.1.7 variant. WGS and quasi-species analysis could provide further confirmation.

The tested melting point genotyping method has limited potential for the analysis of variants in sewage and should only be used in the absence of the other available options.

Recent developments in PCR-based methods for VOC screening in wastewaters offer solutions for the limitation of the currently available techniques. Lee *et al.* (2021) recently published a method for allele-specific RT-qPCR for three genomic loci predictive for B.1.1.7 (HV69/70del, Y144del and A570D), while La Rosa *et al.* (2021) used nested RT-PCR assays for rapid variant screening. These recently published methods seem to be more sensitive for mutations, and the RT-qPCR method is also suitable for quantitative analysis. However, rapid, reliable and quantitative variant analysis in sewage might still require a combination of different techniques (e.g. allele-specific RT-qPCR or RT-ddPCR and WGS).

DATA AVAILABILITY STATEMENT

All relevant data are included in the paper or its Supplementary Information.

REFERENCES

Davies, N. G., Abbott, S., Barnard, R. C., Jarvis, C. I., Kucharski, A. J., Munday, J. D., Pearson, C. A. B., Russell, T. W., Tully, D. C., Washburne, A. D., Wenseleers, T., Gimma, A., Waites, W., Wong, K. L. M., van Zandvoort, K., Silverman, J. D., CMMID COVID-19 Working Group, COVID-19 Genomics UK (COG-UK) ConsortiumDiaz-Ordaz, K., Keogh, R., Eggo, R. M., Funk, S., Jit, M., Atkins, K. E. & Edmunds, W. J. 2021a Estimated transmissibility and impact of SARS-CoV-2 lineage B.1.1.7 in England. *Science* **372**, eabg3055. https://doi.org/10.1126/science.abg3055.

Davies, N. G., Jarvis, C. I., CMMID COVID-19 Working GroupEdmunds, W. J., Jewell, N. P., Diaz-Ordaz, K. & Keogh, R. H. 2021b Increased mortality in community-tested cases of SARS-CoV-2 lineage B.1.1.7. *Nature* **593**, 270–274. https://doi.org/10.1038/s41586-021-03426-1.

ECDC 2021 *European Centre for Disease Prevention and Control: Variants of interest and concern in the EU/EEA*. Available from: https://www.ecdc.europa.eu/en/covid-19/situation-updates/variants-dashboard (accessed 30 July 2021).

European Commission 2021 *Commission Recommendation of 17.3.2021 on a Common Approach to Establish a Systematic Surveillance of SARS-CoV-2 and Its Variants in Wastewaters in the EU*. Available from: https://ec.europa.eu/environment/pdf/water/recommendation_covid19_monitoring_wastewaters.pdf (accessed 30 July 2021).

Heijnen, L., Elsinga, G., de Graaf, M., Molenkamp, R., Koopmans, M. P. G. & Medema, G. 2021 Droplet digital RT-PCR to detect SARS-CoV-2 variants of concern in wastewater. Preprint. Infectious Diseases (Except HIV/AIDS). https://doi.org/10.1101/2021.03.25.21254324.

Jahn, K., Dreifuss, D., Topolsky, I., Kull, A., Ganesanandamoorthy, P., Fernandez-Cassi, X., Bänziger, C., Devaux, A. J., Stachler, E., Caduff, L., Cariti, F., Corzón, A. T., Fuhrmann, L., Chen, C., Jablonski, K. P., Nadeau, S., Feldkamp, M., Beisel, C., Aquino, C., Stadler, T., Ort, C., Kohn, T., Julian, T. R. & Beerenwinkel, N. 2021 Detection and surveillance of SARS-CoV-2 genomic variants in wastewater. Preprint. Infectious Diseases (Except HIV/AIDS). https://doi.org/10.1101/2021.01.08.21249379.

La Rosa, G., Mancini, P., Bonanno Ferraro, G., Veneri, C., Iaconelli, M., Lucentini, L., Bonadonna, L., Brusaferro, S., Brandtner, D., Fasanella, A., Pace, L., Parisi, A., Galante, D. & Suffredini, E. 2021 Rapid screening for SARS-CoV-2 variants of concern in clinical and environmental samples using nested RT-PCR assays targeting key mutations of the spike protein. *Water Research* **197**, 117104. https://doi.org/10.1016/j.watres.2021.117104.

Lee, W. L., Imakaev, M., Armas, F., McElroy, K. A., Gu, X., Duvallet, C., Chandra, F., Chen, H., Leifels, M., Mendola, S., Floyd-O'Sullivan, R., Powell, M. M., Wilson, S. T., Berge, K. L. J., Lim, C. Y. J., Wu, F., Xiao, A., Moniz, K., Ghaeli, N., Matus, M., Thompson, J. & Alm, E. J. 2021 Quantitative SARS-CoV-2 Alpha variant B.1.1.7 tracking in wastewater by allele-specific RT-qPCR. *Environmental Science & Technology Letters*. acs.estlett.1c00375. https://doi.org/10.1021/acs.estlett.1c00375.

Medema, G., Been, F., Heijnen, L. & Petterson, S. 2020 Implementation of environmental surveillance for SARS-CoV-2 virus to support public health decisions: opportunities and challenges. *Current Opinion in Environmental Science & Health* **17**, 49–71. https://doi.org/10.1016/j.coesh.2020.09.006.

Rambaut, A., Loman, N., Pybus, O., Barclay, W., Barrett, J., Carabelli, A., Connor, T., Peacock, T., Robertson, D. L. & Volz, E. 2020 *Preliminary Genomic Characterisation of an Emergent SARS-CoV-2 Lineage in the UK Defined by a Novel Set of Spike Mutations*. Available from: https://virological.org/t/preliminary-genomic-characterisation-of-an-emergent-sars-cov-2-lineage-in-the-uk-defined-by-a-novel-set-of-spike-mutations/563 (accessed 30 July 2021).

Rocchigiani, A. M., Tilocca, M. G., Portanti, O., Vodret, B., Bechere, R., Di Domenico, M., Savini, G., Lorusso, A. & Puggioni, G. 2020 Development of a digital RT-PCR method for absolute quantification of bluetongue virus in field samples. *Frontiers in Veterinary Science* **7**, 170. https://doi.org/10.3389/fvets.2020.00170.

Róka, E., Khayer, B., Kis, Z., Kovács, L. B., Schuler, E., Magyar, N., Málnási, T., Oravecz, O., Pályi, B., Pándics, T. & Vargha, M. 2021 Ahead of the second wave: early warning for COVID-19 by wastewater surveillance in Hungary. *Science of the Total Environment* **786**, 147398. https://doi.org/10.1016/j.scitotenv.2021.147398.

Sims, N. & Kasprzyk-Hordern, B. 2020 Future perspectives of wastewater-based epidemiology: monitoring infectious disease spread and resistance to the community level. *Environment International* **139**, 105689. https://doi.org/10.1016/j.envint.2020.105689.

Walsh, K. A., Jordan, K., Clyne, B., Rohde, D., Drummond, L., Byrne, P., Ahern, S., Carty, P. G., O'Brien, K. K., O'Murchu, E., O'Neill, M., Smith, S. M., Ryan, M. & Harrington, P. 2020 SARS-CoV-2 detection, viral load and infectivity over the course of an infection. *Journal of Infection* **81**, 357–371. https://doi.org/10.1016/j.jinf.2020.06.067.

Wang, P., Nair, M. S., Liu, L., Iketani, S., Luo, Y., Guo, Y., Wang, M., Yu, J., Zhang, B., Kwong, P. D., Graham, B. S., Mascola, J. R., Chang, J. Y., Yin, M. T., Sobieszczyk, M., Kyratsous, C. A., Shapiro, L., Sheng, Z., Huang, Y. & Ho, D. D. 2021a Antibody resistance of SARS-CoV-2 variants B.1.351 and B.1.1.7. *Nature* **593**, 130–135. https://doi.org/10.1038/s41586-021-03398-2.

Wang, P., Casner, R. G., Nair, M. S., Wang, M., Yu, J., Cerutti, G., Liu, L., Kwong, P. D., Huang, Y., Shapiro, L. & Ho, D. D. 2021b Increased resistance of SARS-CoV-2 variant P.1 to antibody neutralization. *Cell Host & Microbe* **29**, 747–751.e4. https://doi.org/10.1016/j.chom.2021.04.007.

Weidhaas, J., Aanderud, Z. T., Roper, D. K., VanDerslice, J., Gaddis, E. B., Ostermiller, J., Hoffman, K., Jamal, R., Heck, P., Zhang, Y., Torgersen, K., Laan, J. V. & LaCross, N. 2021 Correlation of SARS-CoV-2 RNA in wastewater with COVID-19 disease burden in sewersheds. *Science of the Total Environment* **775**, 145790. https://doi.org/10.1016/j.scitotenv.2021.145790.

WHO 2020 *Public Health Surveillance for COVID-19 Interim Guidance 7 August 2020*. Available from: https://www.who.int/publications/i/item/who-2019-nCoV-surveillanceguidance-2020.7 (accessed 30 July 2021).

WHO 2021 *Tracking SARS-CoV-2 Variants*. Available from: https://www.who.int/en/activities/tracking-SARS-CoV-2-variants/ (accessed 30 July 2021).

Wu, F., Zhang, J., Xiao, A., Bonneau, R., Brown, M. A., Bushman, M., Gu, X., Huang, K. H., Xiaoqiong, G., Lee, W. L., Armas, F., Kauffman, K., Hanage, W. P., Matus, M., Ghaeli, N., Endo, N., Duvallet, C., Poyet, M., Moniz, K. S., Washburne, A. D., Erickson, T. B., Chai, P. R., Thompson, J. & Alm, E. J. 2020 SARS-CoV-2 titers in wastewater are higher than expected from clinically confirmed cases. *mSystems* **5** (4), e00614-20. https://dx.doi.org/10.1128%2FmSystems.00614-20.

First received 31 July 2021; accepted in revised form 11 January 2022. Available online 31 January 2022

SARS-CoV-2 in Water doi: 10.2166/wh.2022.218

Wastewater surveillance of SARS-CoV-2 in Austria: development, implementation, and operation of the Tyrolean wastewater monitoring program

Beatrice Daleiden[a,†], Harald Niederstätter[a,†], Martin Steinlechner [a], Stefan Wildt[b], Manfred Kaiser[b], Cornelia Lass-Flörl [c], Wilfried Posch [c], Stefan Fuchs [c], Bernhard Pfeifer [d], Andreas Huber[d] and Herbert Oberacher [a,*]

[a] Institute of Legal Medicine and Core Facility Metabolomics, Medical University of Innsbruck, Innsbruck, Austria
[b] Amt der Tiroler Landesregierung, Innsbruck, Austria
[c] Institute of Hygiene and Medical Microbiology, Medical University of Innsbruck, Innsbruck, Austria
[d] State Institute for Integrated Care Tyrol, Innsbruck, Austria
*Corresponding author. E-mail: herbert.oberacher@i-med.ac.at
†The first two authors contributed equally to this work.

MS, 0000-0003-1622-1572; CL-F, 0000-0002-2946-7785; WP, 0000-0001-8955-7654; SF, 0000-0001-8519-0701; BP, 0000-0003-1507-1015; HO, 0000-0002-0963-8268

ABSTRACT

Wastewater-based epidemiology (WBE) is an effective approach for tracking information on spatial distribution and temporal trends of severe acute respiratory syndrome coronavirus 2 (SARS-CoV-2) at the community level. Herein, the development, implementation, and operation of the wastewater monitoring program serving Tyrol – a federal province of Austria – are described. The development of this program was initiated by Tyrolean health authorities at the end of the first phase of the Coronavirus disease 2019 (COVID-19) pandemic (May 2020). In close co-operation with the water sector and academic institutions, efficient and effective workflows and processes for wastewater surveillance were established. The monitoring program went into operation in November 2020. By the end of July 2021, a total of 5,270 wastewater influent samples collected at 43 sites were analyzed. The monitoring program provided valuable insights into the development of the pandemic situation in Tyrol and fulfilled several tasks that are of importance in different phases of the pandemic. It represented an early-warning system, provided independent confirmation of temporal trends in COVID-19 prevalence, enabled the assessment of the effectiveness of measures, alerted about bursts of disease activity, and provided evidence for the absence of COVID-19. These findings underline the importance of establishing national wastewater monitoring programs as a complementary source of information for efficient and effective pandemic management.

Key words: COVID-19, SARS-CoV-2, sewage, Tyrol, wastewater-based epidemiology, wastewater-based surveillance

HIGHLIGHTS

- The development, implementation, and operation of the Tyrolean SARS-CoV-2 wastewater monitoring program are described.
- Health authorities are receiving regular updates on the pandemic situation in 43 Tyrolean regions.
- Turnaround times are <36 hours.
- Lead times of 3–7 days were observed.
- Ten positive cases per 120,000 persons were detectable.

GRAPHICAL ABSTRACT

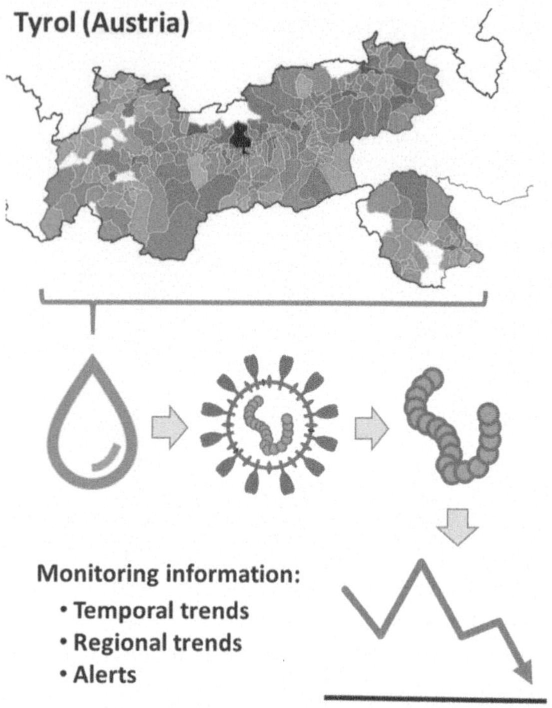

INTRODUCTION

Wastewater-based epidemiology (WBE) is a competent and effective approach for tracking information on spatial distribution and temporal trends of biological and chemical agents at the community level (Wigginton *et al.* 2015; Choi *et al.* 2018; Mao *et al.* 2020; Medema *et al.* 2020a; Boogaerts *et al.* 2021a). WBE involves quantitative analysis of excreted biomarkers in population-pooled wastewater samples collected at or upstream of the influent to a wastewater treatment plant (WWTP). Measured loads are converted into information on exposure, use, or spread of agents in the monitored community providing an unbiased reflection of key aspects of public health.

WBE has successfully been used to detect polioviruses and inform eradication for several decades (Asghar *et al.* 2014). In the current pandemic of Coronavirus disease 2019 (COVID-19), the potential use of environmental surveillance in detecting severe acute respiratory syndrome coronavirus 2 (SARS-CoV-2) circulation in communities has repeatedly been demonstrated (Ahmed *et al.* 2020a; Medema *et al.* 2020b; Wu *et al.* 2020). A summary of relevant publications is provided by a global collaborative at https://www.covid19wbec.org/publication-map (Bivins *et al.* 2020). The strengths of wastewater surveillance can be summarized as follows: (1) wastewater is a pooled sample providing information on virus circulation in the whole community served by the WWTP; (2) infected individuals, including symptomatic, presymptomatic, paucisymptomatic, and asymptomatic carriers, contribute to the total load of SARS-CoV-2 in wastewaters by shedding viruses or their RNA in stool, urine, or respiratory secretions (Jones *et al.* 2020; Weiss *et al.* 2020; Wölfel *et al.* 2020); (3) WBE provides information on spatial and temporal trends; (4) as infected individuals may shed the virus in stool before the infection is detected by a diagnostic test, wastewater signals precede the rise and fall of prevalences, hospitalizations, and intensive care unit admissions (Hillary *et al.* 2021; Wu *et al.* 2021; Rusinol *et al.* 2021); and (5) WBE is more inexpensive and less resource-intensive than diagnostic testing of individuals at a large scale (Thompson *et al.* 2020). Accordingly, wastewater surveillance may complement other well-established techniques for COVID-19 surveillance, such as clinical-based surveillance, hospital admission data, and mortality and morbidity rates.

Health authorities, such as the World Health Organization (WHO 2020), the United States Centers for Disease Control and Prevention (CDC 2021), and the European Commission (EC 2021), have recognized the potential usefulness of WBE as an instrument for pandemic management. Accordingly, all around the world, monitoring programs have been initiated. One of the first monitoring programs in operation was established in Tyrol and is presented here.

Tyrol is a federal province of Austria located in the center of Europe and has a population of 760 thousand. Austria, and in due consequence Tyrol, operates a highly developed healthcare system, which includes an epidemiological surveillance program. Tyrol is one of the leading holiday destinations for winter sports. Approximately 5 million tourists have spent their winter holidays in Tyrol throughout the 2019/2020 season. During the first phase of the pandemic in Tyrol (February–May 2020), large number of tourists in combination with limited experience, information, and measures led to an increased risk of viral spreading in skiing resorts. Thus, Tyrol has potentially acted as a superspreading transmission hub (Popa *et al.* 2020) being responsible for cases exported to different European countries, including Germany (Correa-Martinez *et al.* 2020). A due consequence of these early events was an expansion of available surveillance and response capacities. One important activity initiated by the Tyrolean health authorities in May 2020 was the establishment of a wastewater monitoring program.

In this report, concepts, structure, and results of the Tyrolean wastewater monitoring program are presented. The health and water sector jointly worked together in the conception, development, and operation of this program. Further support was provided by academic institutions. The discovery phase started in May 2020. Feasibility of the approach was demonstrated by the end of August 2020. A first regional monitoring program involving nine WWTPs was launched in September 2020. The monitoring program involving 43 WWTPs was put into operation in mid-November 2020. By the end of July 2021, SARS-CoV-2 concentrations were determined in 5,270 Tyrolean wastewater samples. The existing data support the important role of SARS-CoV-2 wastewater monitoring in complementing clinical disease surveillance. WBE acts as an early-warning system indicating the resurgence of COVID-19 cases, provides independent confirmation of temporal trends in COVID-19 prevalence, and alerts about bursts of disease activity.

METHODS

The monitoring program is accompanied by research and development programs for providing new and advanced features with immediate application in the operational work. Consequently, the workflow was revised several times between April 2020 and March 2021. Important technological milestones included (1) the automation of RNA extraction from precipitates in May 2020, including a platform change in September 2020, (2) an increase in the processed wastewater volume from 50 to 250 mL by the end of July 2020, (3) the implementation of an internal amplification positive control (IPC) by the end of October 2020, (4) the implementation of a Pepper Mild Mottle Virus (PMMoV) target as a process control in mid-February 2021; (5) pipetting samples and reagents by a robotic system in mid-February 2021, and (6) quantification with linear RNA standards and reference material. In the following sections, only the final setup is described.

Sampling sites, wastewater sampling, and transport

Influent samples were collected from 43 Tyrolean WWTPs (Supplementary Material, Figure S1 and Table S1). Samples were collected as 24-hour composite samples using cooled autosamplers (4 °C) operated in a volume proportional mode. The daily flows of wastewater were measured online with sensors. The chemical oxygen demands (CODs) were directly determined on-site. About 500 mL aliquots were filled in polypropylene bottles, which were stored at 4 °C until transportation. The bottles were labeled with the name of the sampling location and the sampling date. The transport of the collected samples to the laboratory was accomplished by a courier service. Samples were transported in coolers (2–8 °C).

Wastewater concentration and automated nucleic acid extraction

Wastewater samples were processed within 24 h of arrival. They were stored at 4 °C until analysis. The protocols for wastewater sample pretreatment and PEG-NaCl-based precipitation of the virus fraction were based on the methods described by Ye *et al.* (2016) and Wu *et al.* (2020). For centrifugation, we used an Eppendorf 5910 R centrifuge equipped with a FA-6×250 fixed angle rotor (both Eppendorf, Hamburg, Germany) and 250 mL polycarbonate centrifuge bottles (TFS; Thermo Fisher Scientific, Waltham, MA, USA). The temperature was set at 4 °C, and the brake was deactivated unless stated otherwise. In a first step, 250 mL aliquots of the wastewater samples were cleared by centrifugation at 4,500×g for 30 min. In 225 mL of the resulting supernatant, 22.5 g of PEG 8000 and 5.1 g of NaCl (both Sigma-Aldrich, St. Louis, MO, USA) were dissolved by careful swirling. These solutions were allowed to sit at 4 °C for 15 min before they were spun at 12,000 × g for 50 min to pellet the precipitated virus fraction. After carefully discarding the supernatant, the pellets were re-centrifuged at 12,000 × g for 5 min (brake set at level 4 on a scale from 0 to 9) and the residual liquid was removed with a pipette. Finally, the pellets were resuspended in 700 μL of buffer MTL (Qiagen, Hilden, Germany).

Four-hundred microliters of pellet suspension were supplemented with 60 μL of a solution comprising 60 ng of carrier RNA per microliter AVL buffer (both Qiagen). Subsequently, samples were subjected to an automated RNA extraction on a Biorobot EZ1 Advanced XL instrument (Qiagen) using the EZ1 Virus Mini kit (v2.0, Qiagen) according to the manufacturer's recommendations. Total RNA was eluted in 60 μL AVE buffer (Qiagen) and typically used immediately in quantitative reverse transcription real-time polymerase chain reaction (RT-qPCR) experiments. For long-term storage, the RNA extracts were kept at −80 °C.

Recoveries were determined by spiking chemically modified viral particles (NATtrol SARS-CoV-2 External Run Controls, 50,000 copies/mL, ZeptoMetrix, Buffalo, NY, USA) into wastewater samples.

Quantitative reverse transcription real-time polymerase chain reaction

Our *triplex* RT-qPCR approach amplified two previously published viral quantification targets – a 72-nt sequence stretch on the SARS-CoV-2 nucleocapsid gene (NC_045512.2:28,287–28,358) (Lu *et al.* 2020) and a 68-nt subsequence of the PMMoV replicase gene (AY859497.1:1878–1945) (Zhang *et al.* 2006; Haramoto *et al.* 2013) – as well as a 78-bp long, in-house developed synthetic IPC target. Single-tube reverse transcription and multiplexed quantitative real-time PCR amplification were performed with the Luna Universal Probe One-Step RT-qPCR kit (NEB; New England Biolabs, Ipswich, MA, USA). The 20-μL RT-qPCR cocktails contained 250 ng/μL of non-acetylated bovine serum albumin (Sigma-Aldrich), 1× of Luna Universal Probe One-Step Reaction Mix, 1× of Luna WarmStart RT Enzyme Mix (both NEB), and 6 μL of target RNA. Information regarding nucleotide sequences, fluorescent labels, and assay concentrations of PCR primers, TaqMan hybridization probes, and the synthetic PCR targets are provided in Supplementary Material, Table S2. To avoid potential contamination issues, we ordered the amplification primers and TaqMan hybridization probes from a different manufacturer (Microsynth, Balgach, Switzerland) than the synthetic PMMoV and IPC targets (Integrated DNA Technologies, Leuven, Belgium).

We used a Freedom EVO 100 robotic platform (Tecan, Männedorf, Switzerland) for an automated reaction setup in 96-well polypropylene PCR plates and two Applied Biosystems 7500 Fast Real-Time PCR System (all TFS) instruments for RT-qPCR amplifications.

The thermocycler protocol consisted of two initial holds at 55 °C for 10 min (reverse transcription) and 95 °C for 1 min (reverse transcriptase inactivation and initial template denaturation) and 40 PCR cycles of 95 °C for 15 s, 55 °C for 30 s, and 60 °C for 30 s. The speed of temperature transitions was set to 'standard'. For raw data acquisition and analysis, we used the 7500 Software (v2.0.6 or v2.3, both TFS). After automatic baseline correction and normalization to the ROX signal, the following fluorescence thresholds were used for Cq determination: 0.15 for SARS-CoV-2 N1 (FAM) and 0.05 for PMMoV (HEX/VIC) and the IPC system (Cy5). All samples, standards, and controls were run in duplicate. The difference of the Cq-values between the replicates was always < 0.5.

For RT-qPCR-based SARS-CoV-2 quantification, a linearized double-stranded plasmid DNA SARS-CoV-2 standard (Promega, Fitchburg, WI, USA) was used initially as reference standard to prepare calibration standards and quality control (QC) standards. This standard was subsequently replaced by an *in vitro* transcribed single-stranded partial RNA sequence of the SARS-CoV-2 N gene (IVT-T, Supplementary Material, Table S2). Furthermore, we used the National Institute of Standards and Technology (NIST)'s research grade test material 10169 – fragment 1 (NIST RGTM 10169, National Institute of Standards and Technology, Gaithersburg, MD, USA; https://bit.ly/3lSFGC6) as reference material. Fragment 1 is a synthetic sense strand RNA molecule corresponding to 3,985 nucleotides on the isolate USA-WA1/2020 SARS-CoV-2 genome (positions MN985325.1:25,949–29,698) (Harcourt *et al.* 2020).

As a quantification standard for the PMMoV genome, we used a synthetically prepared cDNA sequence (PMMoV-T, Supplementary Material, Table S2) that included nucleotide positions 1875–1948 of GenBank entry AY859497.1.

Calibration standards were prepared by diluting a stock solution that contained IVT-T RNA (16,667 copies/μL) and PMMoV-T (500,000 copies/μL; Supplementary Material, Table S2) in TLE buffer (pH 7.5) supplemented with 50 ng/μL yeast total RNA (Baoutina *et al.* 2019) within the range of 0.5–5,000 copies/μL for SARS-CoV-2 and 8×10^2–8×10^6 for PMMoV (in 1 mM Tris–HCl, 0.01 mM Na_2EDTA, pH 7.5). Initially, the calibration was included in every RT-qPCR run. After demonstrating its reproducibility, a master standard curve compiled from 12 independent experiments was used.

In RT-qPCR positive control reactions, 6 μL aliquots of stock dilutions (100- and 2,000-fold in 1 mM Tris–HCl, 0.01 mM Na_2EDTA, pH 7.5) were amplified in parallel to unknown samples.

The limit of detection (LOD) and limit of quantification (LOQ) of the SARS-CoV-2 quantification assay were determined by analyzing NIST RGTM 10,169 dilutions in 60 replicates. The LOD was defined as the lowest concentration where all replicates were positive. The LOQ was defined as the lowest concentration where the relative standard deviation was <30%.

Two negative controls (PCR-grade nuclease-free water) were included in each run. Extraction blanks were processed at least two times a week.

Data analysis and visualization

The SARS-CoV-2 concentration in wastewater was calculated based on the concentration factor, accounting for the effective volume analyzed in the RT-qPCR assay. These concentrations as well as key parameters of the influent wastewaters (CODs, daily flows of wastewater) were electronically submitted to the Tyrolean wastewater monitoring database. COD values were used to estimate the individual WWTP catchment populations. Per capita equivalents were calculated using 120 g/d/person.

Daily mass loads of SARS-CoV-2 were calculated by multiplying the measured concentrations in the 24-h composite samples with the corresponding daily flows of wastewater. Mass loads were converted into 'estimated numbers of positively tested persons' using an estimated average number of viruses that an infected person will contribute to the total load of viruses detectable in wastewater samples as divisor (Gerrity *et al.* 2021; Hasan *et al.* 2021; Saththasivam *et al.* 2021). The conversion factor (3.0×10^9 copies per infected person per day) was determined experimentally by fitting wastewater data to clinical test data.

The database also contained region-specific data on the actual numbers of confirmed COVID-19 cases obtained from clinical test results submitted to the epidemiological reporting system.

The two epidemiological parameters served as input for data visualization in the Tyrolean wastewater monitoring dashboard. The dashboard was created with ArcGIS (ESRI, Redlands, CA, USA).

RESULTS AND DISCUSSION

Vision and mission of the Tyrolean wastewater monitoring program

The monitoring program was designed to provide timely, comprehensive, exhaustive, and reliable information on the pandemic situation and trends in Tyrolean regions. To accomplish this task, samples from influents of Tyrolean WWTPs are regularly collected and analyzed. Less than 36 h after sampling, transformed SARS-CoV-2 loads in combination with other epidemiological parameters (i.e. region-specific prevalence) are made available to health authorities to provide guidance in the decision-making process.

Overview on the selected sampling sites

Tyrol has a population of 760,000 and consists of 279 municipalities. The most populous town is Innsbruck (132,000 inhabitants). There are nine further municipalities with >10,000 inhabitants, namely Kufstein, Telfs, Wörgl, Hall in Tirol, Schwaz, Lienz, and Imst. During the site selection process, utmost coverage of the Tyrolean population was the primary selection criterion.

Tyrol is a traditional tourism destination. In 2019, >12 million people spent their holidays in this country. Nine destinations reported >1 million overnight stays per year. These were Sölden, Innsbruck, Ischgl, Mayrhofen, Neustift im Stubaital, Serfaus, St. Anton am Arlberg, Seefeld in Tirol, and Eben am Achensee. The majority of them are internationally known ski resorts. On average, >100,000 visitors were present per day in Tyrol. During peak times, such as Christmas holidays, even more. Accordingly, tourism was the other factor that was considered during the program development.

There are 53 municipal WWTPs that serve populations ranging from 350 to 174,674 people (Supplementary Material, Figure S1). The WWTPs have a total capacity of 2.2 million population equivalents. The monitoring program covers all WWTPs with capacities >5,000 population equivalents (Supplementary Material, Table S1). With these 43 WWTPs, the coverage of 98% of the Tyrolean population and >90% of potential visitors was accomplished. Figure 1(a) summarizes the individual catchment population sizes. Innsbruck is the most populous region. Nauders can be found at the other end of the scale with 1,531 inhabitants. Figure 1(b) provides an overview on the overnight stays in January 2020 in the monitored regions. Strass reported >1.2 million and Niederndorf only 5,297. With the exception of Strass, where, for instance, Mayrhofen and Eben am Achensee drain their wastewater, the most populous regions are not found among the top touristic destinations and vice versa. Therefore, to cover both, residents and tourists, even rather small WWTPs were monitored.

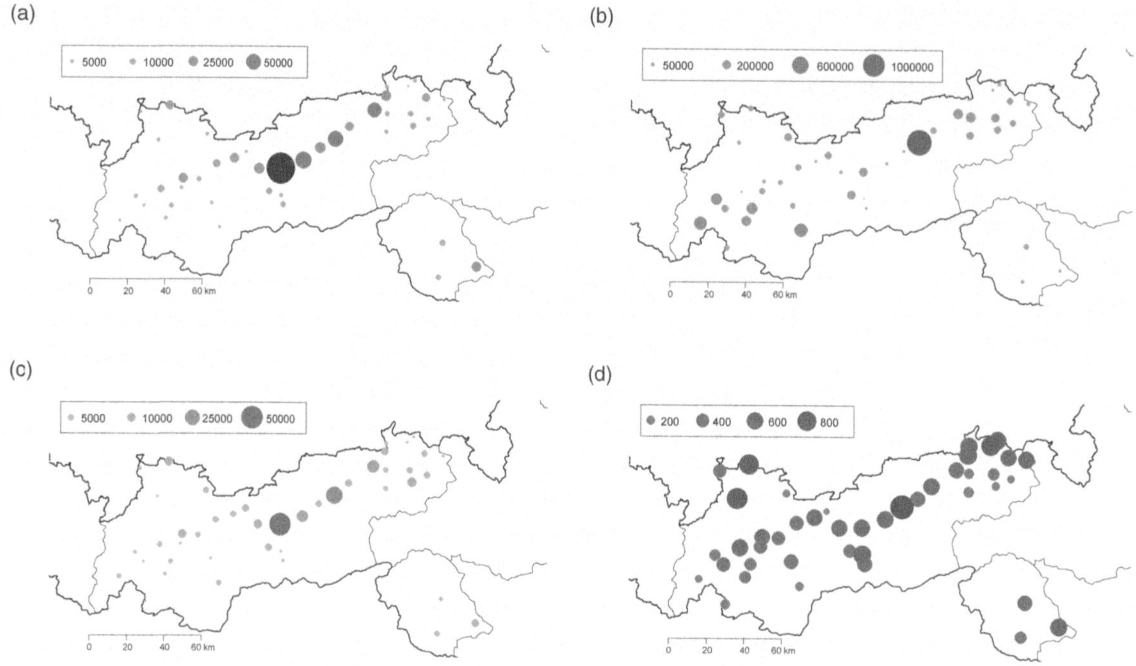

Figure 1 | Geospatial maps of Tyrol showing (a) the number of inhabitants living in the catchment areas of the 43 WWTPs, (b) overnight stays in the catchment areas in January 2020, (c) average daily flow rates (m^3/day) of the monitored WWTPs, and (d) average CODs measured in the collected 24-h composite samples (mg/L).

The project went through three phases with different numbers of WWTPs monitored (Supplementary Material, Figures S2 and S3). In the discovery phase (May–August 2020), Innsbruck was the primary target. By monitoring the region of Innsbruck, 23.1% of the total population and 3% of potential visitors were covered. The monitoring was extended to nine WWTPs in September 2020 (development phase). By monitoring Innsbruck, Fritzens, Schwaz, Strass, Radfeld, Kirchbichl, Kufstein, Niederndorf, and Ischgl, 55.4% of the total population and 33% of potential visitors were covered. The monitoring program with 43 WWTPs was put into operation in mid-November 2020 (delivery phase).

Developed workflow and processes

The workflows and processes established for the Tyrolean wastewater monitoring program are summarized in Figure 2.

Sample collection

In total, 43 Tyrolean WWTPs were included in the monitoring program (Figure 2(a); Supplementary Material, Table S1). Based on the results obtained during the first two phases of the project, we decided that the minimal sampling frequency for a single site should be two samples per week. This decision is in full agreement with published EC recommendations (EC 2021). Based on the available budget and laboratory capacity, however, it was possible to sample a restricted number of the 43 targeted WWTPs with a higher frequency. Here, the primary selection criterion was utmost coverage of the Tyrolean population.

Innsbruck was sampled daily. Twelve other WWTPs provided five samples per week (Sunday–Thursday). In the remaining WWTPs, two samples per week were collected (either Sunday/Tuesday or Monday/Wednesday). Samples were collected as 24-h composite samples using cooled autosamplers (4 °C) operated in a volume proportional mode.

Additionally, CODs and daily influent flows were provided by the WWTPs. Average daily flow rates and average CODs of the 43 WWTPs are summarized in Figure 1(c) and 1(d). The region with the highest average COD caused by a high content of industrial wastewaters was Schwaz. As a considerable number of samples from Schwaz showed inhibition, there might be a link between high COD values and PCR inhibition. Accordingly, the daily COD values were used to check plausibility of (negative) results. Lowest average COD values were observed for touristic destinations in lockdown (e.g. Ischgl, Sölden, and Nauders) and WWTPs with infiltration (e.g. Seefeld and Kitzbühel).

The hydrochemical parameters were electronically transferred to the Tyrolean wastewater database. They were used to calculate epidemiological parameters. Furthermore, they were used as plausibility checks for the analytical results obtained.

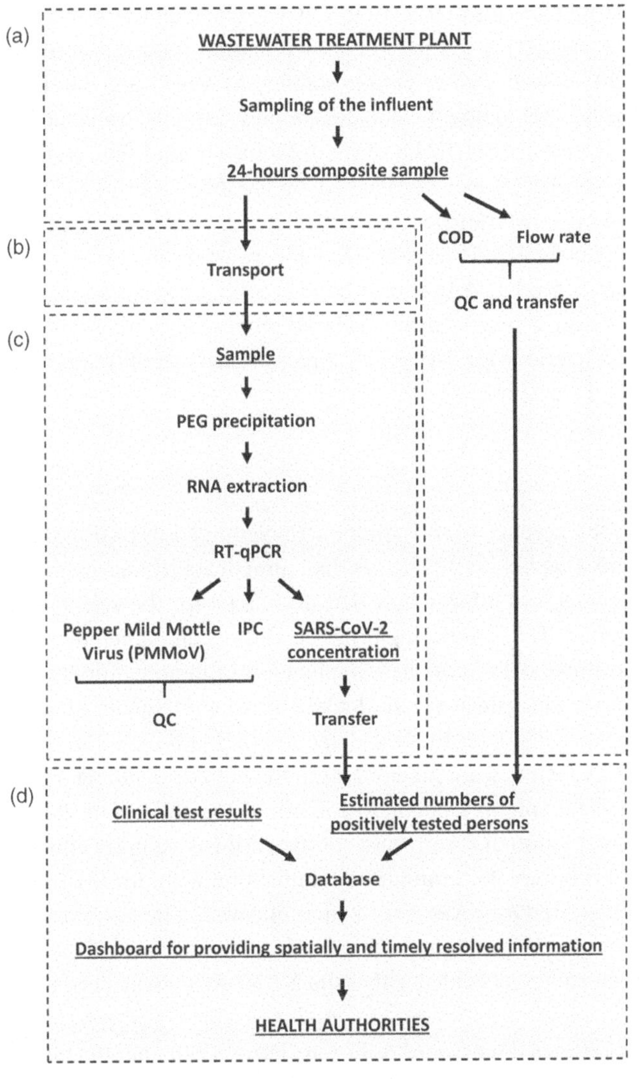

Figure 2 | Workflow established for monitoring SARS-CoV-2 loads in 43 Tyrolean WWTPs. The workflow involves (a) sample collection and measurement of hydrochemical parameters (flow rate and COD) at the WWTPs, (b) transport of the samples to the laboratory, (c) quantification of SARS-CoV-2 with a validated method, and (d) amalgamation of data sets in a single database with consecutive processing, visualization, and transfer of information to health authorities.

High CODs were often found as reasonable explanations for observed PCR inhibitions and, thus, for low and negative results. Increased daily influent flows also helped to determine additional outliers. Increased daily influent flows were usually measured on rainy days. Depending on the precipitation quantity and the sewer system characteristics (i.e. separate or combined sewer systems), sample composition might be altered by stormwater in a way that the measured SARS-CoV-2 concentration does not reflect the actual pandemic situation.

Beside the COD, several other hydrochemical parameters are regularly monitored in influent wastewaters. These include the biological oxygen demand, ammonia-nitrogen (NH_4-N), total-nitrogen (N_{tot}), and total-phosphorus (P_{tot}). However, particularly by small- and medium-sized WWTPs, only parts of these parameters are determined on a weekly or daily basis. The most frequently monitored parameter is the COD. Thus, to avoid extra work and to increase the willingness to co-operate, we decided to ask for the COD alone.

Sample transport and storage

During transport and storage, samples were kept at temperatures between 2 and 8 °C (Figure 2(b)). Storage and transport conditions can affect the detection and quantification of SARS-CoV-2 in wastewater samples because of stability. Various studies provided sufficient evidence that SARS-CoV-2 is persistent in wastewater samples for several days if samples are stored at 4 °C (Ahmed *et al.* 2020b; Boogaerts *et al.* 2021b). As part of the validation process, we have evaluated sample stability. We observed that the N1-gene fragment showed high in-sample stability in wastewater samples for up to 7 days of storage.

We used thermally insulated boxes for shipping the cooled samples. Initially, transport was outsourced to parcel services. With the involvement of external services, at least 24 h of the total turnaround time were spent for transportation. The situation was unsatisfactory, because Innsbruck can be reached by car from any of the 43 monitored WWTPs within 3 h. Accordingly, a faster and more cost-effective transport concept was elaborated during the development phase. This included six collection routes with fixed pick-up and handover times. Sample transportation was organized by the administration of the province Tyrol using their drivers and carpool. With this internal courier service, samples were delivered to the laboratory on the same day of sampling usually before 2 pm. Typically, sample processing started within 1–3 h upon arrival.

We experienced that freezing is deleterious for SARS-CoV-2 recoveries from untreated wastewater samples. Therefore, this option was ruled out for storage and transport.

SARS-CoV-2 quantification

The rapid onset of COVID-19 and the urgency of the pandemic have forced us to provide quantitative data already in early phases of the monitoring program (Figure 2(c)). Factors like practicability, accessibility to instrumentation, supply of materials, and availability of knowhow and information had an impact on the selection of methods that were integrated into the initial and modified workflows. The monitoring program was accompanied by research and development programs for providing new and advanced features with immediate application in the operational work. Published recommendations represented further triggers for innovations (Bustin *et al.* 2009; Ahmed *et al.* 2020c). Important milestones included (1) the principle setup of the workflow in April 2020, (2) the first detection of SARS-CoV-2 in a wastewater sample collected on 16 April 2020, (3) automation of RNA extraction from precipitates in May 2020, including a platform change in September 2020, (4) increase of the processed wastewater volume from 50 to 250 mL by the end of July 2020, (5) implementation of an IPC by the end of October 2020, (6) implementation of the process control PMMoV in mid-February 2021; (7) pipetting samples and reagents by a robotic system in mid-February 2021, and (8) quantification with linear RNA standards and reference material.

The US CDC N1 RT-PCR assay was used for SARS-CoV-2 quantification. The assay was developed for clinical testing. Several studies have evaluated its performance for wastewater analysis (Ahmed *et al.* 2020a; Medema *et al.* 2020b; Gerrity *et al.* 2021; Pecson *et al.* 2021). So, there is sufficient evidence that the N1 RT-PCR assay performs equally or even better than other targets.

As reported by several other laboratories (Fomsgaard & Rosenstierne 2020), manufacturer shortages forced us to modify or exchange methods to accommodate supply issues. In May 2020, ultrafiltration devices were hardly available; hence, we decided to implement PEG-NaCl-based precipitation. In September 2020, reagents for running the initial RNA extraction could not be supplied. Therefore, the extraction platform was changed.

The laboratory responsible for SARS-CoV-2 quantification was accredited for genetic analysis according to ISO 17025. Accredited laboratories have to meet specific process and management system requirements. Process requirements are describing the activities to ensure that results are based on accepted science and aimed at technical validity. Management system requirements are those steps taken by the organization to give itself quality management system tools to support the work of its people in the production of technically valid results.

An important obligation for the analytical laboratory is validation of newly developed methods and workflows. We have performed a panel of experiments to verify that the developed SARS-CoV-2 quantification workflow is fit for the intended purpose. We evaluated the calibration model (assay efficiency of 100%, with an R^2 value of 0.996), determined the LOQ (27 copies per reaction) and LOD (nine copies per reaction), assessed recoveries ($20.6 \pm 12.2\%$), checked the intraday and interday reproducibilities (RSD $< 30\%$), and verified the stability of wastewater samples over several days at 4 °C. A publication summarizing these method validation efforts is in preparation.

Metrological traceability would require use of a certified reference material. Such material is currently not available. NIST RGTM 10169 is probably the best alternative. It differs from a NIST standard reference material in that it is not as highly characterized or traceable to the International System of Units. Nevertheless, it is homogeneous and undergoes continual stability testing. It contains synthetic fragments of the SARS-CoV-2 virus RNA, which is the target of molecular diagnostic tests. The RNA fragments were characterized for concentration using digital PCR methods.

A common way of checking the validity of test and calibration results is participating in appropriate proficiency testing. Inter-laboratory exercises are well established in the framework of WBE of illicit drugs (van Nuijs *et al.* 2018). They helped to improve the performance of participating laboratories and triggered harmonization and standardization of analytical methods. Despite their importance, inter-laboratory studies focusing on SARS-CoV-2 quantification in wastewater are not institutionalized so far (Chik *et al.* 2021; Pecson *et al.* 2021). Therefore, we have organized two such studies with Austrian and German colleagues. Obtained results clearly revealed a lack of standardization rendering a direct comparison of quantitative results from different laboratories nearly impossible (publication in preparation).

Importantly, the laboratory was designed with a one-way path from sample receipt to RNA extraction to PCR preparation to PCR analysis. Each step was accomplished in separate rooms. Owing to the strict separation, sample contamination indicated by positively tested blanks (i.e. no-template controls and extraction controls) was not observed.

Known PCR inhibitors found in wastewater, including bile salts, urea, phenol, ethanol, polysaccharides, sodium dodecyl sulfate (SDS), humic acids, tannic acid, melanin as well as different proteins, can lead to false-negative results if they are not effectively removed during sample preparation (Schrader *et al.* 2012). Their concentrations may vary within a single WWTP and between WWTPs. Furthermore, precipitation methods may enrich not only the target but also certain RT-PCR inhibitors. Accordingly, the presence of inhibitors in wastewater samples needs to be investigated. For this purpose, two controls were implemented in the process. With the IPC, the performance of the RT-qPCR assay is assessed. A sample was flagged if its IPC Cq was substantially delayed relative to that of the simultaneously amplified positive controls ($\Delta Cq > 2$). With PMMoV quantification, the overall process is monitored. A sample was flagged if the determined PMMoV load fell outside of the WWTP-specific average load $\pm$ standard deviation. If one of the controls indicates quality issues, either samples, precipitates, or extracts are reprocessed. Furthermore, dilution of precipitates and extracts was found to be a useful strategy to reduce inhibition. Overall, <2.5% of samples were reanalyzed.

At this moment, there is no ideal external control standard with the same properties of SARS-CoV-2 for quantification available (Boogaerts *et al.* 2021b; Chik *et al.* 2021). Therefore, no spike-in control was in use.

Owing to the optimization, automation, and parallelization of processes, we were able to analyze up to 42 samples per day. This capacity is sufficiently high to do service not only for the Tyrolean monitoring program, but also the Austrian monitoring program that is under-developed.

Turnaround times, defined as the number of hours between sample collection and the electronic transfer of quantitative results, ranged between 5 and 28 h. Including shipping times, results were reported to health authorities within 36 h.

Database, dashboard, and information transfer

All data from the WWTPs (i.e. daily influent flows and CODs) and the analyzing laboratory (SARS-CoV-2 concentrations) were collected from the Tyrolean wastewater monitoring database (Figure 2(d)). Transfer was accomplished electronically. The number of reported COVID-19 cases in the monitored regions represented additional entries to the database.

The wastewater data were used to provide estimated numbers of positively tested persons per region. A mass balance approach was used for this purpose (Gerrity *et al.* 2021; Hasan *et al.* 2021; Saththasivam *et al.* 2021). Individual viral loads were converted into estimated numbers of positively tested persons using an estimated number of viruses that an infected person will contribute on an average to the total load of viruses detectable in wastewater samples as a divisor. The divisor (3×10^9 copies per infected person per day) was determined by fitting data sets available in the discovery phase of this project (May–August 2020). It corresponds well to experimentally and theoretically determined values of virus shedding that were reported in other studies (6×10^8–2×10^{11} copies per infected person per day) (Gerrity *et al.* 2021; Hasan *et al.* 2021; Saththasivam *et al.* 2021).

Prevalence estimates offer several following advantages in comparison to the reporting of concentrations: (1) with these numeric values provided, even unexperienced users will learn very quickly the interpretation of wastewater data; (2) results from clinical and environmental surveillance can directly be compared with each other, which facilitates the assessment of the pandemic situation; (3) different regions can directly be compared with each other if population normalized loads are used; and (4) data from different regions can be aggregated to get representative situation pictures for territories and nations.

The applied mass balance approach has several following limitations: (1) fitting was accomplished without knowing the real numbers of cases; (2) this simple model neglected the fact that viral shedding is variable over time and between patients; and (3) the model does not correct for WWTP-specific parameters, such as in-sewer transportation and stability.

Clearly, with the developed mass balance approach, only a rough estimation of the actual number of infected persons can be provided. Nevertheless, based on the feedback provided by the health authorities, the above-mentioned advantages seem to outweigh the limitations.

The available epidemiological data served as an input for the Tyrolean wastewater monitoring dashboard. With this dashboard, the available temporal and geographic information was visualized to enable health authorities monitoring the pandemic situation in Tyrol as a whole and in each region individually. Multiple visualizations were integrated on a single screen. The dashboard included maps with region-specific prevalence numbers. Furthermore, time series were provided to see potential trends, including warnings and confirmation of absence.

The dashboard was built for internal use by health authorities only. Persons with access included decision-makers at the strategic and operational levels. The available information was used to assess the actual pandemic situation and its development, and in due consequence to provide justification for measures.

On a weekly basis, disclosure of the most relevant information to the general public was accomplished with short reports communicated via www.tirol.gv.at/covid-abwasser and press releases.

Representative monitoring results

Between June 2020 and July 2021, 5,270 samples were analyzed and the corresponding data were transferred to the Tyrolean wastewater monitoring database. The database and the corresponding dashboard represented effective and efficient tools for assessing the pandemic situation in Tyrol. In the following, representative monitoring results are discussed to demonstrate the vast potential and some limitations of the environmental surveillance approach.

Monitoring results from Innsbruck obtained between 2020-06-01 and 2020-09-27

In the discovery phase of the Tyrolean wastewater monitoring program, we have provided the proof-of-concept that WBE can be used to monitor viral circulation in the community (Figure 3(a)). Particularly, we have demonstrated its potential use as an early-warning system. As an early-warning system, wastewater surveillance should signal an initial outbreak or the recurrence of the virus, providing time for health authorities to prepare for the event and to minimize the impact.

Innsbruck was the first Tyrolean region with a regular SARS-CoV-2 monitoring. Monitoring started on 2020-06-01, right after the end of the first wave of COVID-19 in Tyrol. In June 2020, life was returning to relative normality, except for the mandatory wearing of a face mask and maintaining a safe social distance. Between 2020-06-15 and 2020-06-22, Innsbruck had no COVID-19 cases. In July 2020, the average number of cases was 14. Beginning with 2020-08-11, a rapid increase of positive cases was observed. At the end of August 2020, around 150 persons were positively tested.

In the first 2 weeks of June 2020, all tests of wastewater samples were negative. A first signal for resurgence of SARS-CoV-2 transmission in the community was obtained on 15 June 2020. This was 7 days ahead of reporting positive clinical tests. Until the beginning of August 2020, in all but one wastewater sample viral RNA was detected. The average of the estimated number of positively tested persons was 4.8. Taking in consideration that in July 2020, the average number of patients with confirmed SARS-CoV-2 infection was 14, the minimal number of clinical cases detectable by wastewater monitoring was 10 per 120,000 inhabitants. On 8 August 2020, a rapid increase of the viral load was observed. This 'burst' leading to 150 positively tested persons was observed 3 days earlier than the increase of positive clinical tests.

For the wastewater sample collected on 11 July 2020, a negative test result was obtained. We believe that high loads of stormwater entering Innsbruck's combined sewer system on that particular day were responsible for sample dilution and a negative test result. This hypothesis was supported by the fact that the PMMoV load determined by reanalyzing the extract of the sample collected on 11 July 2020 was 60–70% lower than the loads observed on rain-free days in July 2020.

Comparison of the graphs representing the wastewater and clinical testing results (Figure 3(a)) revealed that (1) wastewater monitoring complements established epidemiological monitoring tools, (2) wastewater leads clinical cases by 3–7 days, and (3) the minimum number of SARS-CoV-2-infected cases associated with a positive detection in Innsbruck's wastewater is 10 per 120,000 inhabitants.

The herein reported lead time and detection sensitivity are in full agreement with already published information (D'Aoust *et al.* 2021; Larsen *et al.* 2021; Olesen *et al.* 2021; Wu *et al.* 2021). However, the lead time of WBE is dependent on several factors, including diagnostic testing availability and reporting within the community, the sewage collection and conveyance systems, and the capacity and turnaround time of the WBE laboratory (Bibby *et al.* 2021). As resources allocated to diagnostic

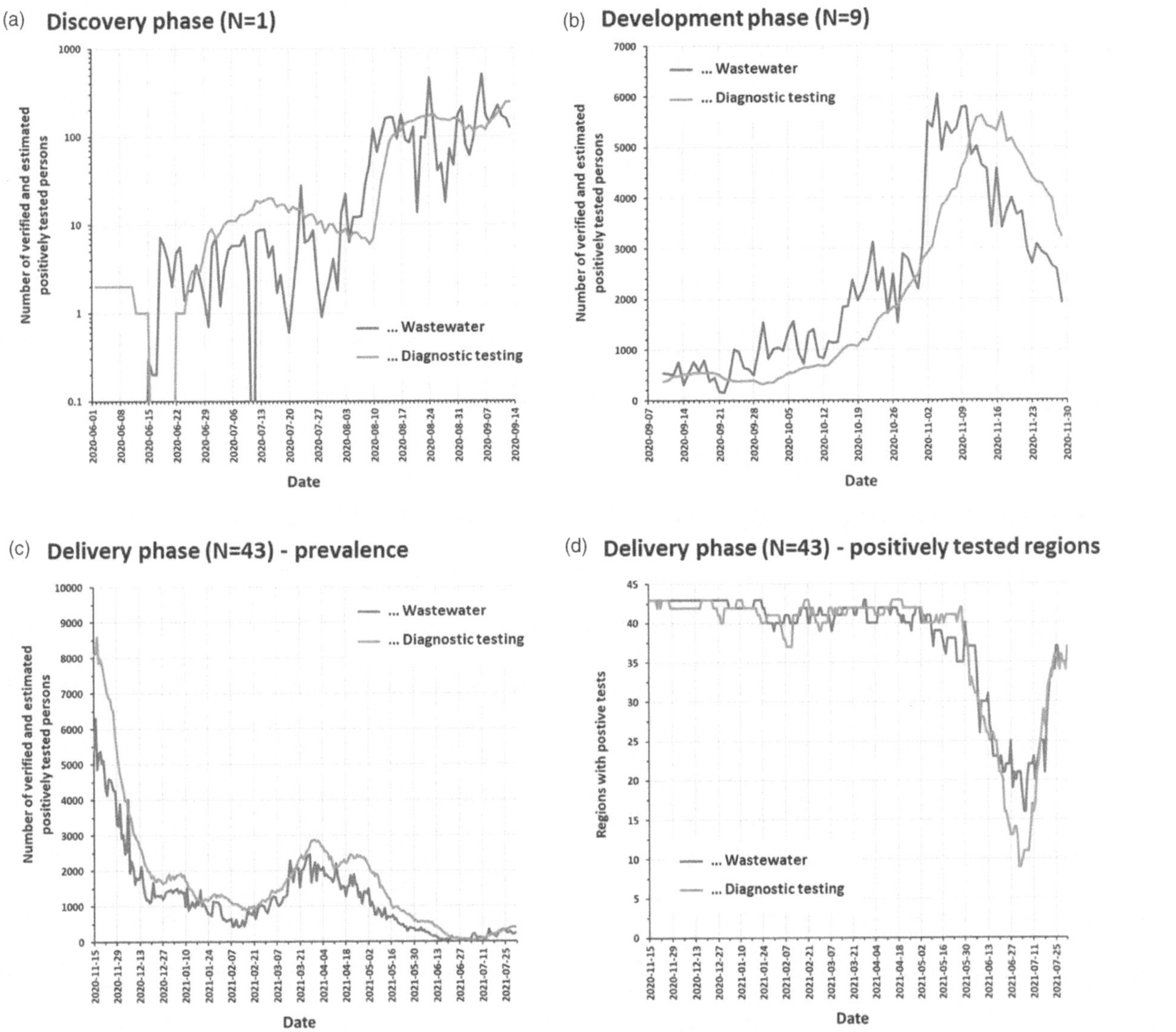

Figure 3 | A summary of important monitoring results obtained in (a) the discovery phase, (b) the development phase, and (c, d) the delivery phase of the project. For the delivery phase, temporal information on (c) the prevalence and (d) the number of positively tested regions is provided. In the discovery phase, only Innsbruck was monitored. In the development phase, Innsbruck, Fritzens, Schwaz, Strass, Radfeld, Kirchbichl, Kufstein, Niederndorf, and Ischgl were covered. In the delivery phase, 43 WWTPs were monitored. Accordingly, aggregated data are provided in (b) and (c).

testing and reporting are continuously expanding during this pandemic, it is possible that lead times might decrease accordingly.

Monitoring results from nine regions obtained between 10 September 2020 and 29 November 2020

After positive evaluation of the project, a monitoring program involving nine WWTPs was started in September 2020 (Figure 3(b)). The monitored WWTPs included Innsbruck, Fritzens, Schwaz, Strass, Radfeld, Kirchbichl, Kufstein, Niederndorf, and Ischgl. The monitoring program was intended to provide information on viral circulation in the community after returning from summer holidays.

On 10 September 2020, the number of positively tested persons was 374 in the nine monitored areas. The number of cases started to rise at the beginning of October 2020. The maximum number of cases was reached in mid-November 2020 (18 November 2020: 5,671 positively tested persons). After that, numbers started to decline.

To counteract the spread of the disease in the community, a number of non-pharmaceutical interventions were undertaken. The intention was to slow down the pandemic by restricting mobility and thus to preserve the capacity of the health systems

(Haug *et al.* 2020). Important measures included (1) limited opening hours (5 am to 10 pm) for restaurants, pubs, and bars (25 September 2020), (2) cancellation of public events, guest registration, and closure of clubhouses (16 October 2020), (3) night-time curfew, closing of restaurants and hotels (3 November 2020), and (4) stay at home requirements, closing of schools and non-essential businesses (17 November 2020).

The clinical testing provided evidence for a steady increase of case numbers in autumn 2020. The wastewater monitoring results seem to indicate a different basis for the COVID-19 spread in the monitored communities at that time. We observed a stepwise increase of the total viral load. Rapid increases of viral loads in wastewater occurred on 24 September 2020, 16 October 2020, and 2 February 2020 (see also Supplementary Material, Figure S4). Interestingly, the three rises coincided with the implementation of reinforced measures. In each case, the stipulated restrictions were communicated to the general public several days before coming into force. Therefore, we speculate that a considerable part of the population used the days between announcement and reinforcement to cultivate social contacts, and this counteracted the effect of the measures taken.

Comparison of the graphs representing the wastewater and clinical testing results (Figure 3(b)) further indicated that already the lockdown measures from 3 November 2020 would have been effective for inducing a decline of positive cases. As the corresponding measures mainly targeted the gastronomy, we speculate that this sector represented an important driver for viral circulation in the monitored communities.

Monitoring results from 43 regions obtained between 15 November 2020 and 31 July 2021

The nationwide monitoring started at the peak of the second wave of COVID-19 infections in Tyrol (Figure 3(c)). The maximum number of COVID-19 cases in Tyrol were reported on 17 November 2020 ($n = 8,589$). Owing to the implemented lockdown measures, there was a steep decline in the number of cases until 20 December 2020 ($N = 1,793$). Owing to a partial withdrawal of measures before and during Christmas holidays as well as reduced acceptance, a slower decrease of cases was observed in the following weeks. After reaching a minimum number of cases on 17 February 2021 ($N = 895$), Tyrol entered into the third wave of infections. The peak number of infections in spring 2021 was observed on 1 April 2021 ($N = 2,845$). The lowest number of positive cases in 2021 were observed at the beginning of July ($N = 35$).

For the total viral load in Tyrolean wastewaters, similar trends were observed, which again substantiates that wastewater monitoring provides meaningful information on disease prevalence. In our opinion, wastewater monitoring and diagnostic testing complement each other. Wastewater analysis is particularly useful in verifying community-wide trends observed with clinical testing. Furthermore, the approach might be helpful in forecasting a potential rise and fall of case numbers. With wastewater monitoring trend reversals might be detected earlier than with clinical testing. For instance, the viral loads reached their minimum in February and June 2021 several days earlier than the clinical case numbers did (Figure 3(c)).

Another important feature of the wastewater monitoring is its ability to provide evidence for the absence of COVID-19 (Figure 3(d)). At the beginning of May 2021, the number of positively tested WWTPs started to decline. At the beginning of July 2021, SARS-CoV-2 was detected in <20 WWTPs. In all other regions, SARS-CoV-2 was not detectable in the collected wastewater samples. During July 2021, however, a turnaround was observed. Viral loads and the number of positively tested WWTPs started to rise again. Eventually, this is the beginning of the fourth pandemic wave.

CONCLUSIONS

In this report, we summarized concepts, structure, and results of the Tyrolean wastewater monitoring program. The development of the program was started after the first phase of the COVID-19 pandemic in Tyrol (February–May 2020) to expand the available surveillance and response capacities. The Tyrolean health and water sector jointly worked together in the conception, development, and operation of this program. The monitoring in Innsbruck started in June 2020. Eight more WWTPs were included in September 2020. The monitoring program involving 43 WWTPs was put in operation in mid-November 2020. By the end of July 2021, SARS-CoV-2 concentrations were determined in 5,270 Tyrolean wastewater samples.

The 24-h composite samples were collected at the influents of the monitored WWTPs 2–7 times a week. The samples were shipped at 4 °C by a courier service to the analytical laboratory in Innsbruck within 5 h upon availability. SARS-CoV-2 quantification was accomplished with a validated workflow, involving PEG-NaCl-based precipitation, automated lysis, and RNA extraction, as well as RT-qPCR of the N1 target. QCs included an IPC, quantification of PMMoV and positive control samples, as well as processing and quantification of blanks. The obtained results were submitted electronically to the Tyrolean wastewater monitoring database. Typically, results were available for review by health authorities in <36 h after finishing the sampling process. The wastewater data together with the clinical test data served as an input for the Tyrolean wastewater monitoring dashboard. With this dashboard, the available temporal and geographic information was visualized to enable health authorities monitoring the pandemic situation in Tyrol as a whole and in each region individually.

The existing data support the important role of SARS-CoV-2 wastewater monitoring in complementing clinical disease surveillance. The monitoring program represents an early-warning system, provides independent confirmation of temporal trends in COVID-19 prevalence, enables the assessment of the effectiveness of measures, alerts about bursts of disease activity, and finally provides evidence for the absence of COVID-19.

ACKNOWLEDGEMENTS

The authors thank Sandra Haslacher, Stefan Sieberer, and Janine Kraker for technical assistance. Furthermore, we thank the management and staff of the participating WWTPs for their support. Financial, personnel, and material support by the Land Tirol is acknowledged. This project received funding by the Austrian Federal Ministry of Education, Science and Research, the Austrian Federal Ministry of Agriculture, Regions and Tourism, the Austrian Städtebund, and several Austrian states as part of the 'Coron-A' project. Finally, we thank the NIST for providing RGTM 10169.

AUTHOR CONTRIBUTIONS

B.D. and H.N.: methodology, validation, investigation, writing – original draft, visualization, writing – review and editing. M.S.: methodology, validation, investigation, writing – review and editing, supervision, project administration. S.W. and M.K.: conceptualization, data curation, writing – review and editing, project administration. C.L.-F., W.P., and S.F.: methodology, resources, writing – review and editing. B.P.: methodology, formal analysis, data curation, writing – review and editing. A.H.: conceptualization, data curation, writing – review and editing, project administration. H.O.: conceptualization, methodology, validation, formal analysis, data curation, writing – original draft, writing – review and editing, visualization, supervision, project administration, funding acquisition.

DATA AVAILABILITY STATEMENT

Data cannot be made publicly available; readers should contact the corresponding author for details.

REFERENCES

Ahmed, W., Angel, N., Edson, J., Bibby, K., Bivins, A., O'Brien, J. W., Choi, P. M., Kitajima, M., Simpson, S. L., Li, J., Tscharke, B., Verhagen, R., Smith, W. J. M., Zaugg, J., Dierens, L., Hugenholtz, P., Thomas, K. V. & Mueller, J. F. 2020a First confirmed detection of SARS-CoV-2 in untreated wastewater in Australia: a proof of concept for the wastewater surveillance of COVID-19 in the community. *Science of the Total Environment* **728**, 138764.

Ahmed, W., Bertsch, P. M., Bibby, K., Haramoto, E., Hewitt, J., Huygens, F., Gyawali, P., Korajkic, A., Riddell, S., Sherchan, S. P., Simpson, S. L., Sirikanchana, K., Symonds, E. M., Verhagen, R., Vasan, S. S., Kitajima, M. & Bivins, A. 2020b Decay of SARS-CoV-2 and surrogate murine hepatitis virus RNA in untreated wastewater to inform application in wastewater-based epidemiology. *Environmental Research* **191**, 110092.

Ahmed, W., Bivins, A., Bertsch, P. M., Bibby, K., Choi, P. M., Farkas, K., Gyawali, P., Hamilton, K. A., Haramoto, E., Kitajima, M., Simpson, S. L., Tandukar, S., Thomas, K. & Mueller, J. F. 2020c Surveillance of SARS-CoV-2 RNA in wastewater: methods optimisation and quality control are crucial for generating reliable public health information. *Current Opinion in Environmental Science and Health* **17**, 82–93. https://www.sciencedirect.com/science/article/pii/S246858442030060X.

Asghar, H., Diop, O. M., Weldegebriel, G., Malik, F., Shetty, S., El Bassioni, L., Akande, A. O., Al Maamoun, E., Zaidi, S., Adeniji, A. J., Burns, C. C., Deshpande, J., Oberste, M. S. & Lowther, S. A. 2014 Environmental surveillance for polioviruses in the Global Polio Eradication Initiative. *Journal of Infectious Diseases* **210**, S294–S303.

Baoutina, A., Bhat, S., Partis, L. & Emslie, K. R. 2019 Storage stability of solutions of DNA standards. *Analytical Chemistry* **91** (19), 12268–12274.

Bibby, K., Bivins, A., Wu, Z. & North, D. 2021 Making waves: plausible lead time for wastewater based epidemiology as an early warning system for COVID-19. *Water Research* **202**, 117438.

Bivins, A., North, D., Ahmad, A., Ahmed, W., Alm, E., Been, F., Bhattacharya, P., Bijlsma, L., Boehm, A. B., Brown, J., Buttiglieri, G., Calabro, V., Carducci, A., Castiglioni, S., Cetecioglu Gurol, Z., Chakraborty, S., Costa, F., Curcio, S., de Los Reyes, F. L., Delgado Vela, J., Farkas, K., Fernandez-Casi, X., Gerba, C., Gerrity, D., Girones, R., Gonzalez, R., Haramoto, E., Harris, A., Holden, P. A., Islam, M. T., Jones, D. L., Kasprzyk-Hordern, B., Kitajima, M., Kotlarz, N., Kumar, M., Kuroda, K., La Rosa, G., Malpei, F., Mautus, M., McLellan, S. L., Medema, G., Meschke, J. S., Mueller, J., Newton, R. J., Nilsson, D., Noble, R. T., van Nuijs, A., Peccia, J., Perkins, T. A., Pickering, A. J., Rose, J., Sanchez, G., Smith, A., Stadler, L., Stauber, C., Thomas, K., van der Voorn, T., Wigginton, K., Zhu, K. & Bibby, K. 2020

Wastewater-based epidemiology: global collaborative to maximize contributions in the fight against COVID-19. *Environmental Science and Technology* **54** (13), 7754–7757.

Boogaerts, T., Ahmed, F., Choi, P. M., Tscharke, B., O'Brien, J., De Loof, H., Gao, J., Thai, P., Thomas, K., Mueller, J. F., Hall, W., Covaci, A. & van Nuijs, A. L. N. 2021a Current and future perspectives for wastewater-based epidemiology as a monitoring tool for pharmaceutical use. *Science of the Total Environment* **789**, 148047.

Boogaerts, T., Jacobs, L., De Roeck, N., Van den Bogaert, S., Aertgeerts, B., Lahousse, L., van Nuijs, A. L. N. & Delputte, P. 2021b An alternative approach for bioanalytical assay optimization for wastewater-based epidemiology of SARS-CoV-2. *Science of the Total Environment* **789**, 148043.

Bustin, S. A., Benes, V., Garson, J. A., Hellemans, J., Huggett, J., Kubista, M., Mueller, R., Nolan, T., Pfaffl, M. W., Shipley, G. L., Vandesompele, J. & Wittwer, C. T. 2009 The MIQE guidelines: minimum information for publication of quantitative real-time PCR experiments. *Clinical Chemistry* **55** (4), 611–622.

Centers for Disease Control and Prevention. 2021 *National Wastewater Surveillance System (NWSS)*. Available from: https://www.cdc.gov/healthywater/surveillance/wastewater-surveillance/wastewater-surveillance.html (accessed 18 December 2021)

Chik, A. H. S., Glier, M. B., Servos, M., Mangat, C. S., Pang, X.-L., Qiu, Y., D'Aoust, P. M., Burnet, J.-B., Delatolla, R., Dorner, S., Geng, Q., Giesy Jr., J. P., McKay, R. M., Mulvey, M. R., Prystajecky, N., Srikanthan, N., Xie, Y., Conant, B. & Hrudey, S. E. & Canadian, S. C. I. Laboratory C 2021 Comparison of approaches to quantify SARS-CoV-2 in wastewater using RT-qPCR: results and implications from a collaborative inter-laboratory study in Canada. *Journal of Environmental Sciences* **107**, 218–229.

Choi, P. M., Tscharke, B. J., Donner, E., O'Brien, J. W., Grant, S. C., Kaserzon, S. L., Mackie, R., O'Malley, E., Crosbie, N. D., Thomas, K. V. & Mueller, J. F. 2018 Wastewater-based epidemiology biomarkers: past, present and future. *TrAC-Trends in Analytical Chemistry* **105**, 453–469.

Correa-Martinez, C. L., Kampmeier, S., Kumpers, P., Schwierzeck, V., Hennies, M., Hafezi, W., Kuhn, J., Pavenstadt, H., Ludwig, S. & Mellmann, A. 2020 A pandemic in times of global tourism: superspreading and exportation of COVID-19 cases from a ski area in Austria. *Journal of Clinical Microbiology* **58** (6), e00588–20.

D'Aoust, P. M., Mercier, E., Montpetit, D., Jia, J. J., Alexandrov, I., Neault, N., Baig, A. T., Mayne, J., Zhang, X., Alain, T., Langlois, M. A., Servos, M. R., MacKenzie, M., Figeys, D., MacKenzie, A. E., Graber, T. E. & Delatolla, R. 2021 Quantitative analysis of SARS-CoV-2 RNA from wastewater solids in communities with low COVID-19 incidence and prevalence. *Water Research* **188**, 116560.

European commission 2021 *Commission Recommendation of 17.3.2021 on a Common Approach to Establish a Systematic Surveillance of SARS-CoV-2 and its Variants in Wastewaters in the EU*. Available from: https://ec.europa.eu/environment/pdf/water/recommendation_covid19_monitoring_wastewaters.pdf (accessed 18 December 2021).

Fomsgaard, A. S. & Rosenstierne, M. W. 2020 An alternative workflow for molecular detection of SARS-CoV-2 – escape from the NA extraction kit-shortage, Copenhagen, Denmark, March 2020. *Eurosurveillance* **25** (14), 2000398.

Gerrity, D., Papp, K., Stoker, M., Sims, A. & Frehner, W. 2021 Early-pandemic wastewater surveillance of SARS-CoV-2 in Southern Nevada: methodology, occurrence, and incidence/prevalence considerations. *Water Research X* **10**, 100086.

Haramoto, E., Kitajima, M., Kishida, N., Konno, Y., Katayama, H., Asami, M. & Akiba, M. 2013 Occurrence of pepper mild mottle virus in drinking water sources in Japan. *Applied and Environmental Microbiology* **79** (23), 7413–7418.

Harcourt, J., Tamin, A., Lu, X., Kamili, S., Sakthivel, S. K., Murray, J., Queen, K., Tao, Y., Paden, C. R., Zhang, J., Li, Y., Uehara, A., Wang, H., Goldsmith, C., Bullock, H. A., Wang, L., Whitaker, B., Lynch, B., Gautam, R., Schindewolf, C., Lokugamage, K. G., Scharton, D., Plante, J. A., Mirchandani, D., Widen, S. G., Narayanan, K., Makino, S., Ksiazek, T. G., Plante, K. S., Weaver, S. C., Lindstrom, S., Tong, S., Menachery, V. D. & Thornburg, N. J. 2020 Severe acute respiratory syndrome coronavirus 2 from patient with coronavirus disease, United States. *Emerging Infectious Diseases* **26** (6), 1266–1273.

Hasan, S. W., Ibrahim, Y., Daou, M., Kannout, H., Jan, N., Lopes, A., Alsafar, H. & Yousef, A. F. 2021 Detection and quantification of SARS-CoV-2 RNA in wastewater and treated effluents: surveillance of COVID-19 epidemic in the United Arab Emirates. *Science of the Total Environment* **764**, 142929.

Haug, N., Geyrhofer, L., Londei, A., Dervic, E., Desvars-Larrive, A., Loreto, V., Pinior, B., Thurner, S. & Klimek, P. 2020 Ranking the effectiveness of worldwide COVID-19 government interventions. *Nature Human Behaviour* **4** (12), 1303–1312.

Hillary, L. S., Farkas, K., Maher, K. H., Lucaci, A., Thorpe, J., Distaso, M. A., Gaze, W. H., Paterson, S., Burke, T., Connor, T. R., McDonald, J. E., Malham, S. K. & Jones, D. L. 2021 Monitoring SARS-CoV-2 in municipal wastewater to evaluate the success of lockdown measures for controlling COVID-19 in the UK. *Water Research* **200**, 117214.

Jones, D. L., Baluja, M. Q., Graham, D. W., Corbishley, A., McDonald, J. E., Malham, S. K., Hillary, L. S., Connor, T. R., Gaze, W. H., Moura, I. B., Wilcox, M. H. & Farkas, K. 2020 Shedding of SARS-CoV-2 in feces and urine and its potential role in person-to-person transmission and the environment-based spread of COVID-19. *Science of the Total Environment* **749**, 141364.

Larsen, D. A., Collins, M. B., Du, Q., Hill, D., Insaf, T. Z., Kilaru, P., Kmush, B. L., Middleton, F., Stamm, A., Wilder, M. L., Zeng, T. & Green, H. 2021 Coupling freedom from disease principles and early warning from wastewater surveillance to improve health security. *medRxiv*. https://doi.org/10.1101/2021.06.11.21258797.

Lu, X., Wang, L., Sakthivel, S. K., Whitaker, B., Murray, J., Kamili, S., Lynch, B., Malapati, L., Burke, S. A., Harcourt, J., Tamin, A., Thornburg, N. J., Villanueva, J. M. & Lindstrom, S. 2020 US CDC real-time reverse transcription PCR panel for detection of severe acute respiratory syndrome coronavirus 2. *Emerging Infectious Disease* **26** (8), 1654–1665.

Mao, K., Zhang, K., Du, W., Ali, W., Feng, X. & Zhang, H. 2020 The potential of wastewater-based epidemiology as surveillance and early warning of infectious disease outbreaks. *Current Opinion in Environmental Science and Health* **17**, 1–7.

Medema, G., Been, F., Heijnen, L. & Petterson, S. 2020a Implementation of environmental surveillance for SARS-CoV-2 virus to support public health decisions: opportunities and challenges. *Current Opinion in Environmental Science and Health* **17**, 49–71.

Medema, G., Heijnen, L., Elsinga, G., Italiaander, R. & Brouwer, A. 2020b Presence of SARS-Coronavirus-2 RNA in sewage and correlation with reported COVID-19 prevalence in the early stage of the epidemic in the Netherlands. *Environmental Science and Technology Letters* **7** (7), 511–516.

Olesen, S. W., Imakaev, M. & Duvallet, C. 2021 Making waves: defining the lead time of wastewater-based epidemiology for COVID-19. *Water Research* **202**, 117433.

Pecson, B. M., Darby, E., Haas, C. N., Amha, Y. M., Bartolo, M., Danielson, R., Dearborn, Y., Di Giovanni, G., Ferguson, C., Fevig, S., Gaddis, E., Gray, D., Lukasik, G., Mull, B., Olivas, L., Olivieri, A., Qu, Y. & Consortium, S. A.-C.-I. 2021 Reproducibility and sensitivity of 36 methods to quantify the SARS-CoV-2 genetic signal in raw wastewater: findings from an interlaboratory methods evaluation in the U.S. *Environmental Science: Water Research and Technology* **7**, 504–520.

Popa, A., Genger, J. W., Nicholson, M. D., Penz, T., Schmid, D., Aberle, S. W., Agerer, B., Lercher, A., Endler, L., Colaco, H., Smyth, M., Schuster, M., Grau, M. L., Martinez-Jimenez, F., Pich, O., Borena, W., Pawelka, E., Keszei, Z., Senekowitsch, M., Laine, J., Aberle, J. H., Redlberger-Fritz, M., Karolyi, M., Zoufaly, A., Maritschnik, S., Borkovec, M., Hufnagl, P., Nairz, M., Weiss, G., Wolfinger, M. T., von Laer, D., Superti-Furga, G., Lopez-Bigas, N., Puchhammer-Stockl, E., Allerberger, F., Michor, F., Bock, C. & Bergthaler, A. 2020 Genomic epidemiology of superspreading events in Austria reveals mutational dynamics and transmission properties of SARS-CoV-2 Alexandra. *Science Translational Medicine* **12** (573), eabe2555.

Rusinol, M., Zammit, I., Itarte, M., Fores, E., Martinez-Puchol, S., Girones, R., Borrego, C., Corominas, L. & Bofill-Mas, S. 2021 Monitoring waves of the COVID-19 pandemic: inferences from WWTPs of different sizes. *Science of the Total Environment* **787**, 147463.

Saththasivam, J., El-Malah, S. S., Gomez, T. A., Jabbar, K. A., Remanan, R., Krishnankutty, A. K., Ogunbiyi, O., Rasool, K., Ashhab, S., Rashkeev, S., Bensaad, M., Ahmed, A. A., Mohamoud, Y. A., Malek, J. A., Abu Raddad, L. J., Jeremijenko, A., Abu Halaweh, H. A., Lawler, J. & Mahmoud, K. A. 2021 COVID-19 (SARS-CoV-2) outbreak monitoring using wastewater-based epidemiology in Qatar. *Science of the Total Environment* **774**, 145608.

Schrader, C., Schielke, A., Ellerbroek, L. & Johne, R. 2012 PCR inhibitors – occurrence, properties and removal. *Journal of Applied Microbiology* **113** (5), 1014–1026.

Thompson, J. R., Nancharaiah, Y. V., Gu, X., Lee, W. L., Rajal, V. B., Haines, M. B., Girones, R., Ng, L. C., Alm, E. J. & Wuertz, S. 2020 Making waves: wastewater surveillance of SARS-CoV-2 for population-based health management. *Water Research* **184**, 116181.

van Nuijs, A. L. N., Lai, F. Y., Been, F., Andres-Costa, M. J., Barron, L., Baz-Lomba, J. A., Berset, J.-D., Benaglia, L., Bijlsma, L., Burgard, D., Castiglioni, S., Christophoridis, C., Covaci, A., de Voogt, P., Emke, E., Fatta-Kassinos, D., Fick, J., Hernandez, F., Gerber, C., González-Mariño, I., Grabic, R., Gunnar, T., Kannan, K., Karolak, S., Kasprzyk-Hordern, B., Kokot, Z., Krizman-Matasic, I., Li, A., Li, X., Löve, A. S. C., Lopez de Alda, M., McCall, A.-K., Meyer, M. R., Oberacher, H., O'Brien, J., Quintana, J. B., Reid, M., Schneider, S., Simoes, S. S., Thomaidis, N. S., Thomas, K., Yargeau, V. & Ort, C. 2018 Multi-year inter-laboratory exercises for the analysis of illicit drugs and metabolites in wastewater: development of a quality control system. *TrAC – Trends in Analytical Chemistry* **103**, 34–43.

Weiss, A., Jellingso, M. & Sommer, M. O. A. 2020 Spatial and temporal dynamics of SARS-CoV-2 in COVID-19 patients: a systematic review and meta-analysis. *EBioMedicine* **58**, 102916.

Wigginton, K. R., Ye, Y. & Ellenberg, R. M. 2015 Emerging investigators series: the source and fate of pandemic viruses in the urban water cycle. *Environmental Science: Water Research and Technology* **1** (6), 735–746.

Wölfel, R., Corman, V. M., Guggemos, W., Seilmaier, M., Zange, S., Muller, M. A., Niemeyer, D., Jones, T. C., Vollmar, P., Rothe, C., Hoelscher, M., Bleicker, T., Brunink, S., Schneider, J., Ehmann, R., Zwirglmaier, K., Drosten, C. & Wendtner, C. 2020 Virological assessment of hospitalized patients with COVID-2019. *Nature* **581** (7809), 465–469.

World Health Organization 2020 *Status of Environmental Surveillance for SARS-CoV-2 Virus*. Available from: https://www.who.int/news-room/commentaries/detail/status-of-environmental-surveillance-for-sars-cov-2-virus (accessed 18 December 2021)

Wu, F. Q., Zhang, J. B., Xiao, A., Gu, X. Q., Lee, W. L., Armas, F., Kauffman, K., Hanage, W., Matus, M., Ghaeli, N., Endo, N., Duvallet, C., Poyet, M., Moniz, K., Washburne, A. D., Erickson, T., Chai, P., Thompson, J. & Alm, E. 2020 SARS-CoV-2 titers in wastewater are higher than expected from clinically confirmed cases. *mSystems* **5** (4), e00614–20.

Wu, F., Xiao, A., Zhang, J., Moniz, K., Endo, N., Armas, F., Bushman, M., Chai, P. R., Duvallet, C., Erickson, T. B., Foppe, K., Ghaeli, N., Gu, X., Hanage, W. P., Huang, K. H., Lee, W. L., McElroy, K. A., Rhode, S. F., Matus, M., Wuertz, S., Thompson, J. & Alm, E. J. 2021 Wastewater surveillance of SARS-CoV-2 across 40 U.S. states from February to June 2020. *Water Research* **202**, 117400.

Ye, Y., Ellenberg, R. M., Graham, K. E. & Wigginton, K. R. 2016 Survivability, partitioning, and recovery of enveloped viruses in untreated municipal wastewater. *Environmental Science and Technology* **50** (10), 5077–5085.

Zhang, T., Breitbart, M., Lee, W. H., Run, J. Q., Wei, C. L., Soh, S. W., Hibberd, M. L., Liu, E. T., Rohwer, F. & Ruan, Y. 2006 RNA viral community in human feces: prevalence of plant pathogenic viruses. *PLoS Biology* **4** (1), e3.

First received 5 September 2021; accepted in revised form 26 December 2021. Available online 19 January 2022

SARS-CoV-2 in Water doi: 10.2166/wh.2021.275

Retrospective epidemiological analysis of SARS-CoV-2 wastewater surveillance and case notifications data – New South Wales, Australia, 2020

H. S. Camphor[a,*], S. Nielsen[b], Z. Bradford-Hartke[a], K. Wall[a] and R. Broome[a]

[a] Health Protection New South Wales, NSW Ministry of Health, St Leonards, NSW 2065, Australia
[b] NSW Biostatistics Training Program, Centre for Epidemiology and Evidence, NSW Ministry of Health, St Leonards, NSW 2065, Australia
*Corresponding author. E-mail: hendrik.camphor@health.nsw.gov.au

HSC, 0000-0002-3416-1975; SN, 0000-0001-6855-4227; KW, 0000-0003-3036-1459

ABSTRACT

This epidemiological study analysed SARS-CoV-2 wastewater surveillance and case notifications data to inform evidence-based public health action in NSW. We investigated measures of association between SARS-CoV-2 RNA fragments detected in wastewater samples ($n = 100$) and case notifications ($n = 1,367$, as rates per 100,000 population) within wastewater catchment areas ($n = 6$); and evaluated the performance of wastewater testing as a population-level diagnostic tool. Furthermore, we modelled SARS-CoV-2 RNA fragment detection in wastewater given the case notification rate using logistic regression. The odds of a viral detection in wastewater samples increased by a factor of 5.68 (95% CI: 1.51–32.1, $P = 0.004$) with rates of one or more notified cases within a catchment. The diagnostic specificity of wastewater viral detection results was 0.88 (95% CI: 0.69–0.97); the overall diagnostic sensitivity was 0.44 (95% CI: 0.33–0.56). The probability of a viral detection result in wastewater exceeded 50% (95% CI: 36–64%) once the case rate within a catchment exceeded 10.5. Observed results suggest that in a low prevalence setting, wastewater viral detections are a more reliable indicator of the presence of recent virus shedding cases in a catchment, than non-detect results are of the absence of cases in a catchment.

Key words: case notifications, COVID-19, public health response, SARS-CoV-2 surveillance, viral detection, wastewater epidemiology

HIGHLIGHTS

- Research was undertaken to investigate SARS-CoV-2 wastewater surveillance findings in relation to observed COVID-19 case notifications data in NSW, Australia.
- In a low prevalence setting such as in NSW, wastewater viral detections, particularly in catchments with no recent case notifications should trigger prompt public health action to rule out low-level virus circulation or undetected community transmission.

INTRODUCTION

In New South Wales (NSW), Australia, COVID-19 containment efforts achieved significant early successes in suppressing the epidemic to very low levels of community transmission (McAnulty & Ward 2020). Wastewater collection for SARS-CoV-2 was initiated within 3 months of the first case detection in NSW (January 2020) and has since been utilised as a supplementary surveillance tool to support the state's COVID-19 response efforts (NSW Health Environmental Health Branch 2020). Formal wastewater testing under the NSW Sewage Surveillance Program for SARS-CoV-2 commenced in July 2020 (NSW Health Environmental Health Branch 2020).

In Australia, wastewater-based epidemiological approaches have long been utilised to supplement enterovirus surveillance (Australian Government Department of Health 2020), and its potential has also been recently demonstrated for SARS-CoV-2 (Ahmed *et al.* 2020a). Despite the rapid development of molecular diagnostic assays for wastewater testing, there is still a paucity of evidence to support interpretation and risk-based decision making arising from wastewater-based surveillance for SARS-CoV-2 in metropolitan wastewater catchments in Australia.

This study analysed wastewater surveillance and notifications data collected in the months preceding active commencement of the NSW Sewage Surveillance Program for SARS-CoV-2 and aimed to support the interpretation of initial SARS-CoV-2 wastewater surveillance findings, in relation to observed COVID-19 case notifications data. The study objectives were threefold. Firstly, to investigate measures of association between two comparison groups, namely catchments with, or without viral RNA fragment detections in wastewater samples, relative to the presence of notified SARS-CoV-2 shedding cases per 100,000 population allocated to the catchment. Secondly, to investigate the diagnostic performance of wastewater testing, where viral detection results were considered as a population-level diagnostic tool for the presence of notified SARS-CoV-2 shedding cases allocated to a catchment. Finally, to model the predicted probabilities of SARS-CoV-2 RNA fragment detection in wastewater samples by notified cases per 100,000 population in a catchment.

METHODS

Wastewater sampling period and testing

Sydney's major water utility (Sydney Water) collected one hundred 24-hour composite samples from seven wastewater treatment plant inlet points servicing six catchment areas in greater metropolitan Sydney over a 14-week sampling period, between late-March and July 2020. To support internal validation of wastewater testing protocols, samples were collected across a range of locations and catchment populations on dates where, based on COVID-19 case notifications data received, testing was anticipated to return both negative and positive results (i.e. including areas from zero to large case notification numbers). The largest wastewater catchment by population size (Malabar) included samples collected from two inlet points. These inlet points have a common upstream catchment, but also distinct catchment areas near the inlet. The inlets were considered separately as one inlet includes an industrial and port zone, with potential inputs from foreign vessels and interference from trade waste. Samples were stored frozen until tested for SARS-CoV-2 RNA fragments. Samples were filtered using electronegative membranes and SARS-CoV-2 RNA, or ribonucleic acid extracted using the RNeasy Power Water kit (Qiagen) as per manufacturer's instructions (Griffiths *et al.* 2021). Samples were analysed using an RT-qPCR assay targeting the N1 and N2 regions of the viral genome (US Centers for Disease Control and Prevention 2020; Griffiths *et al.* 2021; Payyappat 2021). For this study, results were expressed as a binary viral RNA target gene fragment detection (detection or non-detection).

Case notification reference period

Peak virus shedding into wastewater was assumed to occur during a 9-day period, starting 2 days prior and lasting until the 7th day from the calculated onset date (Cevik *et al.* 2021). The study, therefore, aimed to account for all notified (i.e. known) cases located in sampled catchments, with an onset corresponding with the wastewater sampling date (day 1), or ranging from 6 days prior (day 2–7 since onset), to 2 days after the wastewater sampling date (day −2, −1 before onset). Notifications data for laboratory-confirmed SARS-CoV-2 cases were extracted from the NSW Ministry of Health's Notifiable Conditions Information Management System (NCIMS) in November 2020. All notified cases with an onset ranging from 6 days prior to the first wastewater sample date, to 2 days beyond the last sample date were included, to correspond with the wastewater sampling period. These data were collated with additional data relevant to cases managed in the NSW hotel quarantine program (NSW Government 2020a), including accommodation location and dates of arrival, onset, testing and notification.

Allocating case location to wastewater catchments

Case locations were determined through a manual review of NCIMS case records. Unless found to have spent time in hotel quarantine and/or a healthcare facility, cases were assumed to have been residing at their primary residential address for the entirety of their 9-day viral shedding period. Quarantine hotel and healthcare facilities were manually assigned point coordinates using facility address and Google Maps (Google Maps 2020). Residential locations were geocoded using the Geocoded National Address File (GNAF) database and the NSW Health GOPHER server to obtain point coordinates. Cases were allocated to one or more catchments (one catchment, per day) by determining whether their point coordinates and 9-day period fell within sampled catchment boundaries and sampling dates, respectively. Geographic dispersal of case locations from one another within catchments by sampling date were analysed by calculating pairwise Euclidean distances between point coordinates and then case location distance to the group median centroid.

Data integration and analysis

Data cleaning were performed in R software (version 4.0.3) (R Core Team 2020), using methods from the packages 'base', 'stats' (R Core Team 2020), 'tidyverse' (Wickham *et al.* 2019), 'sf' (Pebesma 2018) and 'epitools' (Aragon 2020). Categorical data were compared using Fisher's exact test, with two-tailed P-values < 0.05 considered statistically significant. For univariable analysis, a case–control study approach was used to investigate associations between wastewater detection results (binary outcome), and the presence of notified cases allocated to catchments during their 9-day viral shedding period (categorical exposure). Contingency tables calculated odds ratios (OR) and 95% confidence intervals (CIs). Wastewater detection results were also considered as a population-level diagnostic tool for the presence of notified cases allocated to a catchment during their 9-day viral shedding period, to calculate diagnostic test performance statistics. Logistic regression was employed to model the log odds that either of two viral RNA gene fragments (N1 or N2) were detected in a wastewater sample, given case numbers and catchment population size. The log odds were converted to probabilities and graphed to facilitate interpretation.

Ethics statement

This study was conducted as part of public health surveillance under the *NSW Public Health Act (2010)*, so ethical approval was not required.

RESULTS

Descriptive epidemiology

There were $n = 1,367$ notified cases with an onset falling within the case notification reference period (24 March–11 July 2020) allocated to catchments from where wastewater samples were collected, within which reside a cumulative estimated population of 3,793,038 (period prevalence: 0.04%) (Table 1). There were $n = 1,450$ unique case locations identified (residence, quarantine accommodation or healthcare facility). There were $n = 2,277$ cases counted on wastewater sampling dates, noting that during their 9-day viral shedding period, cases could be counted more than once across wastewater samples. Over 90% of case counts were allocated to three catchments: Malabar (39.79%), Bondi (36.5%) and North Head (14.8%) (Table 1).

Of the wastewater samples collected between 30 March and 9 July 2020, 36/100 (36%) returned a detection of either the N1 or N2 viral RNA fragment (the 'case' to 'control' ratio of 1:1.8) (Table 2). All sampled catchments returned at least two wastewater results with viral detection, with 27/36 (75%) detections occurring during the first month of sampling in April, and 33/36 (92%) during the first half of the sampling period (Figure 1). Of the 100 wastewater samples tested, 42% returned a non-detect result despite having one or more cases allocated to the catchment; while 3% returned viral detections with no notified cases allocated to the catchment (Table 2).

Of the 75% of wastewater samples with one or more cases allocated to the catchment at the time of sampling, the median ratio of allocated cases per 100,000 population in these catchments was 1.8 (range: 0.05–43.6). Of the 42% of samples that returned non-detect results but with one or more cases allocated to the catchment, the median ratio of cases per 100,000 population in these catchments was 0.5 (range: 0.05–29.5). Of note, seven consecutive non-detect samples collected from the Bondi catchment in the latter half of the sampling period (June–July) had a median ratio of 14.25 allocated cases per 100,000 population (range: 7.2–29.5). Of the 3/100 (3%) wastewater samples which returned viral detections with no notified cases allocated to the catchment, two were from Brooklyn catchment in April, and one from St Mary's catchment in May (Figure 1).

Table 1 | Cumulative case counts on wastewater sampling days and estimated resident population of sampled wastewater catchments in greater metropolitan Sydney

Wastewater catchments	Cumulative case count on wastewater sampling days (%)		Estimated resident population (%)	
Brooklyn	0	0	1,241	(0.03)
Penrith	70	(3.10)	110,114	(2.90)
St Marys	133	(5.84)	163,147	(4.30)
Bondi	831	(36.50)	318,810	(8.41)
North Head	337	(14.80)	1,341,986	(35.38)
Malabar	906	(39.79)	1,857,740	(48.98)
Total	**2,277**	**(100.0)**	**3,793,038**	**(100.0)**

Table 2 | Contingency tables comparing viral detection results in wastewater with COVID-19 case notification rates

		'Case' Catchment day with viral detection	'Control' Catchment day without viral detection	Crude odds ratio (95% CI)	*P*-value
Case notifications allocated to wastewater catchment ('Exposure' presence or absence)	All catchments	$n = 36$	$n = 64$		
	Yes	33	42	5.68 (1.51–32.1)	0.004
	No	3	22		
	Catchments with under 1 notified case per 100,000 population	$n = 8$	$n = 51$		
	Yes	5	29	1.26 (0.22–8.99)	0.79
	No	3	22		
	Catchments with 1–9 notified cases per 100,000 population	$n = 20$	$n = 28$		
	Yes	17	6	19.04 (3.84–136)	<0.001
	No	3	22		
	Catchments with 10–50 notified cases per 100,000 population	$n = 14$	$n = 29$		
	Yes	11	7	10.74 (2.08–77.6)	0.001
	No	3	22		

Case notification rates were further compared with viral RNA detection upon stratification at <1, 1–9 and 10–50 notified cases per 100,000 population.

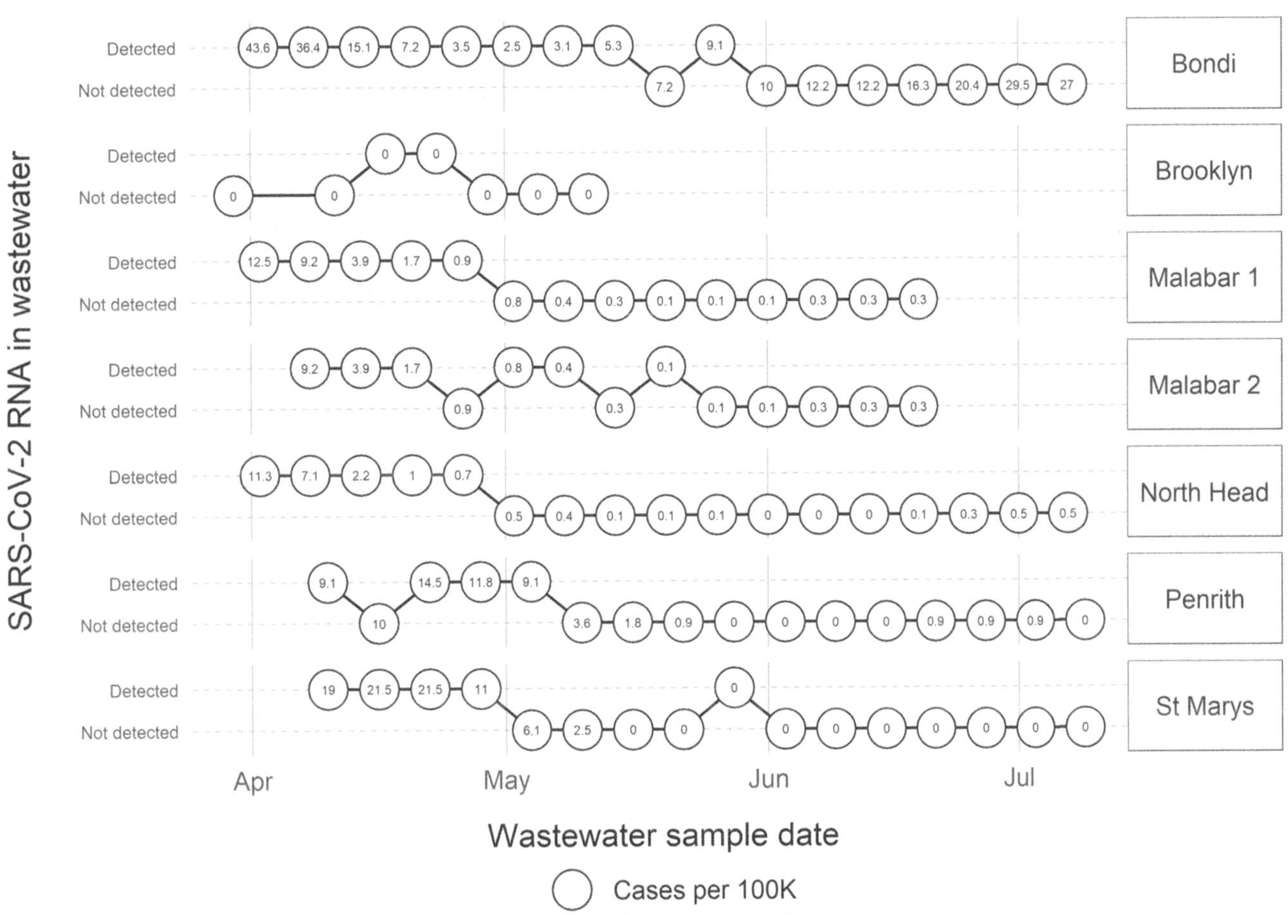

Figure 1 | SARS-CoV-2 RNA fragment detection results and corresponding case notification rates per 100,000 population (within white circles) for 100 samples collected from Sydney wastewater treatment plant inlet points between 30 March and 9 July 2020. Viral detections were either N1 or N2 positive, or both.

A quantitative comparison of the distance between notified case locations per sample date and catchment revealed that wastewater viral detections were observed with wider geographic dispersal of case locations from one another within a catchment, compared with non-detect wastewater results where cases were geographically more clustered. This effect was particularly evident in the Bondi catchment (Figure 2).

Analytical epidemiology

Univariable analysis

Overall, 92% of catchments with viral detection results had one or more cases allocated to the catchment at the time of wastewater sampling, compared with 66% of catchments with non-detect results. The odds of a viral detection result on a given day at a given sampling site were increased in wastewater catchments with one or more cases allocated to the catchment by a factor of 5.68 (95% CI: 1.51–32.1, $P = 0.004$). The OR increased to 19.04 (95% CI: 3.84–136, $P < 0.001$) and 10.74 (95% CI: 2.08–77.6, $P = 0.001$) in wastewater catchments with 1–9 cases, and 10–50 cases per 100,000 population, respectively. In catchments with fewer than 1 case per 100,000 population, there was no statistically significant association between viral detection results and case allocation to the catchment, compared with catchments with no cases (OR = 1.26, 95% CI: 0.22–8.99, $P = 0.79$) (Table 2).

Wastewater viral detection results were also considered as a population-level diagnostic tool for the presence of one or more notified cases allocated to a catchment during their 9-day viral shedding period. The proportion of catchments without cases allocated and corroborated by a non-detect result (specificity) was 0.88 (95% CI: 0.69–0.97). Overall, the proportion of catchments with one or more cases allocated and corroborated by a viral detection result (sensitivity) was 0.44 (95% CI: 0.33–0.56). Sensitivity ranged from 0.15 (95% CI: 0.05–0.31) in catchments with under one case, to 0.74 (95% CI: 0.52–0.9) in catchments with 1–9 cases per 100,000 population (Table 3).

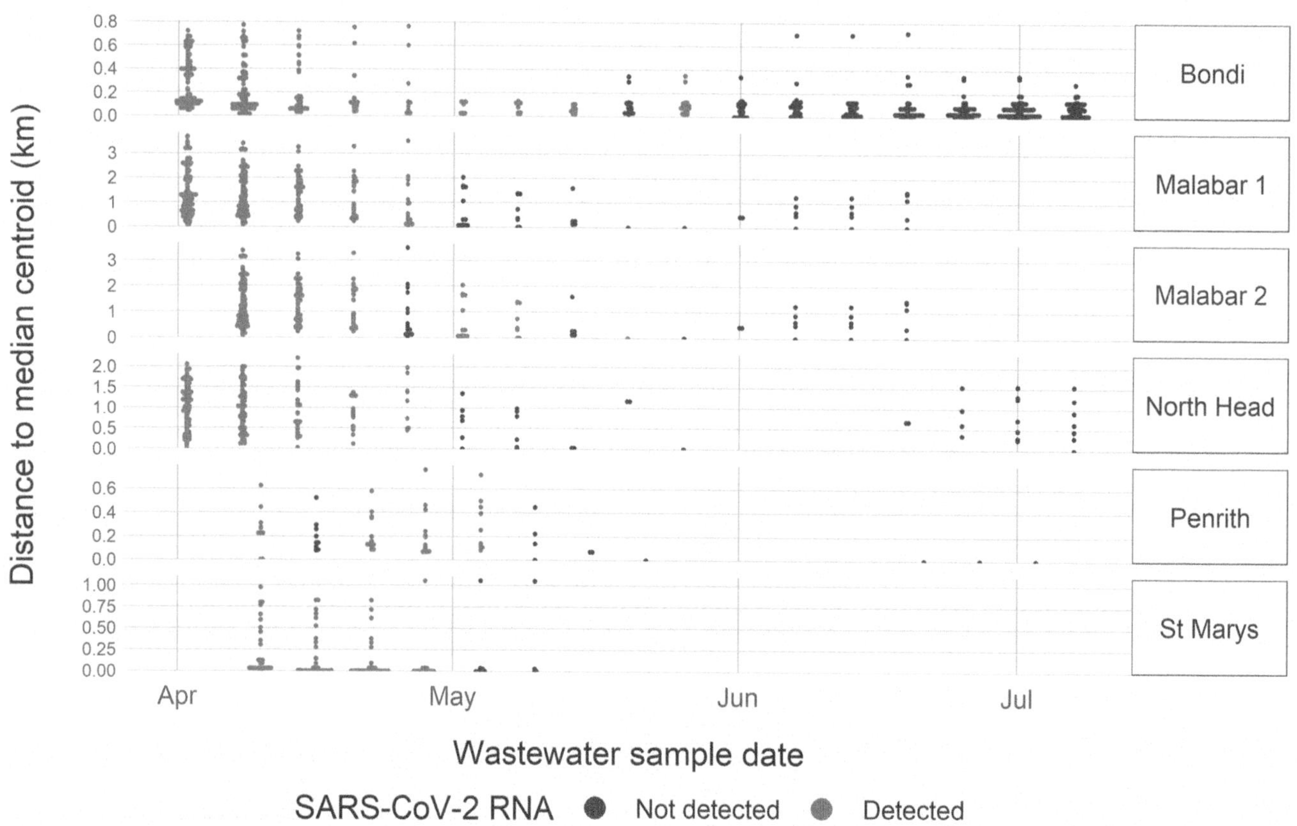

Figure 2 | Dispersion of notified COVID-19 case locations within wastewater catchments on sampling dates, by SARS-CoV-2 RNA fragment detection result. Points represent the case location distance to the group median centroid, calculated by the pairwise Euclidean distances of latitude and longitude of case locations within a catchment and sampling date.

Table 3 | Contingency tables and metrics for the performance of wastewater testing as a population-level diagnostic tool for case allocation to a catchment

		'Test' positive Catchment day with viral detection	'Test' negative Catchment day without viral detection	Diagnostic sensitivity (95% CI)	Diagnostic specificity (95% CI)	Predictive value of a positive (95% CI)	Predictive value of a negative (95% CI)	Positive likelihood ratio (95% CI)	Negative likelihood ratio (95% CI)
Case notifications allocated to wastewater catchment ('Disease' presence or absence)	All catchments								
		$n = 36$	$n = 64$						
	Yes	33	42	0.44 (0.33–0.56)	0.88 (0.69–0.97)	0.92 (0.78–0.98)	0.34 (0.24–0.47)	3.67 (1.72–7.83)	0.64 (0.47–0.86)
	No	3	22						
	Catchments with under 1 notified case per 100,000 population								
		$n = 8$	$n = 51$						
	Yes	5	29	0.15 (0.05–0.31)	0.88 (0.69–0.97)	0.62 (0.24–0.91)	0.43 (0.29–0.58)	1.23 (0.32–4.66)	0.97 (0.79–1.19)
	No	3	22						
	Catchments with 1–9 notified cases per 100,000 population								
		$n = 20$	$n = 28$						
	Yes	17	6	0.74 (0.52–0.90)	0.88 (0.69–0.97)	0.85 (0.64–0.95)	0.79 (0.60–0.90)	6.16 (2.19– 17.36)	0.30 (0.14–0.63)
	No	3	22						
	Catchments with 10–50 notified cases per 100,000 population								
		$n = 14$	$n = 29$						
	Yes	11	7	0.61 (0.36–0.83)	0.88 (0.69–0.97)	0.79 (0.52–0.92)	0.76 (0.58–0.88)	5.09 (1.75–14.85)	0.44 (0.23–0.86)
	No	3	22						

Diagnostic performance was compared with case notification rates stratified at <1, 1–9 and 10–50 notified cases per 100,000 population.

Regression analysis

The number of notified cases per 100,000 population was a statistically significant predictor of wastewater viral detection ($\chi^2 = 12.5$, df = 1, $P < 0.001$). Each additional case per 100,000 population resulted in an increase in the odds of viral detection by a factor 1.11 (95% CI: 1.04–1.19). The predicted probabilities were graphed to aid in interpretation of the model, noting that the relationship is non-linear on the probability scale (Figure 3). At zero, 10.5 and 21.3 notified cases per 100,000 population, the predicted probability of wastewater viral detection was 25, 50 and 75%, respectively. Wide confidence intervals were observed and regression deletion diagnostics (large DFFIT values) associated this with the seven consecutive non-detect results in wastewater samples collected from the Bondi catchment during June and July (median of 14.25 cases per 100,000) (Figure 1).

DISCUSSION

This study showed that during the 14-week wastewater sampling period, SARS-CoV-2 viral RNA fragment detections in wastewater samples showed a positive association at over 1 case per 100,000 population, and provided good predictive power at increasing case notification rates. The probability of a viral detection result exceeded 50% at over 10 cases per 100,000 population. As a population-level diagnostic tool for the presence of recent cases allocated to a catchment, wastewater testing results showed consistently higher specificity than sensitivity at various case notification rates; however, various factors including geographic clustering and distance between case location and wastewater sampling points could have attributed to the lower observed sensitivities.

In 2020, the daily rate of SARS-CoV-2 case notifications in NSW reached its highest peak some 10–14 days prior to the first wastewater sample collection on 30 March (NSW Health Public Health Response Branch 2020). Case notification rates then declined rapidly and remained comparatively much lower during the 14-week wastewater sampling period. This decline was largely attributable to Australia's external border closures on 20 March (Australian Government Media Release 2020), and mandatory hotel quarantine policy for all returning residents and citizens from 27 March onwards (NSW Government 2020a). 75% of wastewater viral detections occurred within the first month (April), and 92% during the first half of the sampling period. This can be explained by the initially much higher case numbers, and subsequent larger volume of residual viral shedding into wastewater from cases with onsets of 7 days and longer prior to the first sampling date. This finding is consistent with the results of an earlier European study which found a significant correlation between the strength of viral assay detection results and COVID-19 case prevalence (Medema *et al.* 2020).

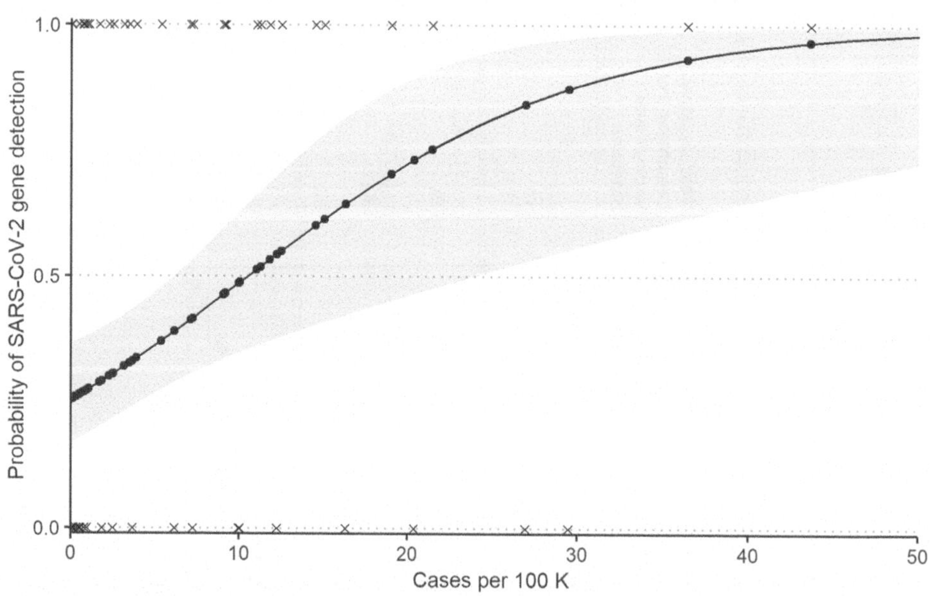

Figure 3 | Predicted probabilities from a logistic regression model of SARS-CoV-2 RNA fragment detection by cases per 100,000 population. The black points are fitted values of the samples used to create the model, and the line and shaded area are the fitted model and 95% confidence interval, respectively. The crosses are the observed viral detections and cases per 100,000 population.

Analysis of the performance of SARS-CoV-2 wastewater surveillance as a population-level diagnostic tool for the presence of cases in a catchment provides invaluable evidence to inform COVID-19 containment efforts. In this study, 56% of catchments with cases allocated on the sampling date returned non-detect wastewater results (false negatives), whereas only 12% of catchments without cases allocated returned viral detections (false positives). Combined with interpretation of the odds ratios (ORs), these findings suggest that SARS-CoV-2 wastewater surveillance may be a highly specific modality for verifying the presence, but not a highly sensitive modality for verifying the absence of recent cases within the metropolitan wastewater catchments sampled. The lower sensitivity suggests that a non-detect wastewater result should not provide a high degree of reassurance that viral shedding cases are indeed not present within the catchment. Conversely, a higher specificity may provide actionable public health intelligence, particularly where wastewater viral detections occur in catchments with few or no notified (i.e. known) cases. Such a finding may point to suboptimal clinical surveillance sensitivity or undetected, low-level virus circulation in the population serviced by the wastewater catchment. This may justify prompt public health action, e.g. issuing public health alerts to maintain high community testing rates, or risk communication targeting vulnerable subpopulations residing in the catchment, e.g. residential aged-care facilities or culturally diverse communities. The influence of disease prevalence on the interpretation of diagnostic test performance metrics is well described (Leeflang *et al.* 2013; Baeyens *et al.* 2019). Considering the very low estimate of SARS-CoV-2 prevalence in sampled catchments during the sampling period (0.04%), the individual observed estimates for sensitivity, specificity, predictive values for detect or non-detect wastewater results, and likelihood ratios should be interpreted with caution. Rather, the inference that wastewater viral detections are a more reliable indicator of the presence of recent virus shedding cases in a catchment, than non-detect results are of the absence of cases in a catchment is based on the consistent trend observed in these estimates across varying case rates per catchment population size, and supported by our interpretation of the positive odds of viral detections in wastewater samples collected from catchments with one or more cases allocated to the catchment.

Study limitations

There are various assumptions and limitations inherent to the methodologies applied in this study. Data captured in NCIMS are dynamic in nature; therefore, the presented data presents a point-in-time analysis of SARS-CoV-2 case notifications and may differ from other reports covering the same period.

We utilised a case–control study approach to investigate measures of association between wastewater viral detection results and case allocation to a catchment. There was an equal probability of incorrect allocation or non-allocation of cases to a catchment (non-differential misclassification). This would be expected to bias the measures of association towards no observable effect. However, the ORs showed a statistically significant and strongly positive association between wastewater viral detections and case allocation to a catchment compared with catchments without allocated case notifications, except where there were fewer than 1 case per 100,000 catchment population. This suggests that the observed results are a true effect. Wide confidence intervals for the observed estimates are likely due to the small sample size; therefore, the trend observed across the variable catchment population sizes is more informative than any one result. Case–control studies require control selection to be independent of exposure status. To validate the molecular diagnostic assay protocol used, historical samples were selected for analysis with knowledge of notifications data. Selection bias may have occurred for those catchments that returned non-detect wastewater results, if these were more likely to have been selected because there were no case notifications reported in the catchment at the time of sampling.

The various possible reasons for misclassification bias (i.e. incorrect allocation or non-allocation of cases to catchments on sampling days) deserve attention. Possible explanations for false negatives observed (i.e. a non-detect wastewater result, despite having one or more cases allocated to the catchment) are many and varied. Upon confirmation of their positive test results, Public Health Order (No 4) 2020 under the *NSW Public Health Act (2010)* required all cases to immediately self-isolate for a period no less than 10 days from onset (NSW Government 2020b). Despite health authorities urging persons to self-isolate upon any symptom development and following SARS-CoV-2 testing, there was no legal requirement to do so until being informed of a positive test result. Unless cases were verified to have been in a quarantine hotel or healthcare facility during their 9-day viral shedding period of interest, we assumed that upon notification of a positive test result, cases would have been self-isolating at their primary residential address recorded in NCIMS. However, we had no readily accessible way of verifying whether the address was correctly recorded, or whether cases were indeed located there during their 9-day viral shedding period of interest. Therefore, cases may have been allocated to a catchment, when in fact they were not present.

Physicochemical and environmental factors, as well as case clustering within a catchment, and cases' geographic location relative to, or distance from the wastewater sampling point could also be important determinants of sensitivity. A combination of these factors could have contributed towards the observed sensitivity of wastewater viral detection results in this study. As enveloped virions, the infectivity of coronaviruses are known to be adversely affected by factors within the sewer network such as elevated temperature, pH, solids, other microorganisms and degradation by detergents or pollutants (Gundy *et al.* 2008; Foladori *et al.* 2020). However, data suggest that the SARS-CoV-2 RNA is likely to remain detectable in wastewater for days to weeks, if temperature is controlled to limit viral degradation once sampled (Ahmed *et al.* 2020b; Bivins *et al.* 2020). Detection in wastewater samples can also be limited by infiltration and ingress of stormwater during high rainfall events causing dilution, inadequate mixing in the network and inhibitory substances in trade waste which can interfere with sample analysis. The impact of such limiting factors may become more pronounced where there is geographic clustering of cases; and further influenced by the distance between case location and the wastewater sampling point. This was particularly evident in the Bondi catchment, where seven consecutive wastewater samples collected during June and July retuned non-detect results, despite case allocations on relevant sampling dates ranging from 7 to 29 cases per 100,000 catchment population. At this stage of the NSW response, most cases allocated to the catchment were confined to hotel quarantine and tightly geographically clustered within the inner-city limits. This effect may be less likely to occur where virus shedding cases are more geographically dispersed across a wastewater catchment area.

There are also various possible explanations for false positives (i.e. a wastewater viral detection, despite having no cases allocated to the catchment). The same uncertainty surrounding cases' true location during their period of mandatory self-isolation that may result in incorrect allocation to a catchment, could also result in cases being non-allocated to a catchment, when they were in fact present. Furthermore, despite very high clinical testing rates in NSW by international comparison (Hasell *et al.* 2020; NSW Health Public Health Response Branch 2020) and an effective contact tracing capability, it is accepted that case notifications represent an undercount of true case numbers. SARS-CoV-2 shedding cases, therefore, may have been present in one or more catchments without the knowledge of health authorities, and therefore not allocated to said catchments. Furthermore, we assumed peak viral shedding occurs during a 9-day period (−2 to 7 days from onset) and restricted our study period of interest. However, it is well-described that SARS-CoV-2 shedding into wastewater may continue for many weeks after clinical recovery (Cevik *et al.* 2021), and well beyond the first week from symptom onset. Therefore, wastewater viral detections could be attributable not only to known cases within their 9-day viral shedding period allocated to the catchment, but also earlier, clinically recovered cases with prolonged virus shedding, or asymptomatic infected persons living, working or visiting a catchment. Of the false-positive wastewater samples, 2/3 (66%) were collected during the NSW school holidays from the Brooklyn catchment, which had the smallest catchment population in this study by far. It is possible that these viral detections were due to one or more non-notified cases, or cases non-allocated for reasons described earlier (e.g. visitors). These findings raise questions whether the sensitivity of wastewater viral detections as an indicator of viral shedding cases may be higher with smaller catchment populations, and/or in higher prevalence settings. This should be an area of further research, particularly considering the potential for wastewater viral detections to provide early warning of undetected cases in non-urban catchment populations, or in vulnerable communities.

CONCLUSION

Field epidemiological approaches are an important tool to integrate and analyse SARS-CoV-2 wastewater surveillance results and case notifications data to inform evidence-based public health action. The results of this study suggest that wastewater testing for SARS-CoV-2 is a valuable supplementary surveillance component to support the public health response in NSW. Considered together, the observed results suggest that in a low prevalence setting, wastewater viral detections are a more reliable indicator of the presence of recent virus shedding cases in a catchment, than non-detect results are of the absence of cases in a wastewater catchment. Wastewater viral detections, particularly in catchments servicing populations with no recent case notifications, should trigger prompt public health investigations to rule out low-level virus circulation or undetected community transmission. This may be supported by the development of upstream sampling methodologies to better localise areas from which wastewater viral detection signals are emanating, thereby allowing health authorities to target public health resources to the highest risk areas or communities. Further research is required to better understand the relationship between prolonged virus shedding and wastewater viral detections. Further research is also needed to explore alternative sampling strategies and testing methodologies to better characterise the sensitivity of wastewater viral detection in low prevalence settings, and in and catchments servicing variable populations sizes. Considering the study assumptions and limitations, the observed results do not exclude the potential of wastewater surveillance as a sensitive modality in certain settings, particularly to provide early warning of undetected cases. This may be particularly useful in smaller or non-urban catchments, vulnerable subpopulations, or in communities that are otherwise less inclined to engage with health authorities or present for clinical testing.

ACKNOWLEDGEMENTS

We thank Water Research Australia, Sydney Water, the NSW Sewage Surveillance Steering Committee, and colleagues in the Public Health Response, Environmental Health and Communicable Diseases branches of NSW Health. This research was funded by NSW COVID-19 Research Grant Funding. Shaun Nielsen was employed as a trainee on the NSW Biostatistics Training Program funded by the NSW Ministry of Health and collaborated on this work while deployed in the Public Health Response Branch (COVID-19).

DATA AVAILABILITY STATEMENT

Data cannot be made publicly available; readers should contact the corresponding author for details.

REFERENCES

Ahmed, W., Angel, N., Edson, J., Bibby, K., Bivins, A., O'Brien, J. W., Choi, P. M., Kitajima, M., Simpson, S. L., Li, J., Tscharke, B., Verhagen, R., Smith, W. J. M., Zaugg, J., Dierens, L., Hugenholtz, P., Thomas, K. V. & Mueller, J. F. 2020a First confirmed detection of SARS-CoV-2 in untreated wastewater in Australia: a proof of concept for the wastewater surveillance of COVID-19 in the community. *Science of The Total Environment* **728**, 138764.

Ahmed, W., Bertsch, P. M., Bibby, K., Haramoto, E., Hewitt, J., Huygens, F., Gyawali, P., Korajkic, A., Riddell, S., Sherchan, S. P., Simpson, S. L., Sirikanchana, K., Symonds, E. M., Verhagen, R., Vasan, S. S., Kitajima, M. & Bivins, A. 2020b Decay of SARS-CoV-2 and surrogate murine hepatitis virus RNA in untreated wastewater to inform application in wastewater-based epidemiology. *Environmental Research* **191**, 110092.

Aragon, T. J. 2020 *epitools: Epidemiology Tools. R Package Version 0.5-10.1*. Available from: https://CRAN.R-project.org/package = epitools (accessed 9 July 2021).

Australian Government 2020 *Office of the Prime Minister, Media Release (19 March 2020)*. Border Restrictions. Available from: https://www.pm.gov.au/media/border-restrictions (accessed 20 December 2020).

Australian Government Department of Health 2020 *Polio – Australian National Enterovirus Reference Laboratory Annual Reports*. Available from: https://www1.health.gov.au/internet/main/publishing.nsf/Content/cda-pubs-annlrpt-polioanrep.htm (accessed 17 December 2020).

Baeyens, J. P., Serrien, B., Goossens, M. & Clijsen, R. 2019 Questioning the 'SPIN and SNOUT' rule in clinical testing. *Archives of Physiotherapy* **9** (1), 4.

Bivins, A., Greaves, J., Fischer, R., Yinda, K. C., Ahmed, W., Kitajima, M., Munster, V. J. & Bibby, K. 2020 Persistence of SARS-CoV-2 in water and wastewater. *Environmental Science & Technology Letters* **7** (12), 937–942.

Cevik, M., Tate, M., Lloyd, O., Maraolo, A. E., Schafers, J. & Ho, A. 2021 SARS-CoV-2, SARS-CoV, and MERS-CoV viral load dynamics, duration of viral shedding, and infectiousness: a systematic review and meta-analysis. *The Lancet Microbe* **2** (1), e13–e22.

Foladori, P., Cutrupi, F., Segata, N., Manara, S., Pinto, F., Malpei, F., Bruni, L. & La Rosa, G. 2020 SARS-CoV-2 from faeces to wastewater treatment: what do we know? A review. *Science of The Total Environment* **743**, 140444.

Google Maps 2020 *Google Maps Version 10.56*. Google. Available from: https://www.google.com/maps (accessed 18 December 2020).

Griffiths, K. R., Emslie, K. R., Burke, D. G., Baoutina, A., Bhat, S., Forbes-Smith, M., Hall, F., Lynch, D., McLaughlin, J., Partis, L., Pinheiro, L. B. & Deere, D. 2021 *ColoSSoS Collaboration on Sewage Surveillance of SARS-CoV-2 Inter-Laboratory Study*. WaterRA Project 2071, National Measurement Institute and Water Research Australia Limited: 73. Available from: https://www.waterra.com.au/_r11479/media/system/attrib/file/2711/WaterRA_Project_2071_FINAL.pdf (accessed 30 September 2021).

Gundy, P. M., Gerba, C. P. & Pepper, I. L. 2008 Survival of coronaviruses in water and wastewater. *Food and Environmental Virology* **1** (1), 10.

Hasell, J., Mathieu, E., Beltekian, D., Macdonald, B., Giattino, C., Ortiz-Ospina, E., Roser, M. & Ritchie, H. 2020 A cross-country database of COVID-19 testing. *Scientific Data* **7** (1), 345.

Leeflang, M. M. G., Rutjes, A. W. S., Reitsma, J. B., Hooft, L. & Bossuyt, P. M. M. 2013 Variation of a test's sensitivity and specificity with disease prevalence. *CMAJ: Canadian Medical Association Journal = Journal de l'Association Medicale Canadienne* **185** (11), E537–E544.

McAnulty, J. M. & Ward, K. 2020 Suppressing the epidemic in New South Wales. *New England Journal of Medicine* **382** (21), e74.

Medema, G., Heijnen, L., Elsinga, G., Italiaander, R. & Brouwer, A. 2020 Presence of SARS-Coronavirus-2 RNA in sewage and correlation with reported COVID-19 prevalence in the early stage of the epidemic in The Netherlands. *Environmental Science & Technology Letters* **7** (7), 511–516.

NSW Government 2020a *COVID-related Legislation*. Available from: https://legislation.nsw.gov.au/information/covid19-legislation (accessed 20 December 2020).

NSW Government 2020b *Public Health (COVID-19 Self-Isolation) Order (No 5) 2020*. Available from: https://www.health.nsw.gov.au/Infectious/covid-19/Documents/self-isolation-order.pdf (accessed 9 July 2021).

NSW Ministry of Health, Health Environmental Health Branch 2020 *COVID-19 Sewage Surveillance Program*. Available from: https://www.health.nsw.gov.au/Infectious/covid-19/Pages/sewage-surveillance.aspx (accessed 17 December 2020).

NSW Ministry of Health, Public Health Response Branch 2020 *COVID-19 Weekly Surveillance Reports*. Available from: https://www.health.nsw.gov.au/Infectious/covid-19/Pages/weekly-reports.aspx (accessed 17 December 2020).

Payyappat, S. 2021 *Development of Sydney Water Method for SARS-CoV-2 Detection in Wastewater* (Manuscript in Preparation).

Pebesma, E. J. 2018 Simple features for R: standardized support for spatial vector data. *The R Journal* **10** (1), 439.

R Core Team 2020 *R: A Language and Environment for Statistical Computing*. Vienna, Australia. Available from: https://www.R-project.org/ (accessed 9 July 2021).

US Centers for Disease Control and Prevention 2020 *The CDC 2019 Novel Coronavirus (2019-nCoV) Real-Time Reverse Transcriptase (RT)–PCR Diagnostic Panel*. Available from: https://www.cdc.gov/coronavirus/2019-ncov/downloads/rt-pcr-panel-primer-probes.pdf (accessed 26 November 2020).

Wickham, H., Averick, M., Bryan, J., Chang, W., McGowan, L. D. A., François, R., Grolemund, G., Hayes, A., Henry, L. & Hester, J. 2019 Welcome to the Tidyverse. *Journal of Open Source Software* **4** (43), 1686.

First received 27 July 2021; accepted in revised form 5 October 2021. Available online 18 October 2021

SARS-CoV-2 in Water doi: 10.2166/wh.2022.183

COVID-19 case prediction via wastewater surveillance in a low-prevalence urban community: a modeling approach

Yifan Zhu [a,†], Wakana Oishi[b,†], Chikako Maruo [b], Sewwandi Bandara [a], Mu Lin[a], Mayuko Saito [c], Masaaki Kitajima [d] and Daisuke Sano [a,b,e,*]

[a] Department of Frontier Sciences for Advanced Environment, Graduate School of Environmental Studies, Tohoku University, Aoba 6-6-06, Aramaki, Aoba-ku, Sendai, Miyagi 980-8597, Japan

[b] Department of Civil and Environmental Engineering, Graduate School of Engineering, Tohoku University, Aoba 6-6-06, Aramaki, Aoba-ku, Sendai, Miyagi 980-8597, Japan

[c] Department of Virology, Graduate School of Medicine, Tohoku University, 2-1 Seiryo-machi, Aoba-ku, Sendai, Miyagi 980-8575, Japan

[d] Division of Environmental Engineering, Faculty of Engineering, Hokkaido University, North 13 West 8, Kita-ku, Sapporo, Hokkaido 060-8628, Japan

[e] Research Institute for Humanity and Nature, 457-4 Kamigamo Motoyama, Kita Ward, Kyoto 603-8047, Japan

*Corresponding author. E-mail: daisuke.sano.e1@tohoku.ac.jp

†These authors contributed equally to this work.

YZ, 0000-0002-6754-3590; CM, 0000-0001-8812-1840; SB, 0000-0001-5669-8602; MS, 0000-0003-4622-7027; DS, 0000-0001-8075-6972

ABSTRACT

Estimating and predicting the epidemic size from wastewater surveillance results remains challenging for the practical implementation of wastewater-based epidemiology (WBE). In this study, by employing a highly sensitive detection method, we documented the time series of SARS-CoV-2 RNA occurrence in the wastewater influent from an urban community with a 360,000 population in Japan, from August 2020 to February 2021. The detection frequency of the viral RNA increased during the outbreak events of COVID-19 and the highest viral RNA concentration was recorded at the beginning of January 2021, amid the most serious outbreak event during the study period. We found that: (1) direct back-calculation still suffers from great uncertainty dominated by inconsistent detection and the varying gap between the observed wastewater viral load and the estimated patient viral load, and (2) the detection frequency correlated well with reported cases and the prediction of the latter can be carried out via data-driven modeling methods. Our results indicate that wastewater virus occurrence can contribute to epidemic surveillance in ways more than back-calculation, which may spawn future wastewater surveillance implementations.

Key words: COVID-19 surveillance, data-driven modeling, epidemic prediction, wastewater-based epidemiology

HIGHLIGHTS

- Time series of SARS-CoV-2 wastewater presence in a low-prevalence community.
- SARS-CoV-2 detection frequency correlates with reported cases.
- Reported cases can be predicted by detection frequency within a period.
- Data-driven methods may facilitate wastewater-based epidemic modeling.

1. INTRODUCTION

The COVID-19 pandemic has caused over 5 million deaths as of 16 November 2021 (WHO 2021) and is still deeply interrupting the global society. In a new era when communities are adapting to revive, there is a substantial need for a surveillance tool that can identify both asymptomatic and symptomatic patients with high efficiency, preferably also having a short turnaround time, wide coverage, non-invasive, anonymous in nature, and with a low implementation cost.

Previous studies have confirmed the persistent shedding of SARS-CoV-2 viral RNA in the feces and urine from patients during the infection course (Wölfel *et al.* 2020) and the viral shedding at the asymptomatic stage (Tang *et al.* 2020; Zhang *et al.* 2020). A review on viral shedding demonstrated that the shedding lasts 14–28 days (Jones *et al.* 2020). Detections of SARS-CoV-2 viral RNA from the wastewater have been reported in a number of studies, including in low-prevalence areas (Haramoto *et al.* 2020; Randazzo *et al.* 2020; D'Aoust *et al.* 2021; Hata *et al.* 2021) and prior to the surge of an outbreak (Peccia *et al.* 2020). These studies provided a clear rationale to monitor the viral components in wastewater to perceive the spread of disease in a given catchment. Improved detection sensitivity had been reported by using the solid fraction of wastewater (Peccia *et al.* 2020; Balboa *et al.* 2021; Kitamura *et al.* 2021), and the continuous detection of viral RNA in the suspended solid rather than the supernatant throughout the surveillance period was reported (Kitamura *et al.* 2021). Pre-amplification of the target sequence is one option to improve sensitivity in RT-PCR assay. A previous study reported a 100-fold improvement of quantification limit in an RT-PCR assay for coronavirus in clinical tests (Lau *et al.* 2003). Therefore, it may also be suitable for wastewater samples that may contain a low amount of target genetic materials.

Although much has already been achieved regarding concentrating and quantifying SARS-CoV-2 from wastewater, wastewater surveillance implementations remain limited, mainly due to a lack of means to interpret the result. The efforts of converting the results into prevalence level, often referred to as 'back-calculation', initiated with the global emergence of detection reports. Being able to quantitatively interpret the wastewater surveillance outcome can greatly facilitate the situation assessment and countermeasure formation (Sims & Kasprzyk-Hordern 2020; Thompson *et al.* 2020). The mass balance function commonly seen in applications regarding drug and pesticide monitoring had been adopted by some recent studies. Briefly, the measured biomarker concentration is multiplied by the wastewater flow rate to obtain the total load (Ahmed *et al.* 2020a; Hart & Halden 2020; Fernandez-Cassi *et al.* 2021; Hasan *et al.* 2021; Saththasivam *et al.* 2021), which can then be divided by an average load to obtain an approximate shedder base. Besides, coefficients such as the degradation factor and the shedding ratio are often included to improve the estimation. However, as research went deeper, it was realized that the shedding profile of infected individuals is rather erratic and cannot be accurately represented by a fixed rate (Walsh *et al.* 2020; Buonerba *et al.* 2021), and other sources of uncertainty may further bring down the credibility of prevalence estimated in this manner (Li *et al.* 2021).

Unquantifiable data points did not receive sufficient consideration either. The wastewater virus concentration sometimes exceeds the limit of detection (LoD) or sensitivity threshold, but not the limit of quantification (LoQ), especially in the early stage of an epidemic or in a low-prevalence region. Data points that fall into this range can only be identified as 'positive' without a quantifiable value, making them binomial (either positive or negative). This creates a grey area of data as the previous mass balance approach fails to apply to data that fall within this range. Thus, it is critical to develop alternative modeling approaches that manage to extract information from these data, yet related studies are scarce.

This study aimed to analyze the relationship between the number of reported COVID-19 cases and the RT-qPCR-based viral RNA detection in wastewater influent. To achieve this, we monitored the time series of SARS-CoV-2 RNA concentrations in wastewater influent from Sendai city, Japan, from August 2020 to February 2021, with a highly sensitive detection method that consists of the recovery of viral RNA from wastewater solids and pre-amplification of cDNA before qPCR assay. Then, using the detection data, we present two metrics that help researchers extract information from the wastewater surveillance results. Specifically, we applied and evaluated two modeling approaches that apply to quantifiable and binomial wastewater virus detections, respectively.

2. METHODS

2.1. Sample collection

A total of 51 influent wastewater samples were collected from a municipal wastewater plant (WWTP) in Sendai, Miyagi, Japan that receives approximately 69% of wastewater generated in the city. Samples were taken twice a week, at 10 a.m., on Tuesday and Thursday from August 2020 to February 2021, from an influent line that serves the major urban area and about 360,000 people. All samples were grab samples (250 mL). Samples were immediately transported to the laboratory after collection and stored at −80 °C until analysis.

2.2. Virus concentration, nucleic acid extraction, and cDNA synthesis

SARS-CoV-2 RNA was recovered from 40 mL influent wastewater sample. The suspended solid was concentrated by centrifugation at 5,000 g for 10 min at 4 °C. After the supernatant was removed, 1 mL of TRIzol reagent (Thermo Fisher Scientific, MA, USA) was added to the concentrated suspended solid, then the suspension was homogenized using a vortex mixer. The total volume of the suspension was less than 3.5 mL. A 140 μL aliquot of the concentrate was processed for viral RNA extraction using the QIAamp Viral RNA Mini Kit (QIAGEN, Hilden, Germany) following the manufacturer's instructions, and the viral RNA was eluted in 60 μL of elution buffer provided in the kit. The whole process recovery of the SARS-CoV-2 RNA was verified based on the concentration of Pepper Mild Mottle Virus (PMMoV) in wastewater samples, and 140 uL aliquot was extracted from samples both before and after concentration. A 10 μL sample of extracted RNA was used to obtain 20 μL of cDNA with the high-capacity cDNA RT Kit (Thermo Fisher Scientific). To synthesize cDNA for SARS-CoV-2, the random primer included in the kit was substituted with CDC nCOV_N1-R Primer (CN 10006831, Integrated DNA Technologies, Inc., Iowa, USA) (10 μM) (CDC 2021).

2.3. SARS-CoV-2 genome pre-amplification and quantification by qPCR assay

The PCR-based pre-amplification method was applied to cDNA prior to the qPCR assay. The pre-amplification was performed with *TaKaRa Ex Taq*® Hot Start Version (Takara Bio, Kusatsu, Japan) and the CDC nCOV_N1 Primers (CN10006830 and CN10006831, Integrated DNA Technologies, Inc.) (CDC 2021). Each 50 μL reaction contained 20 μL of cDNA, 0.25 μL of *TaKaRa Ex Taq HS* (5 U/μL) (Takara Bio), 5 μL of 10×Ex Taq Buffer (Mg^{2+} plus) (20 mM) (Takara Bio), 4 μL of dNTP Mixture (Takara Bio), and 400 nM of forward and reverse primers. The PCR cycling condition was 2 min at 94 °C, followed by 10 cycles of 30 s at 94 °C, 30 s at 55 °C, and 1 min at 72 °C. The pre-amplification step was simultaneously applied to 18 or 20 μL of standard DNA (2.0 to 2.0_{104} copies/μL) created via 10-fold dilution series of 2019-nCoV_N_PositiveControl (CN10006625, Integrated DNA Technologies, Inc.). The pre-amplification step was also applied to 20 μL of TE buffer as the negative control for the pre-amplification and qPCR.

The concentrations of SARS-CoV-2 and PMMoV viral RNA were determined by real-time qPCR on a CFX96 real-time PCR detection system (Bio-Rad, Hercules, CA, USA). The amplification reaction was performed with SsoAdvanced Probes Supermix (Bio-Rad). For SARS-CoV-2 viral RNA, we used the same forward/reverse primers from the previous pre-amplification step with nCOV_N1 Probe Aliquot (CN10006832, Integrated DNA Technologies, Inc.) (CDC 2021), and RT-qPCR was performed only on pre-amplified samples. Each 20 μL reaction contained 5 μL of pre-amplified cDNA, 10 μL of SsoAdvanced Probes Supermix (Bio-Rad), 500 nM of forward and reverse primers, and 200 nM of fluorogenic probe. The PCR cycling condition was 30 s at 95 °C, followed by 40 cycles of 10 s at 95 °C and 30 s at 60 °C. The number of SARS-CoV-2 genome copies was determined by a standard curve generated with the pre-amped standard samples (1.0×10^1 to 1.0×10^5 copies/reaction). Each sample was quantified in triplicate. The amplification efficiency in the real-time PCR was at least 80%.

The concentration of SARS-CoV-2 viral RNA in the influent wastewater sample, $C_{v,w}$ (copies/mL), was determined by the following equation:

$$C_{v,w} = 1,000 \times \frac{V}{V_w} \times \frac{V_{f,ex}}{V_{s,ex}} \times \frac{V_{f,syn}}{V_{s,syn}} \times \frac{V_{f,pre}}{V_{s,pre}} \times C_{qPCR} \qquad (1)$$

where V is the volume of concentrated wastewater suspended with TRIzol regent (mL) (1.1–3.5 mL), C_{qPCR} is the concentration of cDNA applied to qPCR (copies/μL), V_w is the volume of raw wastewater (mL). $V_{f,ex}$, $V_{s,ex}$, $V_{f,syn}$, $V_{s,syn}$, $V_{f,pre}$, and $V_{s,pre}$ are all the volumes of samples in the intermediate steps (μL). Subscripts f and s stand for the final and starting volumes, while subscripts *ex*, *syn*, and *pre* stand for the RNA extraction, the cDNA synthesis, and the pre-amplification, respectively. The LoQ was 118–375 copies/mL based on the quantification limit in a qPCR assay (2 copies/μL of pre-amped standard samples) and Equation (1).

2.4. PMMoV genome quantification by qPCR assay

For the quantification of PMMoV viral RNA, the cDNA was applied to real-time qPCR without pre-amplification. The amplification reaction was performed with SsoAdvanced Probes Supermix (Bio-Rad), the reverse and forward primers, and probe. Each 20 μL reaction contained 5 μL of cDNA, 10 μL of SsoAdvanced Probes Supermix (Bio-Rad), 500 nM of forward and reverse primers, and 200 nM of fluorogenic probe. The PCR cycling condition used for the detection of CDC N1 was also used for PMMoV. The number of PMMoV genome copies of one reaction was determined by a standard curve generated with the standard samples (1.0×10^2–1.0×10^6 copies/reaction). Each sample was quantified in triplicate.

The concentration of PMMoV viral RNA in influent wastewater samples was determined by the following equation:

$$\text{PMMoV, copies/mL} = 1,000 \times \frac{V_{f,ex}}{V_{s,ex}} \times \frac{V_{f,syn}}{V_{s,syn}} \times C_{qPCR} \qquad (2)$$

where C_{qPCR} is the concentration of cDNA applied to qPCR (copies/μL).

2.5. Viral load calculation

For data points with quantified SARS-CoV-2 RNA concentration, we applied a model developed in a prior study (Equation (3)) (Zhu *et al.* 2021). Briefly, patients are assigned different shedding rates according to a pre-defined shedding function that takes into count how many days the patients are into the infection course. The transition of the shedding rate is governed by a matrix whose dimension is determined by the patient report coverage and the shedding duration of infected individuals. Thus, with a consecutive patient number record, the viral load excreted by the active shedder base on a given day can be calculated as follows:

$$L_{p[d]} = \sum_{n=1}^{j} P_{[n,d]} * C_{f,n} * r_f * m_f \tag{3}$$

In this model, $L_{p[d]}$ is the patient viral load on calendar day d, j is the shedding duration, $P_{[n,d]}$ represents the number of patients on the nth day of the infection course on day d and can be called from a matrix used to store and track the infection status of reported cases. $C_{f,n}$ is the fecal shedding rate (copies/g feces) that is calculated from a separate function, r_f is the ratio of patients who develop fecal shedding, and m_f stands for the daily feces load (g/day). More details can be found in Zhu *et al.* (2021).

The number of daily COVID-19 cases in Sendai city was obtained from the Sendai municipal government. The case report covers the entire population of Sendai city (about 1 million) while the catchment area only covers 360,000 people, the patient viral load was thus multiplied by 0.3.

The patient viral load L_p was then compared to the wastewater viral load L_w calculated by Equation (4).

$$L_{w[d]} = C_{v,w[d]} * Q_d * k_1 \tag{4}$$

In this equation, the measured concentration of SARS-CoV-2 RNA in the wastewater $C_{v,w}$ is multiplied by the wastewater flow rate Q and a dilution coefficient k_1 to obtain the wastewater viral load L_w. The daily flow rate was acquired from the treatment plant operator. A dilution coefficient k_1 was introduced to help normalize the wastewater viral load, the reason being that toilet flushing is not evenly distributed throughout the day. Previous studies have reported that a peak of toilet flushing in the morning accounts for about 25% of the total daily flushes (Butler *et al.* 1995; Campisano & Modica 2015). However, due to a lack of in-depth information, k_1 was set to 1 in this study.

2.6. Positive rate modeling

To test whether wastewater surveillance can provide information about how the epidemic may unfold, for data points with positive yet unquantifiable results, a prediction model framework was established (Figure 1). First, the correlation between the reported cases and the positive rate was assessed by Spearman's rank-order correlation and generalized linear model (GLM). Rolling two- and four-week were used to ensure enough data points in a calculation window. The positive rate was calculated as the number of positive signals divided by the total sample number in the given calculation window (rolling two- or four-week). Then, the positive rates from consecutive calculation windows were used as inputs to predict the reported cases in the last calculation window. Both the positive rate and reported cases were assigned to the last week of the calculation window.

Three models, GLM, artificial neural network (ANN), and random forest (RF), were employed to perform the prediction tasks for their ability to solve nonlinear regression problems and learn from available data. For each model, a pre-determined portion of the dataset (80%) was randomly selected for model training; the trained models were then used to perform prediction on the remaining part of the dataset (testing data). This random sampling-training prediction process was repeated 5,000 times. The mean squared error (MSE) of actual value versus predicted value was calculated each time for performance evaluation. The optimal number of inputs was determined through a performance analysis (Supplementary Table S2). The data pretreatment, model configuration, prediction, and statistical analysis were all performed using the R programming language; the related code is provided in the Supplementary material.

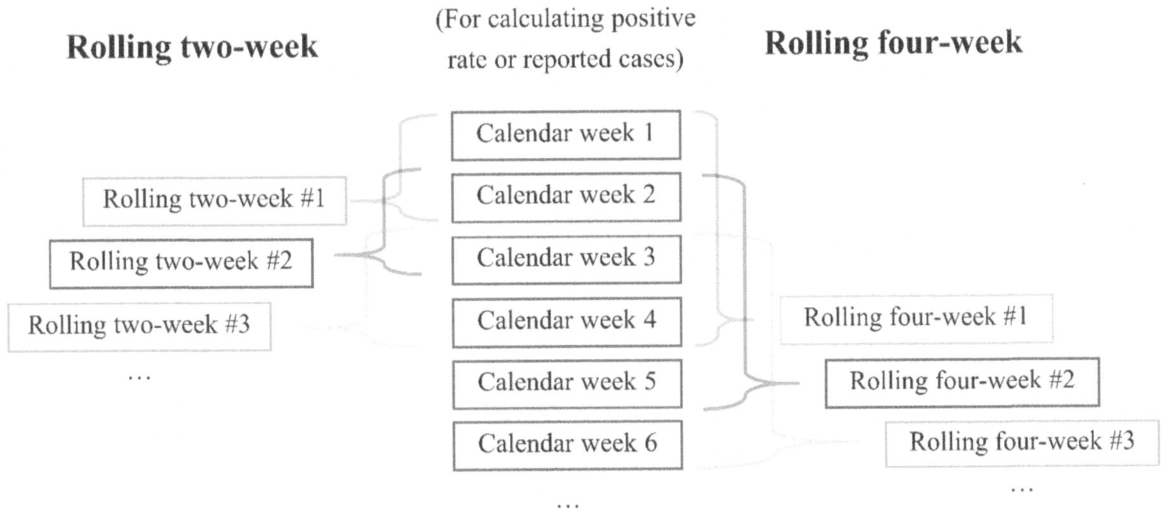

Input 1	Input 2	Output
Positive rate *p* (rolling four-week #1)	*p* (rolling four-week #2)	Reported cases *n* (rolling four-week #3)
p (rolling four-week #2)	*p* (rolling four-week #3)	*n* (rolling four-week #4)
p (rolling four-week #3)	*p* (rolling four-week #4)	*n* (rolling four-week #5)
p (rolling four-week #4)	*p* (rolling four-week #5)	*n* (rolling four-week #6)

Figure 1 | A brief illustration of the model framework. The positive rate and cumulative cases in a rolling week are calculated from a calculation window of two or four weeks, the values are assigned to the last week included in the calculation window. For example, when using rolling two-week, the positive rate during calendar weeks 1 and 2 is denoted as *p* (rolling week #1). In the prediction models, the model inputs are the positive rates in consecutive calculation windows (in chronological order), while the output is the cumulative cases of the following calculation window.

3. RESULTS

3.1. COVID-19 prevalence summary

As a low-prevalence region, Sendai was not severely hit by COVID-19 during the study period. A total of 2,142 cases were reported in Sendai city from August 3, 2020, to February 28, 2021, and can be approximately assigned into two outbreak events (Figure 2). The first one lasted between late October and late November 2020. The peak appeared on October 27, 2020, when 38 patients were reported. The second outbreak event that struck between mid-December and late January was more critical with a higher daily case count. There were 8 days when the daily reported cases exceeded the peak in the first outbreak, and 63 cases were reported on January 14, 2021, marking an all-time high. It is worth mentioning that although the daily reported patient number had a temporal dip during the New Year holiday, it was more likely due to the reduced testing capacity and delayed reporting rather than actual easing of epidemic. Following the second outbreak event, the daily reported cases dropped to a low level and remained that way until the end of the study period.

3.2. Wastewater surveillance summary

The concentration of PMMoV RNA in wastewater influent ranged from 5.2 $\log_{10}$ to 5.8 $\log_{10}$ genome copies/mL, while in concentrated influent samples ranged from 5.4 $\log_{10}$ to 6.4 $\log_{10}$ genome copies/mL. As the occurrence of PMMoV RNA in concentrated wastewater samples was stable, this may suggest there was no significant loss of SARS-CoV-2 RNA in the whole quantification process.

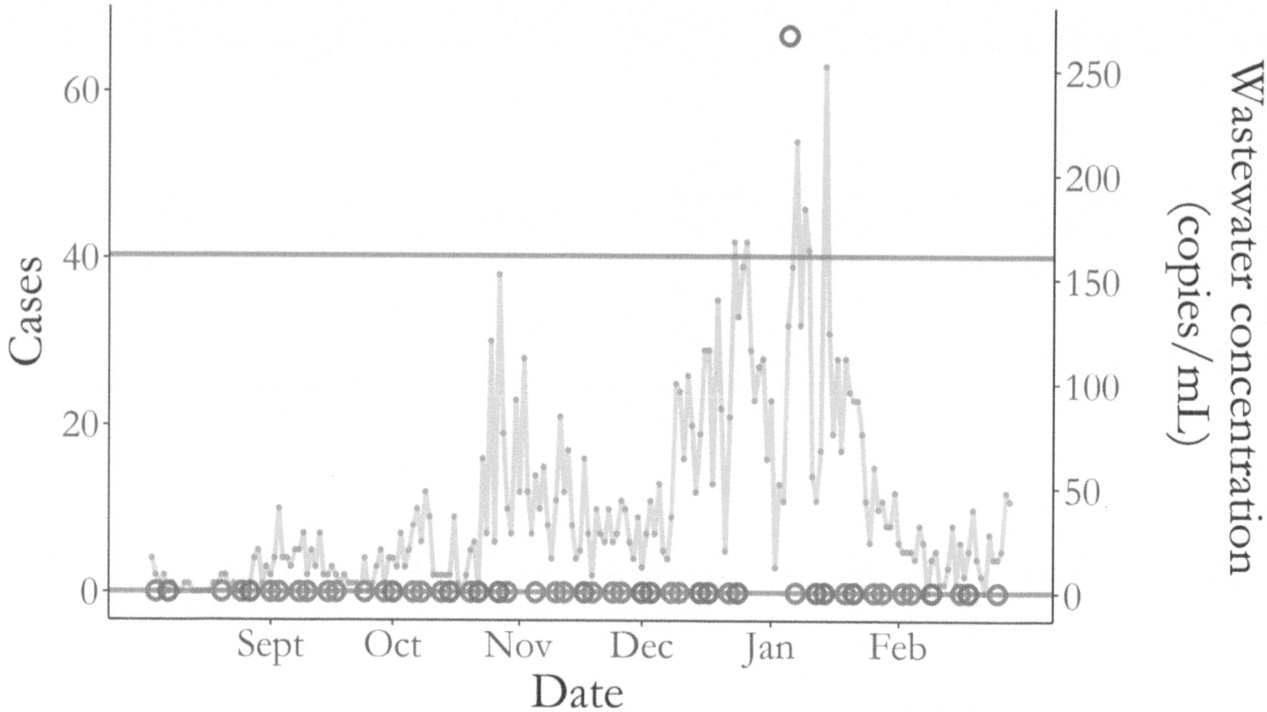

Figure 2 | The time series of the SARS-CoV-2 RNA occurrence in wastewater influent and daily reported cases (red line and points). Positives are shown as blue circles, while negatives are grey. The upper and lower red horizontal lines represent the LoQ (median) and qualitative sensitivity, respectively.

In total, 51 samples were examined throughout the surveillance period, and 33 of them were negative. The genome concentration corresponding to the highest *Ct* value in our assay (referred to as qualitative sensitivity hereafter) was estimated to be 0.025 copies/mL by extrapolating the standard curve. Seventeen samples recorded *Ct* values greater than this, but still lower than the LoQ, ranging from 1.18×10^2 to 3.75×10^2 copies/mL with a median of 1.61×10^2 copies/mL. The amplification efficiency in real-time PCR was from 80 to 120% which was the acceptable range according to the MIQE guidelines (Bustin *et al.* 2009). The coefficient of determination of the standard curves was greater than 0.99 in each assay. No PCR products were detected in negative controls. We considered samples that were tested positive in at least one well in triplicate analysis as positive.

During the study period, the measured viral RNA concentration exceeded the LoQ only once on January 5, 2021 (Figure 2). With a concentration of 2.67×10^2 copies/mL, the daily wastewater viral load estimated from Equation (4) would be 6.74×10^{13} copies. However, the viral load contributed by reported patients on that day, calculated using Equation (3), would be 3.40×10^9 copies with 203 cumulative cases in the 26-day patient viral load calculation window, meaning a L_w/L_p value of 1.98×10^4. Also, this quantifiable signal occurred prior to the peak of reported cases which came 9 days later, on January 14, 2021. The wastewater virus concentration did not exceed the LoQ again despite a higher daily case count reported in the following days.

On the other hand, due to the lack of quantifiable data points, non-quantitative detection gave us a more consistent dataset to work on. The first positive signal occurred on August 7, 2020, with just 21 cumulative cases in the patient viral load calculation window and an estimated patient viral load of 3.42×10^8 copies, which translates into a theoretical wastewater virus concentration of only 2.70×10^{-3} copies/mL, far below the qualitative sensitivity. However, assuming the concentration sits somewhere between the qualitative sensitivity and the LoQ, the wastewater viral load would have a range of 6.32×10^9–4.06×10^{13}, thus a L_w/L_p range of 1.85×10^1–1.19×10^5, covering the L_w/L_p value estimated from the quantifiable detection on January 5, 2021. Over the study period, a total of 18 (35.29%) samples tested positive. Although a positive signal does not directly translate into wastewater viral concentration, consecutive positives may indicate a high viral load with higher confidence. In that sense, two consecutive positives appeared four times and all of them occurred during the two outbreak events.

3.3. Positive rate modeling

A stronger correlation was found between the four-week positive rate and cumulative cases than that between the two-week positive rate and cumulative cases (Figure 3). Therefore, the prediction models were used to predict the cumulative cases using the four-week positive rate. By testing different amounts of input numbers, we found the optimal value was two (Supplementary Table S2). Among the three models (Figure 4), ANN offered the best overall performance (median MSE: 7520.22) followed by the other two (median MSE: 9,038.60 (GLM), 12,021.26 (RF)). For about half of the data points (45.83%, 11 in 24), the actual four-week cumulative cases were within the 95% CI range of the prediction. For the remaining data points, the average error was 17.51%.

4. DISCUSSION

By employing a highly sensitive detection method, we monitored the time series of SARS-CoV-2 RNA occurrence in wastewater influent from an urban community with a population of 360,000. Eighteen out of the 51 influent samples yielded positive signals, and 17 samples had SARS-CoV-2 RNA concentrations lower than the LoQ. By examining the reported cases, we found the positive rate of detection has a strong correlation (four-week rolling window, $\rho=0.7598$, $p<0.05$) with the cumulative cases in the same time frame and established prediction models, hoping to extend the knowledge on wastewater-based epidemiology (WBE) implementation strategies.

In this study, the LoQ was around 1.61×10^2 copies/mL, about 1 log above that reported in the previous studies (Ahmed *et al.* 2020a; Hokajärvi *et al.* 2021). This is due to the decreased volume of samples throughout the analysis in this study. Specifically, the volume of a sample input was approximately 1/10, 1/6, and 1/10 of the total volume of a suspension obtained in a previous step, in RNA extraction, cDNA synthesis, and qPCR, respectively. The LoQ may be improved by using other different extraction kits that use a larger volume of samples for extraction. Nevertheless, the method we employed is a feasible option because the data is obtained within 1 day, and our analysis has not been restricted by the stock shortages of manufacturers.

We used PMMoV as an internal process control for SARS-CoV-2 detection from suspended solids. PMMoV is detected throughout the year and is abundant in wastewater (10^6–10^{10} copies/L) (Kitajima *et al.* 2014; Symonds *et al.* 2018; Ahmed

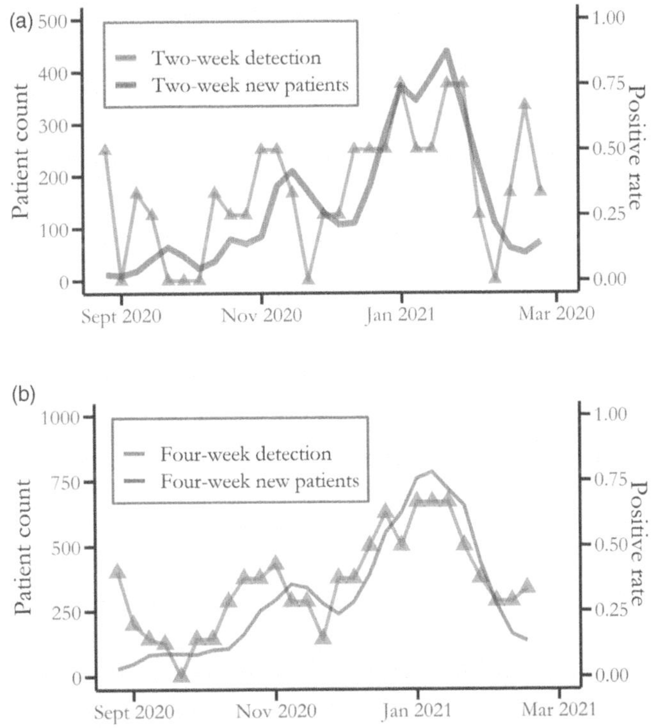

Figure 3 | The time series of positive rate and cumulative cases. (a) The calculation window is rolling two-week. (b) The calculation window is rolling four-week. Spearman's rank-order correlation coefficients: 0.4996 (two-week, $p<0.05$) and 0.7598 (four-week, $p<0.05$).

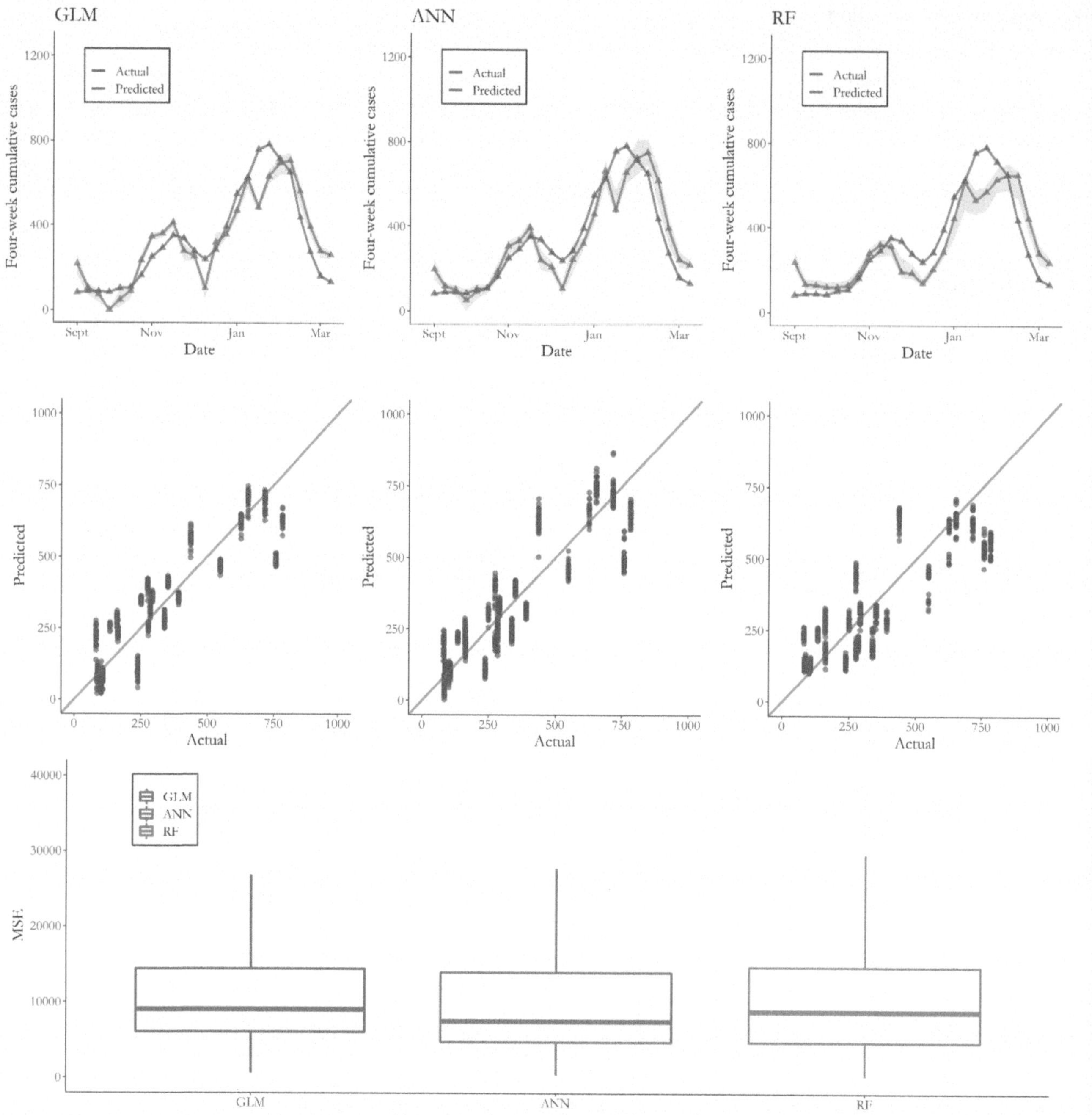

Figure 4 | The cumulative cases predicted by the three models. The inputs were the positive rates in two consecutive rolling four-weeks, while the output was the cumulative cases within the following four-week. Normalization was applied to inputs and output before model training and all data were denormalized back to the original scale once the prediction was conducted. The three rows of figures are the time series of actual versus predicted (median and 95% CI) four-week cumulative cases for each model, the scatter plot of 1,000 randomly selected pairs of actual and predicted values, and the boxplot of MSE distribution from the 5,000 predictions of each model, respectively.

et al. 2020b), which may allow for using it as an internal control of RT-qPCR for a wastewater sample (Haramoto *et al.* 2020). A previous study reported that the recovered load of PMMoV correlated with that of murine hepatitis virus, suggesting that PMMoV is the potential indicator of the efficiency of SARS-CoV-2 (Torii *et al.* 2022). We concluded that there was no significant loss throughout the analysis because the PMMoV concentration was consistent in both raw and concentrated influent samples. Future studies should decide the best whole process control for extraction from suspended solids. The better options are human coronaviruses 229E and HKU1 although their longitudinal concentration has not been reported (Bibby *et al.* 2011; Bibby & Peccia 2013).

The pre-amplification employed in this study increases the number of amplicons in the downstream qPCR. The theoretical qualitative sensitivity should have the same *Ct* value as that obtained from one copy per reaction, which in this study is 32.9. However, this value was surpassed by multiple samples whose *Ct* values reached up to 40.0. A possible explanation for the results was that organic compounds in the influent samples inhibited the amplification efficiency in PCR. We evaluated the effect of inhibitors using a commercially available RNA positive control according to the Gibson *et al.* (2012) study but did not observe lowered efficiency in pre-amplification and qPCR of the positive control RNA, indicating that other reasons were responsible for the greater-than-expected *Ct* values. It may be explained by the different affinity efficiency of primers and polymerase to the target amplicons between the cDNA derived from viral RNA and plasmid DNA used as the positive control.

From the perspective of early warning, getting a positive signal from wastewater can be a solid proof that the virus has started circulating in the community (Randazzo *et al.* 2020; Fernandez-Cassi *et al.* 2021). So far, different studies have reported varied sensitivity. Hong *et al.* (2021) reported that a positive signal in hospital wastewater requires 253–409 positive cases out of 10,000 individuals, while Hata *et al.* (2021) detected the presence of SARS-CoV-2 RNA in municipal wastewater when the number of cases was <1.0 per 100,000 people, and Betancourt *et al.* (2021) reported a positive detection when there were only one symptomatic and two asymptomatic individuals among a total of 311 residents in a student dormitory. In practice, the varied detection sensitivity can be mainly attributed to the different experimental methods used as well as the characteristics of the sewage system from which samples are collected. A standardized method may contribute to the comparison and integration of studies (Weidhaas *et al.* 2021). In this study, when the first positive signal was recorded, the number of active shedders estimated from the clinical reports was only 21 in the catchment area with about 360,000 people. Nevertheless, a bigger active shedder base does not guarantee a positive signal in subsequent detections. Even during the summit of the second outbreak event which enabled the only signal above LoQ, negatives were still recorded. Such inconsistent detection had also been reported in other studies (Fernandez-Cassi *et al.* 2021; Hong *et al.* 2021), adding another layer of complexity.

Despite the strong interest in quantitative wastewater surveillance, a streamlined solution has yet to be formed. Especially, although the experimental side has received substantial attention which led to more sensitive and reliable detection, the analytical side still lacks adequate investigation and verification. There is a noteworthy knowledge gap in how to associate the measured wastewater virus concentration with the epidemic size in the catchment area. So far, most studies had tried directly correlating the abundance of viral RNA with reported cases. However, the following are the findings from our study: (1) a positive signal occurred when the speculated active shedder group was supposedly far from large enough to enable a successful detection, (2) higher reported cases did not translate to high wastewater virus concentration, and (3) there is a significant yet uncertain gap between the observed wastewater viral load and the viral load contributed by the supposed active shedder base, presented as L_w/L_p in this study, all point to a conclusion that at the current stage, the uncertainty associated with the wastewater viral load is still a great hindrance to reliable back-calculation.

The dimensionless metric L_w/L_p largely determines the robustness of back-calculation. Although a stable L_w/L_p is ideal for back-calculation and was therefore assumed in some recent studies, its value seems to be both time- and location-specific due to various factors. For instance, in Sendai, August and September are the rainy season, because the major urban area is served by a combined sewer, which likely aggravated the dilution of viral RNA and led to a lower L_w/L_p. This is supported by the lower bound of L_w/L_p estimated for August 7, 2020 when the first positive signal appeared. The way patient viral load L_p was calculated also implies it can be impacted by societal factors. For instance, a high level of underreporting may occur under limited testing capacity, leading to a smaller speculated active shedder base, thus a smaller L_p and a larger L_w/L_p. Similarly, if the asymptomatic infection ratio increase, a larger L_w/L_p can also be expected. Knowing this, some critical epidemic-related information may be drawn by keeping a close eye on L_w/L_p. Nevertheless, it should be pointed out that our calculations of L_w and L_p were based on a set of assumptions including the shedding profile, which may be further refined once more medical evidence becomes available.

As shown in this study, in wastewater surveillance projects, researchers may obtain positive yet unquantifiable signals, especially in a low-prevalence period/region. As far as we know, no wastewater surveillance study has utilized binominal data other than as occurrence indicator yet. But, the detection frequency, or positive rate, might serve as a suitable indicator of the virus occurrence upon which further analysis can be performed. This indicates that binominal results may also be utilized to help with epidemic surveillance while establishing a precise connection between the wastewater viral load and the prevalence level remains challenging and entails further research. Rolling four-week was used as the calculation window in this study, but it may be shortened by increasing the sampling frequency or the number of samples collected each time, albeit more time- and resource-consuming. It should be noted, though, that using a positive rate as an indicator may only be feasible in a low-prevalence region or at the early stage of an outbreak event. There exists an upper bound of epidemic size beyond which its linear correlation with positive rate becomes invalid, which may explain why the epidemic peaks were not successfully modeled in this study. On the other hand, a higher prevalence level means a higher chance of getting quantifiable signals, and back-calculation models should take over once developed.

When a causal relationship between input and output is difficult to establish, data-driven methods like those used in this study may be employed. However, being data-driven also means training data need to be accumulated to fine-tune the model, and prediction does not always match the reality. With all the uncertainties, it should be reiterated that wastewater surveillance ought not to be a stand-alone tool and its outcome should be interpreted along with other information sources before reaching any conclusion. For instance, in the early stage of an epidemic when clinical testing capacity is often compromised, wastewater detection may be put into action quicker and cover a larger area. Nevertheless, our study shows that a positive rate may be an important indicator, as also recognized by a recent study (Fernandez-Cassi *et al.* 2021). Also, in terms of prediction accuracy, as stated above, environmental and societal factors may affect the detection result; thus, adding explanatory variables into the model may improve the model performance.

Several limitations of this study should be noted. First, the lag between symptom onset and reporting was not included in modeling. Counting in the delay in case reporting may explain the 9-day delay between the quantified virus concentration and the case peak, it may also improve the correlation between the positive rate and cumulative cases, as cases would be assigned to an earlier date. However, existing studies on the delay between symptom onset and hospitalization had varied estimations ranging from 7 days (Huang *et al.* 2020) to a much shorter 1.2 days (Lauer *et al.* 2020). Therefore, without enough information about the local testing and reporting practice, integrating this factor into the model may introduce further error. Second, grab samples are prone to short-term heterogeneity of viral RNA abundance, which may affect the representativeness of samples. While composite samples collected by an autosampler may improve the consistency of detection, the viral RNA may also get highly diluted as toilet flushing mainly occurs during certain times, resulting in false negatives. Designing a sampling strategy that captures the toilet flushing peak, therefore, may be a viable solution as suggested by recent studies (Betancourt *et al.* 2021).

5. CONCLUSIONS

As the world is still under the shadow of COVID-19, on top of timely medical and societal intervention, each and every tool that helps monitor the situation and alerts the society is worth looking into. In this study, using a highly sensitive assay, we (1) monitored the occurrence of SARS-CoV-2 viral RNA in the wastewater of an urban area in Japan for over 7 months and (2) established a model framework to help extend the existing knowledge base about analyzing and interpreting the surveillance results. Particularly, we found that although quantitative epidemic size estimation based on measured virus concentration is still challenging, the positive rate of wastewater virus detection is strongly correlated with reported cases and can be used for its prediction, which may guide toward novel wastewater surveillance strategies. Our findings may not only strengthen the application of wastewater surveillance in the current COVID-19 pandemic but also help the scientific community prepare for other public health challenges.

ACKNOWLEDGEMENTS

This research was supported by the Japan Agency for Medical Research and Development (AMED) under Grant No. JP20wm0125001 and by JST SPRING under Grant Number JPMJSP2114. We appreciate the generous cooperation of the Sendai City Construction Bureau for providing wastewater samples.

REFERENCES

Ahmed, W., Angel, N., Edson, J., Bibby, K., Bivins, A., O'Brien, J. W., Choi, P. M., Kitajima, M., Simpson, S. L., Li, J., Tscharke, B., Verhagen, R., Smith, W. J. M., Zaugg, J., Dierens, L., Hugenholtz, P., Thomas, K. V. & Mueller, J. F. 2020a First confirmed detection of SARS-CoV-2 in untreated wastewater in Australia: a proof of concept for the wastewater surveillance of COVID-19 in the community. *Science of the Total Environment* **728**, 138764. doi:10.1016/j.scitotenv.2020.138764.

Ahmed, W., Kitajima, M., Tandukar, S. & Haramoto, E. 2020b Recycled water safety: current status of traditional and emerging viral indicators. *Current Opinion in Environmental Science and Health* **16**, 62–72. doi:10.1016/j.coesh.2020.02.009.

Balboa, S., Mauricio-Iglesias, M., Rodriguez, S., Martínez-Lamas, L., Vasallo, F. J., Regueiro, B. & Lema, J. M. 2021 The fate of SARS-COV-2 in WWTPS points out the sludge line as a suitable spot for detection of COVID-19. *Science of the Total Environment* **772**, 145268. doi:10.1016/j.scitotenv.2021.145268.

Betancourt, W. Q., Schmitz, B. W., Innes, G. K., Prasek, S. M., Pogreba Brown, K. M., Stark, E. R., Foster, A. R., Sprissler, R. S., Harris, D. T., Sherchan, S. P., Gerba, C. P. & Pepper, I. L. 2021 COVID-19 containment on a college campus via wastewater-based epidemiology, targeted clinical testing and an intervention. *Science of the Total Environment* **779**, 146408. doi:10.1016/j.scitotenv.2021.146408.

Bibby, K. & Peccia, J. 2013 Identification of viral pathogen diversity in sewage sludge by metagenome analysis. *Environmental Science and Technology* **47** (4), 1945–1951. doi:10.1021/es305181x.

Bibby, K., Viau, E. & Peccia, J. 2011 Viral metagenome analysis to guide human pathogen monitoring in environmental samples. *Letters in Applied Microbiology* **52** (4), 386–392. doi:10.1111/j.1472-765X.2011.03014.x.

Buonerba, A., Corpuz, M. V. A., Ballesteros, F., Choo, K. H., Hasan, S. W., Korshin, G. V., Belgiorno, V., Barceló, D. & Naddeo, V. 2021 Coronavirus in water media: analysis, fate, disinfection and epidemiological applications. *Journal of Hazardous Materials* **415**. doi:10.1016/j.jhazmat.2021.125580.

Bustin, S. A., Benes, V., Garson, J. A., Hellemans, J., Huggett, J., Kubista, M., Mueller, R., Nolan, T., Pfaffl, M. W. & Shipley, G. L. 2009 The MIQE Guidelines: Minimum information for publication of quantitative real-time PCR experiments. *Clinical Chemistry* **55**, 611–622.

Butler, D., Friedler, E. & Gatt, K. 1995 Characterising the quantity and quality of domestic wastewater inflows. *Water Science and Technology* **31** (7), 13–24. doi:10.1016/0273-1223(95)00318-H.

Campisano, A. & Modica, C. 2015 Appropriate resolution timescale to evaluate water saving and retention potential of rainwater harvesting for toilet flushing in single houses. *Journal of Hydroinformatics* **17** (3), 331–346. doi:10.2166/hydro.2015.022.

CDC 2021 *Real-Time RT-PCR Primers and Probes for COVID-19*. CDC. Available from: https://www.cdc.gov/coronavirus/2019-ncov/lab/rt-pcr-panel-primer-probes.html (accessed 29 November 2021).

D'Aoust, P. M., Mercier, E., Montpetit, D., Jia, J. J., Alexandrov, I., Neault, N., Baig, A. T., Mayne, J., Zhang, X., Alain, T., Langlois, M. A., Servos, M. R., MacKenzie, M., Figeys, D., MacKenzie, A. E., Graber, T. E. & Delatolla, R. 2021 Quantitative analysis of SARS-CoV-2 RNA from wastewater solids in communities with low COVID-19 incidence and prevalence. *Water Research* **188**, 116560. doi:10.1016/j.watres.2020.116560.

Fernandez-Cassi, X., Scheidegger, A., Bänziger, C., Cariti, F., Tuñas Corzon, A., Ganesanandamoorthy, P., Lemaitre, J. C., Ort, C., Julian, T. R. & Kohn, T. 2021 Wastewater monitoring outperforms case numbers as a tool to track COVID-19 incidence dynamics when test positivity rates are high. *Water Research* **200**, 117252. doi:10.1016/j.watres.2021.117252.

Gibson, K. E., Schwab, K. J., Spencer, S. K. & Borchardt, M. A. 2012 Measuring and mitigating inhibition during quantitative real time PCR analysis of viral nucleic acid extracts from large-volume environmental water samples. *Water Research* **46** (13), 4281–4291. doi:10.1016/j.watres.2012.04.030.

Haramoto, E., Malla, B., Thakali, O. & Kitajima, M. 2020 First environmental surveillance for the presence of SARS-CoV-2 RNA in wastewater and river water in Japan. *Science of the Total Environment* **737**, 140405. doi:10.1016/j.scitotenv.2020.140405.

Hart, O. E. & Halden, R. U. 2020 Computational analysis of SARS-CoV-2/COVID-19 surveillance by wastewater-based epidemiology locally and globally: feasibility, economy, opportunities and challenges. *Science of the Total Environment* **730**, 138875. doi:10.1016/j.scitotenv.2020.138875.

Hasan, S. W., Ibrahim, Y., Daou, M., Kannout, H., Jan, N., Lopes, A., Alsafar, H. & Yousef, A. F. 2021 Detection and quantification of SARS-CoV-2 RNA in wastewater and treated effluents: surveillance of COVID-19 epidemic in the United Arab Emirates. *Science of the Total Environment* **764**, 142929. doi:10.1016/j.scitotenv.2020.142929.

Hata, A., Hara-Yamamura, H., Meuchi, Y., Imai, S. & Honda, R. 2021 Detection of SARS-CoV-2 in wastewater in Japan during a COVID-19 outbreak. *Science of the Total Environment* **758**, 143578. doi:10.1016/j.scitotenv.2020.143578.

Hokajärvi, A. M., Rytkönen, A., Tiwari, A., Kauppinen, A., Oikarinen, S., Lehto, K. M., Kankaanpää, A., Gunnar, T., Al-Hello, H., Blomqvist, S., Miettinen, I. T., Savolainen-Kopra, C. & Pitkänen, T. 2021 The detection and stability of the SARS-CoV-2 RNA biomarkers in wastewater influent in Helsinki, Finland. *Science of the Total Environment* **770**, 145274. doi:10.1016/j.scitotenv.2021.145274.

Hong, P. Y., Rachmadi, A. T., Mantilla-Calderon, D., Alkahtani, M., Bashawri, Y. M., Al Qarni, H., O'Reilly, K. M. & Zhou, J. 2021 Estimating the minimum number of SARS-CoV-2 infected cases needed to detect viral RNA in wastewater: to what extent of the outbreak can surveillance of wastewater tell us? *Environmental Research* **195**, 110748. doi:10.1016/j.envres.2021.110748.

Huang, C., Wang, Y., Li, X., Ren, L., Zhao, J., Hu, Y., Zhang, L., Fan, G., Xu, J., Gu, X., Cheng, Z., Yu, T., Xia, J., Wei, Y., Wu, W., Xie, X., Yin, W., Li, H., Liu, M., Xiao, Y., Gao, H., Guo, L., Xie, J., Wang, G., Jiang, R., Gao, Z., Jin, Q., Wang, J. & Cao, B. 2020 Clinical features of patients infected with 2019 novel coronavirus in Wuhan, China. *The Lancet* **395** (10223), 497–506. doi:10.1016/S0140-6736(20)30183-5.

Jones, D. L., Baluja, M. Q., Graham, D. W., Corbishley, A., McDonald, J. E., Malham, S. K., Hillary, L. S., Connor, T. R., Gaze, W. H., Moura, I. B., Wilcox, M. H. & Farkas, K. 2020 Shedding of SARS-CoV-2 in feces and urine and its potential role in person-to-person transmission and the environment-based spread of COVID-19. *Science of the Total Environment* **749**, 141364. DOI: 10.1016/j.scitotenv.2020.141364

Kitajima, M., Iker, B. C., Pepper, I. L. & Gerba, C. P. 2014 Relative abundance and treatment reduction of viruses during wastewater treatment processes – identification of potential viral indicators. *Science of the Total Environment* **488–489** (1), 290–296. doi:10.1016/j.scitotenv.2014.04.087.

Kitamura, K., Sadamasu, K., Muramatsu, M. & Yoshida, H. 2021 Efficient detection of SARS-CoV-2 RNA in the solid fraction of wastewater. *Science of the Total Environment* **763**, 144587. doi:10.1016/j.scitotenv.2020.144587.

Lau, L. T., Fung, Y. W. W., Wong, F. P. F., Lin, S. S. W., Wang, C. R., Li, H. L., Dillon, N., Collins, R. A., Tam, J. S. L., Chan, P. K. S., Wang, C. G. & Yu, A. C. H. 2003 A real-time PCR for SARS-coronavirus incorporating target gene pre-amplification. *Biochemical and Biophysical Research Communications* **312** (4), 1290–1296. doi:10.1016/j.bbrc.2003.11.064.

Lauer, S. A., Grantz, K. H., Bi, Q., Jones, F. K., Zheng, Q., Meredith, H. R., Azman, A. S., Reich, N. G. & Lessler, J. 2020 The incubation period of coronavirus disease 2019 (CoVID-19) from publicly reported confirmed cases: estimation and application. *Annals of Internal Medicine* **172** (9), 577–582. doi:10.7326/M20-0504.

Li, X., Zhang, S., Shi, J., Luby, S. P. & Jiang, G. 2021 Uncertainties in estimating SARS-CoV-2 prevalence by wastewater-based epidemiology. *Chemical Engineering Journal* **415**, 129039. doi:10.1016/j.cej.2021.129039.

Peccia, J., Zulli, A., Brackney, D. E., Grubaugh, N. D., Kaplan, E. H., Casanovas-Massana, A., Ko, A. I., Malik, A. A., Wang, D., Wang, M., Warren, J. L., Weinberger, D. M., Arnold, W. & Omer, S. B. 2020 Measurement of SARS-CoV-2 RNA in wastewater tracks community infection dynamics. *Nature Biotechnology* **38**, 1164–1167. doi:10.1038/s41587-020-0684-z.

Randazzo, W., Truchado, P., Cuevas-Ferrando, E., Simón, P., Allende, A. & Sánchez, G. 2020 SARS-CoV-2 RNA in wastewater anticipated COVID-19 occurrence in a low prevalence area. *Water Research* **181**. doi:10.1016/j.watres.2020.115942.

Saththasivam, J., El-Malah, S. S., Gomez, T. A., Jabbar, K. A., Remanan, R., Krishnankutty, A. K., Ogunbiyi, O., Rasool, K., Ashhab, S., Rashkeev, S., Bensaad, M., Ahmed, A. A., Mohamoud, Y. A., Malek, J. A., Abu Raddad, L. J., Jeremijenko, A., Abu Halaweh, H. A., Lawler, J. & Mahmoud, K. A. 2021 COVID-19 (SARS-CoV-2) outbreak monitoring using wastewater-based epidemiology in Qatar. *Science of the Total Environment* **774**, 145608. doi:10.1016/j.scitotenv.2021.145608.

Sims, N. & Kasprzyk-Hordern, B. 2020 Future perspectives of wastewater-based epidemiology: monitoring infectious disease spread and resistance to the community level. *Environment International* **139**, 105689. doi:10.1016/j.envint.2020.105689.

Symonds, E. M., Nguyen, K. H., Harwood, V. J. & Breitbart, M. 2018 Pepper mild mottle virus: a plant pathogen with a greater purpose in (waste)water treatment development and public health management. *Water Research* **144**, 1–12. doi:10.1016/j.watres.2018.06.066.

Tang, A., Tong, Z., Wang, H., Dai, Y., Li, K., Liu, J., Wu, W., Yuan, C., Yu, M., Li, P. & Yan, J. 2020 Detection of novel coronavirus by RT-PCR in stool specimen from asymptomatic child, China. *Emerging Infectious Diseases* **26** (6), 1337–1339. doi:10.3201/eid2606.200301.

Thompson, J. R., Nancharaiah, Y. V., Gu, X., Lee, W. L., Rajal, V. B., Haines, M. B., Girones, R., Ng, L. C., Alm, E. J. & Wuertz, S. 2020 Making waves: wastewater surveillance of SARS-CoV-2 for population-based health management. *Water Research* **184**. doi:10.1016/j.watres.2020.116181.

Torii, S., Oishi, W., Zhu, Y., Thakali, O., Malla, B., Yu, Z., Zhao, B., Arakawa, C., Kitajima, M., Hata, A., Ihara, M., Kyuwa, S., Sano, D., Haramoto, E. & Katayama, H. 2022 Comparison of five polyethylene glycol precipitation procedures for the RT-qPCR based recovery of murine hepatitis virus, bacteriophage phi6, and pepper mild mottle virus as a surrogate for SARS-CoV-2 from wastewater. *Science of the Total Environment* **807**, 150722. doi:10.1016/j.scitotenv.2021.150722.

Walsh, K. A., Jordan, K., Clyne, B., Rohde, D., Drummond, L., Byrne, P., Ahern, S., Carty, P. G., O'Brien, K. K., O'Murchu, E., O'Neill, M., Smith, S. M., Ryan, M. & Harrington, P. 2020 SARS-CoV-2 detection, viral load and infectivity over the course of an infection. *Journal of Infection* **81** (3), 357–371. doi:10.1016/j.jinf.2020.06.067.

Weidhaas, J., Aanderud, Z. T., Roper, D. K., VanDerslice, J., Gaddis, E. B., Ostermiller, J., Hoffman, K., Jamal, R., Heck, P., Zhang, Y., Torgersen, K., Laan, J., Vander & LaCross, N. 2021 Correlation of SARS-CoV-2 RNA in wastewater with COVID-19 disease burden in sewersheds. *Science of the Total Environment* **775**, 145790. doi:10.1016/j.scitotenv.2021.145790.

WHO 2021 *WHO Coronavirus (COVID-19) Dashboard*. WHO. Available from: https://covid19.who.int/ (accessed 16 November 2021).

Wölfel, R., Corman, V. M., Guggemos, W., Seilmaier, M., Zange, S., Müller, M. A., Niemeyer, D., Jones, T. C., Vollmar, P., Rothe, C., Hoelscher, M., Bleicker, T., Brünink, S., Schneider, J., Ehmann, R., Zwirglmaier, K., Drosten, C. & Wendtner, C. 2020 Virological assessment of hospitalized patients with COVID-2019. *Nature* **581** (7809), 465–469. doi:10.1038/s41586-020-2196-x.

Zhang, J. C., Wang, S. B. & Xue, Y. D. 2020 Fecal specimen diagnosis 2019 novel coronavirus–infected pneumonia. *Journal of Medical Virology* **92** (6), 680–682. doi:10.1002/jmv.25742.

Zhu, Y., Oishi, W., Saito, M., Kitajima, M. & Sano, D. 2021 Early warning of COVID-19 in Tokyo via wastewater-based epidemiology: how feasible it really is? *Journal of Water and Environment Technology* **19** (3), 170–183. doi:10.2965/jwet.21-024.

First received 28 July 2021; accepted in revised form 30 January 2022. Available online 7 February 2022

SARS-CoV-2 in Water doi: 10.2166/wh.2022.185

Establishment of local wastewater-based surveillance programmes in response to the spread and infection of COVID-19 – case studies from South Africa, the Netherlands, Turkey and England

M. Tlhagale[a,*], S. Liphadzi[a,b], J. Bhagwan[a], V. Naidoo[a], K. Jonas[a], L. van Vuuren[a], G. Medema[c], L. Andrews[c], F. Béen[c], M. L. Ferreira[c], A. M. Saatci[d], B. Alpaslan Kocamemi[e], F. Hassard[f,g], A. C. Singer[h], J. T. Bunce[i,j], J. M. S. Grimsley[i], M. Brown[i,k] and D. L. Jones[l]

[a] Water Research Commission, 4 Daventry St, Lynnwood Manor, Pretoria, South Africa
[b] Univerisity of Venda, University Rd, Thohoyandou, 0950, South Africa
[c] KWR, Groningenhaven 7, 3433 PE Nieuwegein, Netherlands
[d] Turkish Water Institute (SUEN), Libadiye Cad. 54 Küçükçamlıca Üsküdar 34696, Istanbul, Turkey
[e] Environmental Engineering Department, Marmara University, Kadıkoy 34722, Istanbul, Turkey
[f] Cranfield University, Bedfordshire, MK43 0AL, UK
[g] Institute for Nanotechnology and Water Sustainability, University of South Africa, UNISA Science Campus, 1710 Roodepoort, Johannesburg, South Africa
[h] UK Centre for Ecology and Hydrology, MacLean Building, Benson Ln, Crowmarsh Gifford, Wallingford, OX10 8BB, UK
[i] United Kingdom Health Security Agency, Windsor House, Victoria Street, London, SW1H 0LT, UK
[j] Department for Environment, Food and Rural Affairs, Seacole Building, 2 Marsham Street, London SW1P 4DF, UK
[k] School of Engineering, Newcastle University, Newcastle-upon-Tyne NE1 7RU, UK
[l] Environment Centre Wales, Bangor University, Bangor, LL57 2UW, UK
*Corresponding author. E-mail: mamohlodingt@wrc.org.za

MT, 0000-0002-2827-8798; GM, 0000-0003-0475-6465; MLF, 0000-0002-8264-4496; AMS, 0000-0003-4715-6953; ACS, 0000-0003-4705-6063; DLJ, 0000-0002-1482-4209

ABSTRACT

The COVID-19 pandemic has resulted in over 340 million infection cases (as of 21 January 2022) and more than 5.57 million deaths globally. In reaction, science, technology and innovation communities across the globe have organised themselves to contribute to national responses to COVID-19 disease. A significant contribution has been from the establishment of wastewater-based epidemiological (WBE) surveillance interventions and programmes for monitoring the spread of COVID-19 in at least 55 countries. Here, we examine and share experiences and lessons learnt in establishing such surveillance programmes. We use case studies to highlight testing methods and logistics considerations associated in scaling the implementing of such programmes in South Africa, the Netherlands, Turkey and England. The four countries were selected to represent different regions of the world and the perspective based on the considerable progress made in establishing and implementing their national WBE programmes. The selected countries also represent different climatic zones, economies, and development stages, which influence the implementation of national programmes of this nature and magnitude. In addition, the four countries' programmes offer good experiences and lessons learnt since they are systematic, and cover extensive areas, disseminate knowledge locally and internationally and partnered with authorities (government). The programmes also strengthened working relations and partnerships between and among local and global organisations. This paper shares these experiences and lessons to encourage others in the water and public health sectors on the benefits and value of WBE in tackling SARS-CoV-2 and related future circumstances.

Key words: COVID-19, public health surveillance, SARS-CoV-2, sewage surveillance, wastewater-based epidemiology, wastewater surveillance

HIGHLIGHTS

- Designing and refining methodologies, epidemiological tools and indicators.
- Influencing public health decision-making through research data.
- Experiences on programme design and methodology for WBE surveillance.
- Partnerships strengthening communication and knowledge dissemination.
- Capacity building activities among the science communities, local authorities, industry nationally and internationally.

INTRODUCTION

The COVID-19 pandemic has resulted in at least 340 million infection cases (as of 21 January 2022) and more than 5.57 million deaths globally. The SARS-CoV-2 virus is the newest of the family of coronaviruses associated with human infections that are grouped into the beta-CoV genus, with 79% genetic similarity to SARS-CoV-1 (Pal *et al.* 2020). The outbreak of SARS-CoV-2 was declared a Public Health Emergency of International Concern on 30 January 2020. On 11 February 2020, the World Health Organization (WHO) announced a name for the new coronavirus disease: COVID-19. One month later, the WHO upgraded the status of this coronavirus outbreak from an epidemic to a pandemic.

The emergence of several SARS-CoV-2 variants has become a significant challenge for COVID-19-related policies and vaccination implementation in most countries. In this context, the need to design new and refine existing epidemiological tools, including the use of wastewater-based epidemiology (WBE) to inform decision-making, has increased in many parts of the world. Surveillance programmes have since been set up to monitor the spread of COVID-19 in at least 58 countries (https://ucmerced.maps.arcgis.com/).

The concept of screening municipal wastewater and environmental waters as an epidemiological tool is not new. This approach of using markers of infection has, for instance, been used to help inform broader infectious disease epidemiological surveillance and mitigation efforts such as the Global Polio Eradication Initiative (Hovi *et al.* 2012; Humayun *et al.* 2014). Environmental surveillance has also been used and recommended for the detection of other infections such as typhoid (WHO 2018), hepatitis A and norovirus outbreaks (Hellmér *et al.* 2014), and for antimicrobial resistance (Hendriksen *et al.* 2019). Modelling techniques were used to assist both the design and interpretation of these efforts (Wang *et al.* 2020). WBE is also commonly used to monitor the use of illicit drugs and various chemical contaminants that may impact human health (Choi *et al.* 2018).

Studies have shown the value of WBE surveillance programmes in tracing and monitoring changes in the prevalence of infections in urban populations. This includes, for example, COVID-19 containment on a college campus via wastewater-based epidemiology, targeted clinical testing and an intervention (Corchis-Scott *et al.* 2021). In the Netherlands, test buses were deployed to an area of undertesting (Anon 2022) while in February 2021 in New Zealand, case data was coupled with wastewater data to decide whether to extend a lockdown after three positive cases were detected (Sharara *et al.* 2021). Negative results from wastewater PCR testing at the municipal level allowed government officials to rule out community spread of infections and defined it rather as a contained and traceable cluster of individuals (Sharara *et al.* 2021).

The presence of SARS-CoV-2 in wastewater treatment plant (WWTP) influent, and within the sewer network, can help determine the presence of infection in a community. This is because the detection of SARS-CoV-2 genetic material in wastewater is indicative of the number of COVID-19 cases (dependent on the testing levels of that community) and thus can present a valuable early indicator for local hospitalisations, viral outbreaks or COVID-19 cases (Larsen & Wigginton 2020; Galani *et al.* 2022). This information can be used as an epidemiological indicator even in areas where community testing is not possible, for instance in low-income environments where resources are limited.

WBE surveillance during the pandemic

The objective of this paper is to outline some of the complex conditions under which four countries, namely South Africa, the Netherlands, Turkey and England, found unique ways to navigate their individual paths towards the implementation of WBE surveillance for the tracking of SARS-CoV-2. These methods included the formation of strategic networks and partnerships, the coordination of WBE surveillance efforts at the national and catchment level, the prioritising of actions and resources for tracking SARS-CoV-2, knowledge dissemination and purposeful communication with stakeholders, capacity building, and data collection and analysis. The paper also considers how the data generated were used to inform public health decision-making and ensure future scientific and policy impact.

This section of the paper outlines some of the methods and structures used to establish and coordinate such surveillance programmes, as well as the objectives and strategic intent that informed the efforts. Information was gathered by collecting institutions' own experiences in undertaking WBE surveillance (particularly in the Netherlands, Turkey and South Africa); through discussion groups that shared information on the implementation and challenges of WBE surveillance; and through personal communication with particularly policy- and decision-makers (particularly for the England case study). Due to the qualitative nature of the findings a synthesis as opposed to analysis was undertaken. A concise overview of the implementation of WBE surveillance programmes for SARS-CoV-2 in the studied countries is presented in Table 1.

Table 1 | Overview of implementation of WBE monitoring programmes for SARS-CoV-2 in South Africa, The Netherlands, Turkey and England

	South Africa	The Netherlands	Turkey	England
Programme design, including methodology	• Initial aim: To monitor local and national spread of SARS-CoV-2 in provincial hotspot areas • Evolved into early warning system with pilots in cities and hotspot areas i.e., 38 Wastewater Treatment Plants (WWTP) sites and 21 non-sewered sites) • Focus on sewered and non-sewered communities • Multidisciplinary networks and partnerships, with a leverage model to complement other national surveillance efforts • Three-phased approach to implementation (proof of concept, pilot phase and nationwide rollout) • Variant detection included	• Initial aim: To monitor local and national spread of SARS-CoV-2 and develop methodology for detection of the virus in WWTPs in 6 cities and an airport • Methodology developed, applied and published at early stage of the pandemic • Evolved into a high-resolution analysis of the surveillance pyramid in Rotterdam, alongside a national rollout at all WWTPs in the country • Only sewered systems • Multidisciplinary networks and partnerships implemented to support knowledge exchange and support within the European Union and globally • Variant detection included in the Rotterdam study	• Initial aim: To monitor SARS-CoV-2 distribution nationally • Evolved into early warning system for government, providing routine scanning of 7 geographical regions of Turkey, 21 pilot cities and Istanbul • Focus on sewered systems • Functional interdisciplinary project through involvement of various partners, central governmental and local governmental authorities, governmental entities, and state universities • Systemic methodology, incl. 3 major steps: Planning, Execution and Reporting via governmental dashboards (https://covid19.tarimorman.gov.tr) of several pilots leading to the nationwide surveillance phase • Variant detection included	• Initial aim: To monitor SARS-CoV-2through routine collection and analysis of wastewater samples for SARS-CoV-2 RNA concentrations from WWTPs, in-network sites and near-to-source at critical infrastructure • Evolved into national programme routinely reporting on the detection of variants, providing a prevalence indicator, and case estimates, to inform local response and national strategy • Environmental Monitoring for Health Protection programme (EMHP) monitors SARS-CoV-2 RNA levels in wastewater – covers 70% of the population and major cities • Focus on sewered systems • Pilots focused on determining if WBE could provide a national and regional prevalence indicator and be used effectively to provide local public health insight • In December 2020, EMHP initiated a successful pilot study on the use of whole-genome sequencing to detect SARS-CoV-2 variants in wastewater • Routine reports on the detection of variants in samples from all monitored WWTPs and in-network sites
Programme coordination	• Led by government entities Water Research Commission (WRC), the National Institute of Communicable Diseases (NICD) in partnership with the South African Medical Research Council (SAMRC) • Collaborative engagements with river action groups, community	• RIVM mandated by Ministry of Public Health to start national monitoring program to all WWTPs in The Netherlands. • KWR initiated the programme, also involved experts from Erasmus Medical Centre, Municipal Health Services (GGD) Rotterdam-Rijnmond, the National Institute of Public Health and the Environment	• Led by the Ministry of Agriculture and Forestry and coordinated by the Turkish Water Institute (SUEN)	• Led by Joint Biosecurity Centre within the Department of Health and Social Care (DHSC) now part of the United Kingdom Health Security Agency (UKHSA) and the Department for Environment, Food & Rural Affairs (Defra), the Environment Agency, inputs, and strong

(Continued.)

Table 1 | Continued

	South Africa	The Netherlands	Turkey	England
	leaders, universities and research facilitators for monitoring of non-sewered environments	(RIVM, epidemiology), Partners4UrbanWater, Royal-Haskoning DHV (engineering consultancy), Foundation for Applied Water Research, local water authorities (Hollandse Delta, Hoogheemraadschap van Delfland, Hoogheemraadschap van Schieland en de Krimpenerwaard) and KWR		collaboration with universities and academic partners • Multidisciplinary working groups set up in each region
Networks and partnerships (local and international)	• SACCESS network established on a national level • Strategic partnerships were strengthened (including those with policymakers) • Technical partnerships strengthened • Work closely with GWRC • Knowledge-sharing and learning partnerships strengthened (in the co-development of joint knowledge products with NICD)	• KWR shared findings with Dutch water boards, municipalities and international partners and networks • Knowledge-sharing and learning partnerships strengthened through strategic initiatives such as WHO, IWA, WRF, the European Joint Research Centre and EU vs Virus Setting-up and chairing webinar series on sewage surveillance with the Global Water Research Coalition • Strategic partnerships were strengthened (including those with policymakers) • Technical partnerships strengthened (including those with research, laboratory and municipality networks) with Morocco, Kenya, South Africa and Peru • Working with PATH, Michigan State University and UCMerced on W-SPHERE, a global data center for SARS-CoV-2 in wastewater as part of the Global Water Pathogen Project	• Joint venture with China to build a surveillance system using sewage sludge monitoring from a community health perspective • Thematic group for the 9th World Water Forum to provide a platform for knowledge-sharing	• Partnered approach with public health teams at the local authority levels. Results communicated via weekly reporting • National policy and incident response • Technical and research networks across sectors; focusing on sampling, laboratory methods, analysis, epidemiology, and genomics • Inputs into supra-national data sharing repositories to aid international effort
Communication and knowledge dissemination	• Webinars: Targeted at science community local and international (workshops, dialogues, meetings) • Broadcast (TV and Radio), Print media; and Online news: Targeted at public (expert interviews, press releases, factsheets, radio content for local communities) • Website: Dedicated programme pages	• Webinars: Targeted at technical audiences, policy-makers, NGOs, etc. (knowledge sharing dialogues, strategic meetings) • Broadcast Media (TV, Radio, Podcasts): Targeted at the public • Website and dedicated webpage: Targeted at internal and external stakeholders and mainly to respond to public concerns. (News items, project information)	• Webinars: Targeted at international stakeholders, partners and NGOs • Broadcast (TV), Print Media: Targeted at public (non-scientific TV speeches) • Dashboard to share information and data with the general public by the Ministry of Agriculture and Forestry website (https://covid19.	• Seminar presentations and webinars targeted at government, academic and international audiences • Direct communication: Targeted at local health protection teams (insights and data via operational dashboards) • Weekly reporting to the UK Secretary of State for Health and Social Care and the UK Prime Minister

(*Continued.*)

Table 1 | Continued

	South Africa	The Netherlands	Turkey	England
	targeted at internal and external stakeholders • Direct communication: Targeted at municipal authorities, policy and decision-makers • Social media: Targeted at public and WRC social media community • Dashboard to share information and data with the general public: https://www.nicd.ac.za/diseases-a-z-index/disease-index-covid-19/surveillance-reports/weekly-reports/wastewater-based-epidemiology-for-sars-cov-2-in-south-africa/	• Direct communication: Targeted at health authorities and the national Outbreak Management Team, scientists and water authorities • Targeted, strategic communications (best practices guide for communicating science during a public health crisis) • Social media: Targeted at public and KWR and Watershare social media community (LinkedIn, Twitter) • Dashboards, sampling information and methodologies shared, scientific papers published • Dashboard to share information and data with the general public: Dutch sewage surveillance use case (arcgis.com)	tarimorman.gov.tr) (Dashboards, sampling information and scientific papers) • Direct communication: Targeted at local authorities • Social media: Targeted local and international stakeholders (Twitter, YouTube and LinkedIn) • Dashboard to share information and data with the general public	• Direct communication: Targeted at key stakeholders (weekly reports) • Dashboard to share information and data with the general public: Monthly data is published on the government web page 'Monthly statistics for the Environmental Monitoring for Health Protection (EMHP) wastewater program (England) - GOV.UK (www.gov.uk)' • Publications in peer reviewed journals, and Government public communication forums • Engagement via various media channels
Capacity-building activities	• SACCESS network established (network partners collaborated to standardise the methodology and sampling methods now used in surveillance efforts)	• KWR involved in several international capacity-building projects to develop wastewater testing capacities, design regional and national monitoring programmes, and develop tools for reporting and data sharing • Published best practice guidelines to limit misinformation and enhance knowledge dissemination • Supported global community by setting up the global data centre W-SPHERE (see above) and sharing methodologies and lessons learned via Watershare	• COVID-19 Thematic group for World Water Forum March 2022: COVID-19 platform for all COVID-19 groups (global) for the development of capacity building activities for low- and middle-income countries	• Local and national public health teams supported with wastewater data • Secondments, mentorship, and training to post-graduate scientists • Stimulation of UK water industry, consultants and supply chain for sampling, analytics and end-to-end-delivery of wastewater monitoring for public health • Worked closely with members of UK Research and Innovation research network (NWESP)
Data collection, analysis, sharing and use	• Samples collected from WWTW influent, non-sewered communities (stormwater systems and rivers) • Primary sludge and grab samples collected from industry sewage plants, prisons and hospitals	• Samples collected at the inflow of municipal WWTPs and pumping stations in the sewer network • Collection of 24-hour, high-frequency flow composite samples • Results shared on a weekly basis with the health authorities via direct communication and with the general public through dedicated web page	• Samples collected from influent, effluent, primary and waste activated sludge at WWTPs • Routine sampling covering 40–50% of the Turkey population, 25–95% of megacity Istanbul population • Routine sampling of touristic districts in Aegean and Mediterranean parts of Turkey	• Sampling from WWTPs, the sites providing coverage of around 70% of the population, and sampling from major cities at the community level • Data available on the Secure Research Service and on gov.uk • Whole-genome sequencing to identify lineage-defining mutations

(Continued.)

Table 1 | Continued

	South Africa	The Netherlands	Turkey	England
			• Routine sampling of some military areas	• Variants of concern and variants of interest tested in all WWTP and 50% of in-network sites weekly • Pilot programs collect samples from prisons, schools, managed quarantine facilities, halls of residence, food production factories • Collaboration with Scientific Pandemic Influenza Group to develop English prevalence models Integration with local response teams to sample and provide insights in areas with data gaps
Reach and impact on public health decision-making	• WBE surveillance not yet fully integrated into national COVID-19 surveillance programme • City of Cape Town and the Western Cape Provincial Department of Health incorporated WBE monitoring into their local responses	• WBE surveillance data has been evaluated in combination with conventional health surveillance data • Differences in trends observed via variations in clinical testing and wastewater testing, which was used by the health authorities to confirm or falsify apparent trends in the clinical testing data, the latter more prone to changes in testing behaviour and capacity ('everyone uses the toilet, not everyone uses the test facilities'. Also, city quarters differed in the ratio between reported cases and SARS-CoV-2 concentration in wastewater, suggesting underreporting in some city quarters. The health authority mobilised testing buses to these underreported areas to increase testing	• Early warning system established with routine systematic scanning in geographical regions • National roll-out	• National operational programme established • Wastewater concentration and sequencing data integrated into national situational awareness reporting and local health risk assessment processes • Wastewater database supports public health decision-making • Data frequently triggers public health action such as deployment and prioritisation of resources • Wastewater monitoring used with clinical testing for coordinated outbreak response

Programme design of the four case studies

The aim of establishing the national WBE surveillance programme in all the case studies was to provide scientific evidence regarding the prevalence of SARS-CoV-2 within communities. The programmes were further used as an early warning system to guide the governments' decision-making regarding their national responses which normally centred around allocation of resources and restriction measures to curb the spread of the virus. Once established, the programmes were also used to monitor the emergence of variants of concern in the population and the success of mitigation measures (e.g., targeted vaccination programmes, lockdowns, border control and quarantining).

In South Africa, the initial objective was to establish a national surveillance programme rollout to enhance the ability of the Department of Water and Sanitation to react to water supply and sanitation needs during COVID-19, and to integrate water quality and health data through a dedicated data platform. The programme was further re-designed to track the spread of

COVID-19 in communities, and South Africa is the only country among the four case studies to include the environmental surveillance in the context of non-sewered sanitation. In Turkey, the nationwide surveillance programme was initially established to evaluate the spread and extent of COVID-19 nationally before evolving into an early warning system for the government providing a routine systematic scanning of the different geographical regions. In England, in addition to the surveillance and early warning system the programme further provided public health intelligence and insight to guide the deployment of clinical resources. For the Netherlands, while the National Institute for Public Health and the Environment (RIVM) was focused on establishing a national programme of SARS-CoV-2 sewage surveillance (Lodder & de Roda Husman 2020), KWR focused on the development and finetuning of the analytical methodologies to detect SARS-CoV-2 in wastewater, observing patterns and trends and investigating how sewage surveillance could be used as a sensitive tool to monitor the circulation of the virus in the population, complementing clinical surveillance (Medema *et al.* 2020a).

With the viral mutation giving rise to new variants experienced by all the countries in the case study, the national programmes' scope was expanded to also include detection of signals of potential SARS-CoV-2 variants of concern (VOCs) and variants under investigation (VUIs) in WWTPs. The emerging COVID-19 variants are further reported using the publicly accessible government website and media to ensure public dissemination and awareness of the variants currently circulating in the communities and country. England is further working on detecting variants and mutation presence in near-source sites building on work undertaken by and in collaboration with the Natural Environment Research Council (NERC) Environmental Omics Facility (NEOF) (Hillary *et al.* 2021) and has incorporated co-occurrence analysis into its workflow (Jahn *et al.* 2021).

Programme coordination

Comparing the design and methodology of establishing the national programmes of the countries featured in the case studies, two programmes (England and Turkey) were initiated/owned by a national government department (e.g., health, environment or agriculture) and coordinated by a government entity or a research institute. Table 2 outlines the countries' funding and coordinating partnerships. The inclusion of national government representatives in the core partnership was expected to enable effective feedback and guidance to the national decision-making process. However, integration of WBE and surveillance into a national COVID-19 management strategy by governmental decision makers remains a challenge. The reason for this could be that as seen in Turkey, the overall national COVID-19 decision-making lies within the health ministry. Therefore, efforts should be made to promote understanding of the importance of WBE surveillance studies by health authorities. In South Africa, the programme is owned and coordinated by government entities. The WRC has continuously sought to secure government support through the Department of Water and Sanitation and the Department of Health, which are key for successful implementation of a nationwide rollout of the programme.

In The Netherlands, KWR initiated a monitoring process in six cities and the WWTP collecting wastewater from the Schiphol Airport in early 2020, and then decided to shift its focus towards three cities, namely Amsterdam, Utrecht and Rotterdam. In the course of 2020, the RIVM was mandated by the Ministry of Public Health to expand its national monitoring program to all WWTPs in the Netherlands. To avoid overlap, KWR pursued its research activities with the goal of further refining WBE and improving the understanding about the link between wastewater signals and virus circulation in the

Table 2 | WBE surveillance programme funders and coordinators at national and local level

Country	Coordinator(s)	Funder
England	United Kingdom Health Security Agency (UKHSA)	Department of Health and Social Care (DHSC), Department for Environment, Food & Rural Affairs (Defra)
The Netherlands	RIVM (national programme) KWR + partners (local projects)	Ministry of Health, Welfare and Sport (national programme) Top consortium of Knowledge and Innovation Water Technology and Health Holland, Foundation for Applied Water Research (STOWA), Erasmus Foundation
South Africa	National Institute of Communicable Diseases (NICD) in partnership with the South African Medical Research Council (SAMRC)	Water Research Commission (WRC), MRC and NICD
Turkey	Turkish Water Institute (SUEN)	Minister of Agriculture and Forestry

population (Medema *et al.* 2020a). Together with various partners, KWR set up a study in the city of Rotterdam, which involves the collection of high-resolution wastewater and epidemiological (e.g., swab testing, syndromic surveillance, genomic sequencing) data. At the time of writing, wastewater data was being collected both at the level of the WWTP influents and upstream at pumping stations, which serve specific neighbourhoods. More recently, the study also involves the monitoring of variants of concern through both targeted analyses and genome sequencing.

The multidisciplinary partnership model applied by all countries promoted national coordination, facilitated data sharing and complementarity with other national programmes implemented by the various health and water sector partners. All four countries included a mix of academic, industry and government experts in their partnership model. Although all countries implemented their programmes using a phased approach i.e., feasibility and/or proof of concept phase, pilot phase and nationwide programme phase, there was a distinct difference in the English and Dutch case studies.

In addition to the pilot to determine if WBE could provide a national and regional prevalence indicator which all countries undertook, England took a step further and implemented pilot projects to assess the feasibility of WBE in providing local public health insight in Exeter, and tested critical infrastructure and large buildings near-to-source through pilots monitoring schools and prisons in collaboration with academics from Universities of Middlesex, Cranfield, Newcastle and the UK Centre for Ecology and Hydrology (Gutierrez *et al.* 2021). More recently, SARS-CoV-2 monitoring has been undertaken in England at several managed quarantine facilities (for arrivals into the UK from any Red List country) to screen for variants of concern (VOC) or variants under investigation (VUI). These unique feasibility pilots were only seen in the England case study. On the other hand, The Netherlands, through KWR, was the only country initially which had a strong technical interest and dedicated efforts towards fine-tuning the analytical methodologies and getting a better understanding of the link between wastewater signals and virus circulation in the population to complement the clinical surveillance (Medema *et al.* 2020a). The England programme subsequently undertook a major research programme to optimize the methods to enable the rapid sequencing of VOCs and VUIs in wastewater alongside approaches for automated sample collection and on-site analysis.

Additionally, The Netherlands (through KWR) and England (through key government agencies, the Environment Agency and key academic partners), had a strong technical interest and dedicated efforts towards fine-tuning the analytical methodologies and getting a better understanding of the link between wastewater signals and virus circulation in the population to complement the clinical surveillance (Medema *et al.* 2020a).

In South Africa, a group of laboratories involved in the WBE and clinical surveillance collectively established a forum to communicate challenges and share experiences and lessons learnt. The group was named the South African Collaborative COVID-19 Environmental Surveillance System (SACCESS) network. Webinars were held to share methodologies for concentration, extraction and polymerase chain reaction (PCR) testing for the detection of SARS-CoV-2, and a compendium of methodologies was published (Corman *et al.* 2020). Below are the examples of the South African and Turkish WBE surveillance programme partnerships and networks. A similar network to SACCESS was developed in July 2020 in the UK, the National Wastewater-based Epidemiology Surveillance Programme (NWESP). The UK Natural Environment Research Council (NERC) funded network supported research into WBE sampling, analysis, and epidemiological model development, which directly supported the UKHSA, Defra and the Environment Agency of England, as well as the other devolved governments, as they established their surveillance programmes.

It is clear from all the case studies that strategic networks and partnerships are considered vital to the implementation of WBE surveillance programmes. In South Africa, the WRC categorised partnership needs into three categories, namely strategic, technical and knowledge partnerships. Ultimately, the biggest development with regards to establishing networks for WBE surveillance in South Africa to date has been the establishment of the SACCESS network. This network facilitates knowledge sharing and capacity building amongst its members, who also collaborated to standardise the methodology and sampling methods that are now used in many of the existing surveillance efforts.

Networks and partnerships

Organisations in all four case studies have prioritised the establishment of partnerships, collaborations and networks in their respective countries to ensure that the WBE programmes are national or regional initiatives. A collaborative COVID-19 Environmental Surveillance System (SACCESS) network for South Africa, KWR's international network and local partnership with Erasmus Medical Centre and the Municipal Health Service and others in Rotterdam, a thematic group for the 9th World Water Forum for Turkey, and Technical and Research networks in England were established to advance collaboration and partnerships among institutions and scientists working with COVID-19 surveillance in wastewater. South Africa and The Netherlands forged partnerships not only with scientific and knowledge organisations, which all four countries did, but also reached out to local governments and policymakers to be strategic partners for funding, implementation, knowledge dissemination, and use of the results from their programmes. All the institutions ensured that their programmes remain connected and engaged with the public health organisations so that the results from WBE can be used to save lives and manage the spread of the virus.

Communication and knowledge dissemination

The COVID-19 pandemic brought with it increased limitations to physical engagements, while at the same time catalysing the uptake and adoption of virtual engagement platforms such as video conferencing and online channels like websites, e-mails, broadcast and social media. Webinars are cited in three of the four country case studies as the preferred and dominant platforms used for knowledge sharing between the local and international science community. They enabled high participation and attendance of virtual events and facilitated global collaboration as well as extended reach beyond the science community to other key stakeholder groups specifically for the Netherlands, South Africa and Turkey teams.

The dominant channel for continuous programme updates, directed knowledge dissemination and official programme responses and news updates was website communication and via dashboards. This channel provided teams with a central, internally owned and controlled, publicly accessible repository for internal and external stakeholder engagement. The Turkish team further expanded its reach on website platforms by using its partnership with the Ministry of Agriculture and Forestry to upload dashboards and scientific papers on the ministry's website (https://covid19.tarimorman.gov.tr). KWR used the data collected in six cities to develop an analytical method to detect SARS-CoV-2 in wastewater, which was published in scientific paper in an early stage of the pandemic (Medema *et al.* 2020b). KWR, SUEN and Cranfield university published several scientific papers during 2020, targeting the scientific community.

The Netherlands, South Africa and Turkey teams referenced direct communication with government authorities, policy- and decision-makers, while both the Netherlands and England highlighted direct communication in engaging other scientists, water authorities and project teams.

Engagement with the public was done through broadcast media. Turkey focused specifically print media and television broadcast on prime time, which the team found to be highly effective. The South African context required the use of both television and radio broadcast mediums to extend reach to local communities, while print as a medium of communication was used once the lock-down restrictions were relaxed. Three of the four case studies (the Netherlands, Turkey and South Africa) included social media as a common dissemination tool for public dissemination both locally and internationally. Each case study illustrates that there is no one size fits all solution when it comes to building effective communication and that it is imperative that appropriate channels are used to reach relevant stakeholders.

Capacity building

Capacity building activities mentioned in the case studies predominantly took the form of publications and disseminating knowledge to the science community. The South Africa case study highlights capacity building initiatives embarked on by the SACCESS network and led by the WRC, who used a publication to facilitate the standardisation of sampling methodology and methods in South African surveillance programmes (Corman *et al.* 2020). The Netherlands, through KWR, published several papers on WBE methodologies, a best practice guide aimed at helping local teams and stakeholders to limit the spread of misinformation and support effective knowledge sharing. All country case studies mentioned the use of workshops, strategic meetings and dialogues as effective platforms for the sharing of learnings, experience and knowledge.

Targeted capacity building programmes were outlined in two case studies. KWR (Netherlands) highlighted capacity building programmes targeted at upskilling local authorities in wastewater testing capacities. Turkey outlined future plans by the COVID-19 thematic group for the 9th World Water Forum that aimed to build a global knowledge and experience sharing platform. The platform would focus on capacity building activities for low- and middle-income countries. As surveillance programmes in the various countries mature, there may be a more deliberate focus on capacity building that extends beyond the science community, the current focus is on building standards, developing tools for sampling, reporting and data sharing as well as upskilling local authorities.

The English programme was different in that it provided stimulus to industry to build commercial capacity, and automated sampling and analysis, via an initiative providing funding and collaborative working across sectors. This occurred in the

English (UK) water industry and via consultants and supply chain. These sectors were vital for sampling, analytics, and end-to-end-delivery of WBE for public health as capacity for analysis of SARS-CoV-2 abundance and genotyping of VOC and VUI increased.

Data collection, analysis, sharing and use

In all four countries, WWTP were sampled on a regular basis to test the presence of the COVID-19 virus genome. However, stormwater systems and urban rivers were also sampled in South Africa to determine the prevalence of COVID-19 in communities without sewer sanitation or using non-sewer sanitation, such as pit latrines.

The programmes in all countries targeted samples from most parts of the countries, via the network of researchers at their disposal. While testing for the prevalence and the spread of the virus was the main drive of these programmes, all countries showed interest in developing the early warning system or protocol to ensure the safety of vulnerable communities and to inform decision making by concerned institutions such as public health and transportation sectors. All the countries' programmes showed interest in studying the variants of the COVID-19 virus to advance science and support the development of effective vaccines.

Research and impact on public health decision-making

WBE has been implemented on a national scale in Turkey and the Netherlands. Sewage surveillance alone has not triggered any policy changes in the Netherlands, but has been used in combination with conventional swab testing as a piece of the puzzle for SARS-CoV-2 monitoring. In Turkey, scientific reports were submitted to the Ministry of Agriculture and Forestry before weekly cabinet meetings and the information was used to evaluate the need of local measures for pandemic control (Alpaslan Kocamemi *et al.* 2020). In South Africa, only the City of Cape Town and the Western Cape Provincial Department of Health have incorporated the WBE programme into their local responses. This is helping the City and province understand the emergence and patterns of infections. At the time of writing, however, efforts were underway to get municipalities that are considered COVID-19 hotspots to incorporate this approach into their local actions in managing this pandemic.

The use of real-time dashboard announcements, including simple and easily understandable maps, charts and graphs, combined with the national health interventions increases decision-makers' timely understanding of the findings of WBE studies and management of the disease. More importantly, proving that the early warnings provided by WBE surveillance correlate with data from clinical testing will increase the trust of decision-makers in the programme.

Discussion and recommendations from the four case studies

WBE surveillance for COVID-19 has proven to be an effective means of providing an early warning of the spread and trends of COVID-19 infections. Continued WBE sampling at priority sites will allow for the expansion of pandemic trend monitoring. Results indicated an increase over time in the viral load of the samples tested at surveillance sites, which corresponded to an increase in case numbers in hotspot areas.

Regular samples must, however, be taken over time to establish trends and baselines due to the inherent variability of sampling from smaller populations compared to a regional WWTP. Sampling in defined populations could provide a less invasive means of continuous screening. Where an increase in viral load is detected – signalling an early warning – additional clinical test methods could be rolled out in a timely way.

The lead-lag relationship between wastewater concentrations of SARS-CoV-2 RNA and case rates has not remained constant during the pandemic and more research is needed to understand the drivers of this variance (Wade *et al.* 2021). In England, the quantitative polymerase chain reaction qPCR) data are generated on average within 36 hours of sample collection, while in South Africa the time varies depending on the testing sites and location from labs, ranging between 2 to 4 days. In Turkey, the turn-around time for Istanbul is around 2 days, however, for the other cities in Turkey it takes up to 4–5 days since samples are all analyzed in center labs after all samples have been picked up and transferred. For the Netherlands, the Rotterdam-Rijnmond study showed that the lead time is related to the delay in clinical testing (delay time between disease onset day and day of testing). This delay was on average 6 days in the beginning of the second wave (which peaked in September 2020) and decreased to 1.5 days in December 2020 by the installation of large-scale testing facilities. When the clinical surveillance data of Rotterdam-Rijnmond were corrected for the testing delay, the clinical and sewage surveillance data overlapped, both in the rise and fall of the waves, which is to be expected since virus shedding is highest around onset.

Although wastewater surveillance provides a powerful tool to evaluate disease trends at the community level, its results must be combined with other public health data. It must, for example, be included in campaign-based and randomised testing of individuals (for the presence of pathogens or antibodies), clinical case reporting, mobile-based contact tracking and self-reporting systems (Boulos & Geraghty 2020). It is also important to consider how best to ethically and legally balance public health with civil liberties when handling epidemiological tracking information (Gostin *et al.* 2020).

In most case studies, communicating with public health authorities was mentioned as challenging due to the lack of expertise in the water sector and the different SARS-CoV-2 monitoring methods. In addition, public health officials were initially unfamiliar with the science behind WBE surveillance. As a result, 'translating' the wastewater data and communicating it in such a way that public health authorities can use it have proven taxing to researchers, as indicated in the Netherlands case study. Additionally, communication and knowledge dissemination have been demanding and time-consuming for the researchers and scientists working on the topic in all four case studies, which shows an area to better prepare for and support in the future.

An important outcome of the initial pilot studies in the UK/England and the monitoring of the six cities in the Netherlands is the improved trust in wastewater data to provide a reliable indication of SARS-CoV-2 trends. Comparisons were made with other epidemiological indicators, including large-scale longitudinal surveys and clinical testing data. In addition, work was undertaken to characterise the uncertainties in the data and determine how to account for them in reporting and modelling. Wastewater data were integrated into routine situational awareness reporting and local health risk assessment processes. It complemented other datasets and provided new data in areas where community infection levels were hard to measure. This approach of partnering with public health teams, including field epidemiologists and policymakers, helped ensure that the data provided were appropriate and useful.

The experience of WBE interventions has demonstrated its powerful contribution in complementing other health intervention efforts in managing this pandemic. It has shown:

- Consistent detection SARS-CoV2 RNA in wastewater samples from upstream and downstream WWTW
- Positive gene amplification observed in environmental (NSS) samples – i.e., river water samples
- WBE has proven to be a useful complementary surveillance tool for management of COVID-19
- Wastewater surveillance is a less invasive continuous screening approach
- Offers opportunity for correlation between increase in viral load and increase in case numbers with time
- Allowed the building of a robust collaborative platform of scientists, laboratories and WSIs

The WBE experiences and lessons have set up the potential for effective water quality surveillance of new and emerging contaminants and emerging pathogens, as well future tracking and management of the health of a population.

Some of the limitations of establishing and managing the WBE programme are: the insufficient funding to cover a wide range of the national programme, lack of coordination and buy-in by the key stakeholders limiting dissemination and uptake of the data and information generated during the surveillance programme, availability of test kits and equipment, number of laboratories registered and qualified to do the tests, standard for testing, collection etc.

CONCLUSION

Further work is needed to understand the impacts of network characteristics and population dynamics on SARS-CoV-2 detection in wastewater, especially in low-prevalence times. It is particularly important to standardise the methods for detection and quantifying the virus and other pathogens in wastewater. Based on the experiences in England and the Netherlands, the combination of wastewater monitoring with other epidemiological indicators (such as clinical detection and sequencing data, and longitudinal infection survey data) is recommended and already applied in practice to gain a full understanding of the changes in the prevalence of SARS-CoV-2 and its variants in communities and on a regional and national scale.

Communication efforts should be aimed at transparency and effectiveness to fully convey the strength of the method to those unfamiliar with this type of environmental monitoring. WBE surveillance offers a potential early warning system for the spread of SARS-CoV-2 infections. Surveillance of defined communities can be used to direct community screening efforts and alert medical authorities to potential increases in patient numbers. This may also provide valuable input for dashboards tracking or monitoring the COVID-19 pandemic. Wastewater testing can be useful in monitoring the prevalence of COVID-19 infections across the globe and alludes to the application of such a monitoring system to other factors affecting human health. Furthermore, the manner in which the programmes were coordinated presents useful lessons for national responses to future pandemics and on-going non-COVID related public health issues.

DATA AVAILABILITY STATEMENT

All relevant data are included in the paper or its Supplementary Information.

REFERENCES

Alpaslan Kocamemi, B., Kurt, H., Sait, A., Kadi, H., Sarac, F., Aydin, I., Saatci, A. M. & Pakdemirli, B. 2020 Routine SARS-CoV-2 wastewater surveillance results in Turkey to follow COVID-19 outbreak. *medRxiv Preprint.* DOI: doi.org/10.1101/2020.12.21.20248586.

Anon, Dutch sewage surveillance use case 2022 *Using Wastewater to Identify Areas for Increased Clinical Testing.* Available from: https://storymaps.arcgis.com/stories/8888f5bfb4704180afeda3d476f2aa63 (accessed 21 January 2022).

Boulos, M. N. K. & Geraghty, E. M. 2020 Geographical tracking and mapping of coronavirus disease COVID-19/severe acute respiratory syndrome coronavirus 2 (SARS-CoV-2) epidemic and associated events around the world: how 21st century GIS technologies are supporting the global fight against outbreaks and epidemics. *International Journal of Health Geographics* **19** (1), 8. DOI: doi.org/10.1186/s12942-020-00202-8.

Choi, P. M., Tscharke, B. J., Donner, E., O'Brien, J. W., Grant, S. C., Kaserzon, S. L., Mackie, R., O'Malley, E., Crosbie, N. D., Thomas, K. V. & Mueller, J. F. 2018 Wastewater-based epidemiology biomarkers: past, present and future. *Trends in Analytical Chemistry (TrAC)* **105**, 453–469.

Corchis-Scott, R., Geng, Q., Seth, R., Ray, R., Beg, M., Biswas, N., Charron, L., Drouillard, K. D., D'Souza, R., Heath, D. D., Houser, C., Lawal, F., McGinlay, J., Menard, S. L., Porter, L. A., Rawlings, D., Scholl, M. L., Siu, K. W. M., Tong, Y., Weisener, C. G., Wilhelm, S. W. & McKay, R. M. L. 2021 Oct 31 Averting an outbreak of SARS-CoV-2 in a university residence Hall through wastewater surveillance. *Microbiol Spectr.* **9** (2), e0079221. doi: 10.1128/Spectrum.00792-21. Epub 2021 Oct 6. PMID: 34612693; PMCID: PMC8510253.

Corman, V. M., Landt, O., Kaiser, M., Molenkamp, R., Meijer, A., Chu, D. K. W., Bleicker, T., Brünink, S., Schneider, J., Schmidt, M. L., Mulders, D. G. J. C., Haagmans, B. L., van der Veer, B., van den Brink, S., Wijsman, L., Goderski, G., Romette, J.-L., Ellis, J., Zambon, M., Peiris, M., Goossens, H., Reusken, C., Koopmans, M. P. G. & Drosten, C. 2020 Detection of 2019 novel coronavirus (2019-nCoV) by real time RT-PCR. *Eurosurveillance* **25** (3), 2000045.

Galani, A., Aalizadeh, R., Kostakis, M., Markou, A., Alygizakis, N., Lytras, T., Adamopoulos, P. G., Peccia, J., Thompson, D. C., Kontou, A., Karagiannidis, A., Lianidou, E. S., Avgeris, M., Paraskevis, D., Tsiodras, S., Scorilas, A., Vasiliou, V., Dimopoulos, M.-A. & Thomaidis, N. S. 2022 SARS-CoV-2 wastewater surveillance data can predict hospitalizations and ICU admissions. *Science of The Total Environment* **804**, 150151. https://doi.org/10.1016/j.scitotenv.2021.150151.

Gostin, L. O., Friedman, E. A. & Wetter, S. A. 2020 Responding to COVID-19: how to navigate a public health emergency legally and ethically. *Hastings Center Report* **50**, 8–12. doi: doi.org/10.1002/hast.1090.

Gutierrez, V. C., Hassard, F., Vu, M., Leitao, R., Burczynska, B., Wildeboer, D., Stanton, I., Rahimzadeh, S., Baio, G., Garelick, H., Hofman, J., Kasprzyk-Hordern, B., Kwiatkowska, R., Majeed, A., Priest, S., Grimsley, J., Lundy, L., Singer, A. C. & Di Cesare, M. 2021 Monitoring occurrence of SARS-CoV-2 in school populations: a wastewater-based approach. *medRxiv Preprint.* DOI: doi.org/10.1101/2021.03.25.21254231.

Hellmér, M., Paxéus, N., Magnius, L., Enache, L., Arnholm, B., Johansson, A., Bergström, T. & Norder, H. 2014 Detection of pathogenic viruses in sewage provided early warnings of Hepatitis A and Norovirus outbreaks. *Applied and Environmental Microbiology* **80** (21), 6771–6781. doi:10.1128/AEM.01981-14.

Hendriksen, R. S., Munk, P., Njage, P., van Bunnik, B., McNally, L., Lukjancenko, O., Röder, T., Nieuwenhuijse, D., Pedersen, S. K., Kjeldgaard, J., Kaas, R. S., Clausen, P. T. L. C., Vogt, J. K., Leekitcharoenphon, P., van de Schans, M. G. M., Zuidema, T., de Roda Husman, A. M., Rasmussen, S., Petersen, B., Global Sewage Surveillance project consortiumAmid, C., Cochrane, G., Sicheritz-Ponten, T., Schmitt, H., Alvarez, J. R. M., Aidara-Kane, A., Pamp, S. J., Lund, O., Hald, T., Woolhouse, M., Koopmans, M. P., Vigre, H., Nordahl Petersen, T. & Aarestrup, F. M. 2019 Global monitoring of antimicrobial resistance based on metagenomics analyses of urban sewage. *Nature Communications* **10** (1), 1124.

Hillary, L. S., Farkas, K., Maher, K. H., Lucaci, A., Thorpe, J., Distaso, M. A., Gaze, W. H., Paterson, S., Burke, T., Connor, T. R., McDonald, J. E., Malham, S. K. & Jones, D. L. 2021 Monitoring SARS-CoV-2 in municipal wastewater to evaluate the success of lockdown measures for controlling COVID-19 in the UK. *Water Research* **200**, 117214. DOI: doi.org/10.1016/j.watres.2021.117214.

Hovi, T., Shulman, L. M., van der Avoort, H., Deshpande, J., Roivainen, M. & de Gourville, E. M. 2012 Role of environmental poliovirus surveillance in global polio eradication and beyond. *Epidemiology & Infection* **140** (1), 1–13. doi:10.1017/S095026881000316X.

Humayun, A., Diop, O. M., Weldegebriel, G., Malik, F., Shetty, S., El Bassioni, L., Akande, A. O., Al Maamoun, E., Zaidi, S., Adeniji, A. J., Burns, C. C., Deshpande, J., Oberste, M. S. & Lowther, S. A. 2014 Environmental surveillance for polioviruses in the global polio eradication initiative. *The Journal of Infectious Diseases* **210** (Suppl 1), S294–S303.

Jahn, K., Dreifuss, D., Topolsky, I., Kull, A., Ganesanandamoorthy, P., Fernandez-Cassi, X., Bänziger, C., Stachler, E., Fuhrmann, L., Jablonski, K. P., Chen, C., Aquino, C., Stadler, T., Ort, C., Kohn, T., Julian, T. R. & Beerenwinkel, N. 2021 Detection of SARS-CoV-2 variants in Switzerland by genomic analysis of wastewater samples. *medRxiv Preprint.* DOI: doi.org/10.1101/2021.01.08.21249379.

Larsen, D. A. & Wigginton, K. R. 2020 Tracking COVID-19 with wastewater. *Nat Biotechnol* **38**, 1151–1153. https://doi.org/10.1038/s41587-020-0690-1.

Lodder, W. & de Roda Husman, A. M. 2020 SARS-CoV-2 in wastewater: potential health risk, but also data source. *The Lancet Gastroenterology & Hepatology* **5** (6), 533–534. DOI: doi.org/10.1016/S2468-1253(20)30087-X.

Medema, G., Been, F., Heijnen, L. & Petterson, S. 2020a Implementation of environmental surveillance for SARS-CoV-2 virus to support public health decisions: opportunities and challenges. *Current Opinion in Environmental Science & Health* **17**, 49–71.

Medema, G., Heijnen, L., Elsinga, G. & Italiaander, R. 2020b Presence of SARS-Coronavirus-2 RNA in sewage and correlation with reported COVID-19 prevalence in the early stage of the epidemic in the Netherlands. *Environmental Science & Technology Letters* **7** (7), 511–516. doi:10.1021/acs.estlett.0c00357.

Pal, M., Berhanu, G., Desalegn, C. & Kandi, V. 2020 Severe acute respiratory syndrome coronavirus-2 (SARS-CoV-2): an update. *Cureus*. **12** (3), e7423. Published 2020 Mar 26. doi:10.7759/cureus.7423.

Sharara, N., Endo, N., Duvallet, C., Ghaeli, N., Matus, M., Heussner, J., Olesen, S. W., Alm, E. J., Chai, P. R. & Erickson, T. B. 2021 Wastewater network infrastructure in public health: applications and learnings from the COVID-19 pandemic. *PLOS Global Public Health* **1** (12), e0000061. https://doi.org/10.1371/journal.pgph.0000061.

Wade, M., Lo Jacomo, A., Armenise, E., Brown, M., Bunce, J., Cameron, G., Fang, Z., Farkas, K., Gilpin, D., Graham, D., Grimsley, J., Hart, A., Hoffmann, T., Jackson, K., Jones, D., Lilley, C., McGrath, J., McKinley, J., McSparron, C., Nejad, B. F., Morvan, M., Quintela-Baluja, M., Roberts, A., Singer, A., Souque, C., Speight, V., Sweetapple, C., Watts, G., Weightman, A. & Kasprzyk-Hordern, B. 2021 *Understanding and Managing Uncertainty and Variability for Wastewater Monitoring Beyond the Pandemic: Lessons Learned From the United Kingdom National COVID-19 Surveillance Programmes*. doi:10.1002/essoar.10507606.2.

Wang, W., Xu, Y., Gao, R., Lu, R., Han, K., Wu, G. & Tan, W. 2020 Detection of SARS-CoV-2 in different types of clinical specimens. *JAMA* **323** (18), 1843–1844.

World Health Organization. 2018 *Typhoid and Other Invasive Salmonellosis*. World Health Organization Press, Geneva, Switzerland.

First received 2 August 2021; accepted in revised form 30 January 2022. Available online 7 February 2022

IWA Publishing's authorised EU representative for General Product Safety Regulations is Diane D'Arras, 15 rue Duret, 75116 Paris, France, e-mail: safety@iwap.co.uk.

Printed and bound by CPI Group (UK) Ltd, Croydon, CR0 4YY
12/05/2026
02108879-0004